Preface

When the temperature of a gas is not too high and the density of a gas is not too low, the transfer of heat by radiation is usually negligibly small in comparison with that by conduction and convection. However, in the hypersonic flow of space flight, particularly in the re-entry of a space vehicle, and in the flow problem involving nuclear reaction such as in the blast wave of nuclear bomb or in the peaceful use of the controlled fusion reaction, the temperature of the gas may be very high and the density of the gas may be very low. As a result, thermal radiation becomes a very important mode of heat transfer. A complete analysis of such high temperature flow fields should be based upon a study of the gasdynamic field and the radiation field simultaneously. Hence during the last few years, considerable efforts have been made to study such interaction problems between gasdynamic field and radiation field and a new title, Radiation Gasdynamics, has been suggested for this subject.

Even though radiative transfer has been studied for a long time by astrophysicists, the interaction between the radiation field and the gadsynamic field has been only extensively studied recently. Since the knowledge of gasdynamics and those of radiative transfer are reported usually in unrelated sources, it is desirable to write a book which would furnish the readers some basic elements of both radative transfer and gasdynamics and their interactions which would be very useful to scientists and engineers who are interested in the high temperature flow problems, and who are not familiar with both subjects. It is the author's hope that this book will furnish the readers the basic elements of this new subject so that it is useful for further study and research in this new field.

After the introduction of radiation gasdynamics, the author reviews the fundamentals of radiative transfer in chapters II to IV and the gasdynamics in chapters V and VI with special emphasics on the coupling terms between the radiation terms and the gasdynamics. In chapter VII, the important parameters of radiation gasdynamics are discussed.

In chapters VIII and IX, the flow problems of radiation gasdynamics based on the continuum point of view are discussed with particular emphasis on the wave motion, shock waves and heat transfer. In chapter X the kinetic theory of radiating gas will be discussed. Because the photon gas moves with a speed of light, relativistic effect must be considered. A brief discussion of relativistic mechanics is included. In chapter X, we also treat some problems of rarefied radiating gas, particularly the free molecule flow. Finally in chapter XI, the radiative properties of high temperature gases will be briefly discussed with special emphasis on the absorption coefficient of high temperature air and hydrogen.

A large portion of the materials in this book was given by the author in a seminar on Plasma Dynamics during the academic year 1962—1963 in the

Institute for Fluid Dynamics and Applied Mathematics, University of Maryland. This seminar was joinly conducted by the author with Professors J. M. BURGERS and T. D. WILKERSON. The author would like to express his appreciation to Professors BURGERS and WILKERSON for many interesting discussions on this subject; and to Professor M. H. MARTIN for his interest and encouragement. In conclusion, the author takes this occasion to thank his wife, ALICE YEN-LAN WANG PAI, for her constant encouragement and help in the proof-reading during the preparation of the manuscript.

College Park, Maryland, U. S. A.
January 15, 1966.

Shih-I Pai

RADIATION GAS DYNAMICS

BY

SHIH-I PAI

RESEARCH PROFESSOR
INSTITUTE FOR FLUID DYNAMICS AND APPLIED MATHEMATICS
UNIVERSITY OF MARYLAND, COLLEGE PARK, MARYLAND, U.S.A.

WITH 76 FIGURES

1966

SPRINGER-VERLAG NEW YORK INC.

TITLE-NO. 9145
ISBN-13: 978-3-7091-5733-6   e-ISBN-13: 978-3-7091-5730-5
DOI: 10.1007/ 978-3-7091-5730-5

TO

ALICE, STEPHEN, SUE, ROBERT, LOU

Table of Contents

Page

Chapter I. **Introduction** 1

   1. Radiation gasdynamics 1
   2. Thermal radiation effects 2
   3. Some thermal radiation phenomena 4
   References 6

Chapter II. **Fundamentals of Radiative Transfer** 8

   1. Specific intensity 8
   2. The flux of radiation 9
   3. Energy density of radiation...................................... 11
   4. The stress tensor of radiation 12
   References 14

Chapter III. **Equation of Transfer of Radiation** 15

   1. Introduction 15
   2. Absorption coefficient 15
   3. Emission coefficient 20
   4. The equation of radiative transfer 21
   5. A solution of the equation of radiative transfer 23
   References 23

Chapter IV. **Radiative Equilibrium** 25

   1. Introduction 25
   2. Kirchhoff's law of radiation 25
   3. Wien's displacement law... 27
   4. Planck's radiation law.. 29
   5. Stefan-Boltzmann's law of radiation 32
   6. Adiabatic changes in an inclosure containing matter and radiation..... 32
   7. Local thermodynamic equilibrium 34
   References 34

Chapter V. **Fundamental Equations of Radiation Gasdynamics** 35

   1. Introduction 35
   2. Equation of state 36
   3. Equation of continuity 36
   4. Equations of motion 36
   5. Equations of energy 37
   6. Equation of radiative transfer 37
   7. General remarks on the fundamental equations 38
   8. Case of small mean free path of radiation 39
   9. Case of finite mean free path of radiation 41
   10. One dimensional radiative transfer 45
   11. The exponential integrals 47
   References 49

Chapter VI. **Boundary Conditions of Radiation Gasdynamics** 51

   1. Introduction 51
   2. Boundary conditions of gasdynamic field......................... 51
   3. Boundary conditions of radiation field 52
   4. Smooth surface 53

Page

    5. Rough surface 55
    6. Radiative transfer between two opaque parallel plates 59
    7. Emissivity of a constant temperature gas layer.................... 61
    8. Radiation slip at finite mean free path of radiation 63
    References 64

Chapter VII. **Similarity Parameters of Radiation Gasdynamics** 66
    1. Introduction 66
    2. Dimensional analysis and π-theorem 66
    3. Non-dimensional equations of radiation gasdynamics 69
    4. Important parameters of radiation gasdynamics 71
    5. Some further remarks for the non-dimensional parameters.......... 75
    References 80

Chapter VIII. **Waves and Shock Waves in Radiation Gasdynamics** 82
    1. Introduction 82
    2. Wave of small amplitude in an optically thick medium 82
    3. Wave of small amplitude in an radiating gas of finite mean free path
       of radiation 92
    4. Shock waves in an optically thick medium 98
    5. Shock wave structure in an optically thick medium 103
    6. Shock wave in a medium of finite mean free path of radiation 106
    7. Flow field behind shock waves................................. 112
    References 120

Chapter IX. **Heat Transfer in Radiation Gasdynamics** 122
    1. Introduction 122
    2. Radiative heat transfer in a non-absorbing medium 123
    3. Radiative heat transfer in an absorbing medium 125
    4. Heat transfer by simultaneous heat conduction and radiation in an ab-
       sorbing medium 126
    5. Radiative processes in the atmosphere 130
    6. Flow between two parallel plates in radiation magnetogasdynamics... 133
    7. Boundary layer flow in radiation gasdynamics.................... 141
    8. Stagnation point heat transfer in radiation gasdynamics 149
    9. Miscellaneous problems of heat transfer in radiation gasdynamics 161
    References 162

Chapter X. **Kinetic Theory of Radiating Gases** 164
    1. Introduction 164
    2. Molecular velocity and molecular distribution functions............. 165
    3. Relativistic mechanics 166
    4. Boltzmann aquation for material particles....................... 169
    5. Boltzmann equation for photons 174
    6. Conservation equations 175
    7. Radiation stresses and radiation energy density 180
    8. Local thermodynamic equilibrium 184
    9. Rarefied radiation gasdynamics 186
    10. Free molecule flow 191
    References 196

Chapter XI. **Radiative Properties of High Temperature Gases** 197
    1. Introduction 197
    2. Classical theory of absorption and emission of radiation............ 198
    3. The quantum theory of radiation 202
    4. Spectroscopy of high temperature gas 206
    5. The absorption coefficient of high temperature gases.............. 208
    6. Scattering coefficient of radiation 211
    7. Planck and Rosseland mean absorption coefficients of air and hydrogen 214
    8. Some experimental investigations of opacity of gases 217
    9. Non-equilibrium radiation.................................... 217
    References 218

A List of Important Symbols 220

Author Index 224

Subject Index 226

Chapter I
Introduction

1. Radiation Gasdynamics. For the flow of a compressible fluid, we have to study the fluid mechanics simultaneously with the heat transfer problem. There are three basic modes of heat transfer:

   (i) The transfer of heat by convection in fluids in a state of motion,

   (ii) The transfer of heat by conduction in solids or fluids, and

   (iii) The transfer of heat by radiation which takes place with no material carrier.

In general, all these three modes of heat transfer occur simultaneously. If the temperature is not too high and the density of the fluid is not too low, the heat transfer by radiation is usually negligible in comparison with the heat transfer by conduction or by convection. Hence in ordinary gasdynamics, the thermal radiation effects are always neglected. In the present space age, we are concerned with many technological developments in hypersonic flight, gas cooled nuclear reactors, power plants for space exploration needs, fission and fusion reactions in which the temperature is very high and the density is rather low. As a result, the thermal radiation becomes an important mode of heat transfer. A complete analysis of very high temperature flow field should be based upon a study of both the gasdynamic field and the thermal radiation field simultaneously. We use the term "Radiation Gasdynamics" for such a new branch of fluid mechanics (7, 11, 19[1]).

The study of radiation in high temperature gases has been made by physicists for a long time. At the turn of the present century, Planck found the correct theory of radiation (12). It is not the intention of the author to discuss the physics of radiation but only the influence of thermal radiation on the flow field of high temperature gases, i.e., the radiative heat transfer. The radiative heat transfer has been extensively studied by astrophysicists (2, 3, 5, 15, 17) because the spectral distribution of the radiation from stars, planets etc. is the main experimental verification of astrophysical analyses. However in most of the radiative transfer problems studied by the astrophysicists, the interaction between the gasdynamic field and the radiation is negligibly small. Hence we may assume that the temperature distribution is independent of the radiative heat transfer and we study the radiative transfer under the known distribution of temperature. This procedure is not accurate if the rate of heat transfer by radiation is of the same order of magnitude as those by convection and conduction. On the other hand, if the rate of heat transfer by radiation is small,

---

[1] The number refers the number of references at the end of each chapter.

we may estimate the thermal radiation from the temperature distribution in the flow field without radiation effect. Since many basic concepts of radiative transfer are not familiar to many engineers and scientists in the field of aerodynamics, we shall discuss these fundamental concepts of radiation in chapters II to IV.

The modern trend of aerodynamics is toward high speed and high temperature as well as low density and high altitude (1, 6). One of the extensive reviews of the state of art of aerodynamics was given by the late Professor Theodore von KÁRMÁN in 1961 (18). Even as late as 1961, the thermal radiation effects are not important in many practical problems of hypersonic flight. For an ICBM, the maximum radiative heat transfer is only $1/_{10}$ of that of aerodynamic heat transfer. Hence the interaction between the aerodynamic field and the radiation field is not important. As we shall see later, at higher reentry speeds such as that for a Mars probe (8), the interaction of aerodynamic field and radiation is no longer negligible. It is the main purpose of this book to discuss the effects of thermal radiation in a very high temperature gas flow.

2. Thermal radiation effects. There are three different thermal radiation effects on the flow field of a high temperature gas which are

   (i) Radiation stresses,
  (ii) Radiation energy density and
 (iii) Heat flux by radiation.

The exact expression for these radiation terms are very complicated and will be derived in later chapters. In order to show the relative importance of heat transfer by radiation and that by convection, we consider the case of sufficient opacity that the radiation can be considered as being trapped in the fluid and it is close to the radiative equilibrium condition. Under this condition, the radiation terms are as follows:

(i) Radiation pressure (cf. chapter II, § 4). The only components of the radiation stresses which differ from zero are the radiation pressure p_R which may be expressed as

$$p_R = \frac{1}{3} a_R T^4 \qquad\qquad (1.1)$$

where T is the temperature of the gas in °K and a_R is known as the Stefan-Boltzmann constant which is 7.67×10^{-15} erg-cm^{-3} — °K^{-4}. This radiation pressure should be added to the gas pressure in order to get the total pressure at each point of the flow field.

(ii) Radiation energy density (cf. chapter II, § 3). The radiation energy density per unit mass of the fluid is

$$E_R/\rho = a_R T^4/\rho \qquad\qquad (1.2)$$

where ρ is the density of the fluid. We should add E_R/ρ to the internal energy $U_m = c_v T$ of the fluid where c_v is the specific heat at constant volume of the fluid.

In order to show the relative magnitude of radiation energy density and the internal energy, we calculate these values for air at an altitude of 72 km or

45 miles. The density of the air behind a normal shock at this altitude in a hypersonic flow is about $\rho = 1.23 \times 10^{-6}$ gr/cm^3, i.e., 10^{-3} of the density at the standard sea level value. We take $c_v = 7 \times 10^6$ erg gr-°K which is the value for a diatomic gas such as air. The radiation energy density and the internal energy at various temperatures are shown in Table 1.

Table 1. *Radiation energy density vs internal energy*

T °K	10^3	10^4	10^5	10^6
E_R/ρ, erg gr	6×10^3	6×10^7	6×10^{11}	6×10^{15}
$c_v T$ erg gr	7×10^9	7×10^{10}	7×10^{11}	7×10^{12}

It should be noted that the estimate given in table 1 may be called the astro-physical estimate in which if two quantities are within one or two order of magnitude, they are not negligible with respect to one another. Only when the difference between them is of several order of magnitude, the smaller one is negligible. Hence from table 1, we see that when the temperature is less than 10^4 °K, the radiation energy density is negligibly small in comparison with the internal energy under this given density while for high temperature, the radiation energy density may be of the same order of magnitude or even larger than the ordinary internal energy of the gas. If we increase the density of the fluid, the temperature above which the radiation energy becomes important increases too. It is easy to show that whenever the radiation energy density is not negligible, the radiation pressure is also not negligible. Most of the flow problems of current interest such as reentry problem have the conditions that the maximum temperature is of the order of 10^4 °K and the density of the fluid is higher than 1.23×10^{-6} gr. cc. Hence the radiation pressure and the radiation energy density are still negligible. However, for the cases of higher temperature such as in the fusion research where 10^6 °K is a relative low temperature and for the cases of lower density such as in the outer space, the radiation pressure and radiation energy density should be considered in the analysis of the flow problem.

(iii) Flux of radiation. For the radiation equilibrium condition, the radiation flux is given by the formula (cf. chapter II, § 2)

$$q_R = \frac{c}{4} \frac{E_R}{\rho} = \frac{\sigma T^4}{\rho} \tag{1.3}$$

where c is the velocity of light which is 3×10^{10} cm/sec. in vacuum. The factor $\sigma = c\, a_R/4 = 5.75 \times 10^{-5}$ erg-cm^{-2} — sec^{-1} — °K^{-0} is also referred to as the Stefan-Boltzmann constant which is used quite often in the literature to show the radiative heat transfer. The term q_R should be compared with the heat flux by convection, q_v, i.e.,

$$q_v = U\, c_v\, T \tag{1.4}$$

where U is a typical flow velocity. If we take the same conditions used in table 1 and assume a mean flow velocity $U = 10^4$ m/sec. which is an average speed of a satellite, we obtain the values in table 2 as follows:

1*

Table 2. *Radiation heat flux vs heat convection flux*

T °K	10^3	10^4	10^5	10^6
q_R erg-cm/gr-sec.	4.5×10^{13}	4.5×10^{17}	4.5×10^{21}	4.5×10^{25}
q_v erg-cm/gr-sec.	7×10^{15}	7×10^{16}	7×10^{17}	7×10^{38}

From the values in table 2, we see that at the temperature $T = 10^4$ °K, the heat transfer by radiation is about the same order of magnitude as that by convection. Hence the radiation heat transfer should be considered in the analysis of the flow problem under this condition even though the radiation pressure and radiation energy density are still negligible.

In many practical cases, the mean free path of radiation [cf. eq. (5.12)] is not small and we have the cases of thermal radiation of optically thin medium. The formulas for the radiation flux and radiation stresses for optically thin cases are different from those given by equations (1.1) and (1.3) (cf. chapter II). In general, these formulas should be expressed in terms of integrals which will be derived in chapter II. For the case of optically very thin medium, a simple formula may also be obtained. Near the stagnation point, the radiation heat flux may be expressed by the following formula (8) (cf. chapter IX):

$$q_R = I \frac{R_b \, \rho_\infty}{2 \, \rho_0} \tag{1.5}$$

where I is the integrated intensity of radiation, R_b is the nose radius of the blunt body which is proportional to the detached distance of the shock wave. At the stagnation point, ρ_∞ and ρ_0 are respectively the density in front and behind the normal shock. The intensity I is computed according to statistical mechanics for thermodynamic equilibrium and is a function of the stagnation temperature and density. In reference 8, the comparison of the aerodynamic heating and radiative heating based on equation (1.5) for an ICBM and a Mars probe are given. The results are shown respectively in Figs. 1.1 and 1.2. They show that for an ICBM, the radiative heating is no longer negligibly small in comparison with the aerodynamic heating and for a Mars probe, the radiative heating is much larger than the aerodynamic heating.

Both of the above estimations give only the order of magnitude of the thermal radiation and are not very accurate because the interaction between the radiation field and the flow field is not taken into account. Whenever the effect of the radiation terms is not small, we have to take the interaction between the radiation field and the flow field into account in order to have accurate results (7, 9, 10, 11, 13, 14, 16). It is the main purpose of this book to investigate the interaction phenomena between the radiation field and the flow field.

3. Some thermal radiation phenomena. Generally speaking, the thermal radiation is a far more complicated phenomenon than the conduction of heat or the convection of heat. Heat convection depends on the velocity field of the fluid and heat conduction depends essentially on the temperature distribution of the medium. Thermal radiation depends on the temperature of the medium in a very complicated manner. There are cases in which the radiation of heat is apparently independent of the temperature of the medium through which it passes. For instance, we may concentrate the solar rays at a focus by passing them through a converging lens of ice which remains at a constant temperature of 0 °C and ignite an inflammable body.

Radiation of heat is a mode of transfer of heat by photons or waves. The creation and destruction of photons or radiation rays depend on the interaction

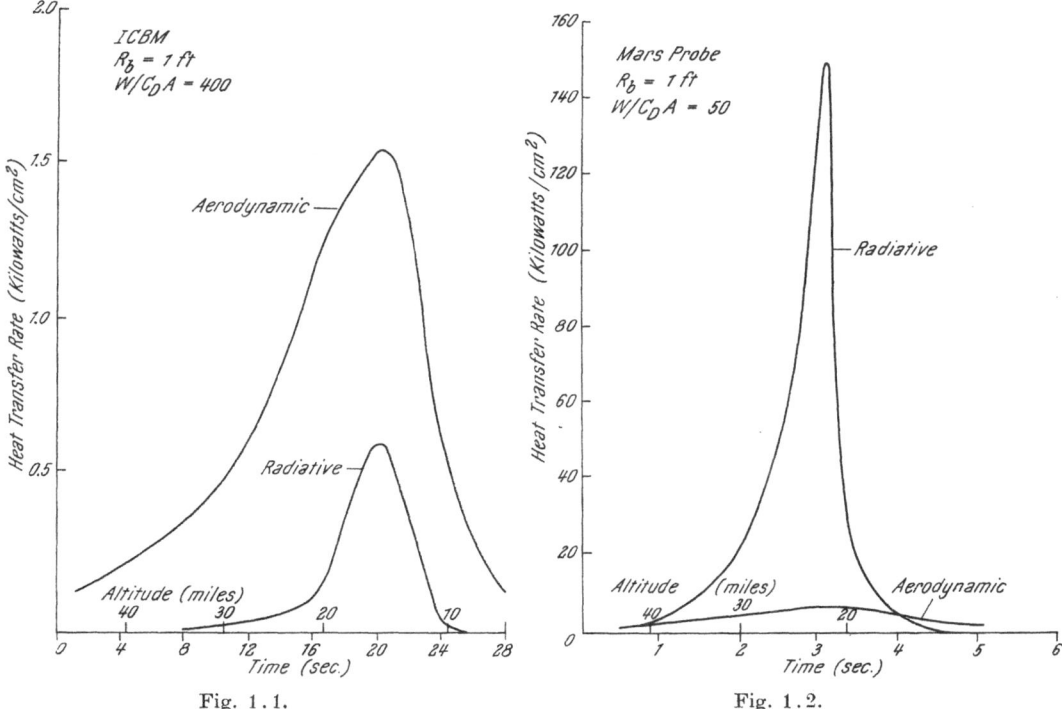

Fig. 1.1.   Fig. 1.2.

Fig. 1.1. Comparison of aerodynamic and radiative heat transfer during re-entry for a typical ICBM. (Fig. 5 of reference 8 by Bennett KIVEL, Courtesy of AIAA and AVCO)

Fig. 1.2. Comparison of radiative and aerodynamic heating during re-entry for a Mars probe. (Fig. 6 of reference 8 by Bennett KIVEL, Courtesy of AIAA and AVCO)

between the energy and matter of the gas particles which should be studied from microscopic point of view by quantum mechanics. Basically there is no difference between the heat wave, electromagnetic wave and light wave. Hence we may study the radiation of heat by considering the radiation of electromagnetic energy from the atoms of gas particles by solving the Maxwell equation of electromagnetic fields. Even though such a study is very useful in the basic understanding of the radiation phenomena, it is too detailed to be useful in the treatment of the flow problems. For flow problems of engineering practice, we are interested in the macroscopic point of view. In other words, we are not interested in the detailed nature of the wave pattern in flow field but the overall effects of thermal radiation on the gross variables such as temperature, velocity etc. of the fluid. One way of description of the radiation of heat from the macroscopic point of view is to represent the radiation of heat by heat rays according to the concept of geometrical optics (*7, 11*). Each ray travels in a specific direction. In a homogeneous isotropic medium, the rays travel in straight lines with the same velocity, the velocity of light, in all directions. In order to specify completely the state of radiation, the intensity of rays of radiation must be known in all directions, infinite in number, which pass through the point considered. The resultant effect is due to the integral of all the rays. As we shall see in chapter II, the general expression of radiation term is usually a complicated integral. In macroscopic treatment, the creation and destruction of radiation rays may be expressed in terms of coefficient of emission and coefficient of extinction or absorption which are functions of state of the gas. These coefficients of emission and extinction of radiation are similar to those coefficients of viscosity and of heat conductivity in ordinary gasdynamics. One of the objects of this book is to formulate the radiation field in terms of these radiation coefficients and then to solve them simultaneously with other gasdynamic variables.

The coefficient of emission and that of absorption of radiation should be determined from the microscopic treatment of quantum mechanics. It is not the intention of the author to give the detailed description of these treatments. However, a brief review of the essential features as well as the main results for high temperature gases (*4*) will be given in chapter XI. In general such a theoretical calculation is very complicated and many assumptions should be made as to the important processes in the interaction of energy and matter of the gas particles. These radiation coefficients may be determined experimentally. Many useful information of emission and absorption of radiation may be obtained by experiments of high temperature gases. We shall also discuss some of the results in chapter XI.

There are a great number of radiation phenomena such as fluorescence, Phosphorescence, etc. besides the thermal radiation. We shall, however, consider only the thermal radiation for which the emission coefficient of radiation depends, apart from the frequency ν and the nature of the medium, only on the temperature T of the medium.

References

1. ALLEN, R. A., P. H. ROSE, and J. C. CAMM: Non-equilibrium radiation at super-satellite re-entry velocities. I. A. S. paper 63–77, Institute of the Aero. Sci. 1963.

2. ALLER, L. H.: Astrophysics: The atmospheres of the sun and stars. Ronald Press Co., New York, second edition, 1963.

3. AMBARTSUMYAN, V. A.: Theoretical Astrophysics. Pergamon Press, New York, 1958.

4. ARMSTRONG, B. H., SOKOLOFF, R. W. NICHOLLS, H. D. HOLLAND, and R. E. MEYEROTT: Radiative properties of high temperature air. Jour. Quan. Spect. Radi. Transfer, Vol. 1, pp. 143–162, Pergamon Press, 1960.

5. CHANDRASEKHAR, S.: Stellar Structures. Dover Publications, Inc., New York, 1957.

6. FAY, J. A., W. C. MOFFATT, and R. F. PROBSTEIN: An Analytical study of meteor entry. AIAA Jour. vol. 2, No. 5, pp. 845–854, 1964.

7. GOULARD, R.: Fundamental equations of radiation gas dynamics. Report A & ES 62-4, School of Aero. & Eng. Sci., Purdue Univ., 1962.

8. KIVEL, B.: Radiation from hot air and its effect on stagnation point heating. Jour. Aero. Sci. vol. 28, No. 2, pp. 96–102, Feb. 1961.

9. MAGEE, J. L., and J. D. HIRSCHEFELDER: Thermal radiation phenomena. Chap. III of Blast Wave. Los Alamos Lab. report LA-2000, 1958.

10. MAGHREBLIAN, R. V.: Thermal Radiation in gaseous fission reactors for propulsion. Tech. report No. 32-139, Jet Propulsion Lab., 1961.

11. PAI, S. I.: Thermal radiation effects on hypersonic flow field. Proc. of Non-linear prob. in Eng. Academic Press, pp. 163–183, 1964.

12. PLANCK, M.: The Theory of Heat Radiation. Dover Publications, Inc., New York, 1959.

13. POMERANTZ, J.: The influence of the absorption of radiation in shock tube phenomena. NAVORD Report 6136, U. S. N. O. L., 1958.

14. ROSE, P. H., and J. D. TEARE: On chemical effects and radiation in hypersonic aerodynamics, AMP 72, AVCO Res. Lab., 1962.

15. ROSSELAND, S.: Theoretical Astrophysics. Oxford Press, 1936.

16. SCALA, S. M., and D. H. SAMPSON: Heat Transfer in hypersonic flow with radiation and chemical reaction. Techn. Inf. series R 63 SD 46, Space Sci. Lab. G. E., 1963.
Also, Supersonic Flow, Chemical Processes and Radiative Transfer, Pergamon Press, pp. 319−354, 1964

17. ÜNSOLD, A.: Physik der Sternatmosphären. Springer Verlag, Berlin, 1938.

18. VON KÁRMÁN, T.: From low Speed Aerodynamics to Astronautics. Pergamon Press, New York, 1963.

19. ZHIGULEV, V. N., YE. A. ROMISHEVSKII, and V. K. VERTUSHKIN: Role of radiation in modern Gasdynamics. AIAA Jour. vol. 1, No. 6, pp. 1473-1485, 1963.

Chapter II
Fundamentals of Radiative Transfer

1. Specific intensity (*2, 3*). We shall consider the thermal radiation pheno-mena from macroscopic point of view. Hence our linear dimension in the flow field is larger compared with the wave length of the radiation rays and our time scale is larger compared with the period of all frequencies contained in the radia-tion rays. In analogy of the theory of gasdynamics, our present theory of thermal radiation is similar to the continuum theory of gasdynamics instead of the kinetic theory of gas. Without further discussion of the foundation of this theory of thermal radiation, we shall represent the radiation of heat by heat rays in the following chapters. For those readers who would like to know more the limitation of the scales of length and time in the present theory, the classical book on the theory of heat radiation by PLANCK (reference *8*) should be referred to.

The heat rays may be specified by a specific intensity I_ν. Let $d\sigma_0$ be an arbitrarily oriented small area, P be a point on this area and $\vec{n}$ be the normal of the area at point P. At a given instant of time, there will be heat rays, trav-ersing this element in all directions. Let us consider a specific direction along which we draw a line L which makes an angle θ to the normal $\vec{n}$. We take L as the axis of an elementary cone of solid angle $d\omega$ (Fig. 2.1). Through each point of $d\sigma_0$ we construct cones having axes parallel to the line L with solid angle at the apex all equal to $d\omega$. These cones define a truncated semi-infinite cone $d\Omega$, whose cross sectional area perpendicular to L at the point P will be $d\sigma_0 \cos\theta$. Let $dE\nu$ be the total amount of energy passing through the area $d\sigma_0$ inside the cone $d\Omega$ in the time dt and in the frequency range between ν and $\nu + d\nu$. The specific intensity of radiation or simply the intensity is de-fined as

$$I_\nu = \lim_{d\sigma_0, d\omega, dt, d\nu \to 0} \left(\frac{dE\nu}{d\sigma_0 \cos\theta \cdot d\omega \cdot dt \cdot d\nu} \right) \tag{2.1}$$

This limit is in general a function of the position P, the direction L, the time t and the frequency ν. One of the main objects of radiative transfer study is to find the specific intensity I_ν for a given physical problem. If I_ν is independent of the direction L, the radiation field is said to be isotropic, while if I_ν is in-dependent of both the position and the direction L, the radiation field is said to be homogeneous and isotropic.

If we know the intensity I_ν, the amount of energy flowing through the area $d\sigma_0$ in the frequency range (ν and $\nu + d\nu$) and in the direction L within the elementary solid angle $d\omega$ and in the time interval dt is

$$dE_\nu = I_\nu \cos\theta \, d\sigma_0 \, d\omega \, d\nu \, dt \tag{2.2}$$

If we define an integrated intensity I such that

$$I = \int_0^\infty I_\nu \, d\nu \qquad (2.3)$$

the total amount of energy radiated over the whole spectrum is

$$dE = \int_0^\infty \left(\frac{dE_\nu}{d\nu}\right) d\nu = I \cos\theta \, d\sigma_0 \, d\omega \, dt \qquad (2.4)$$

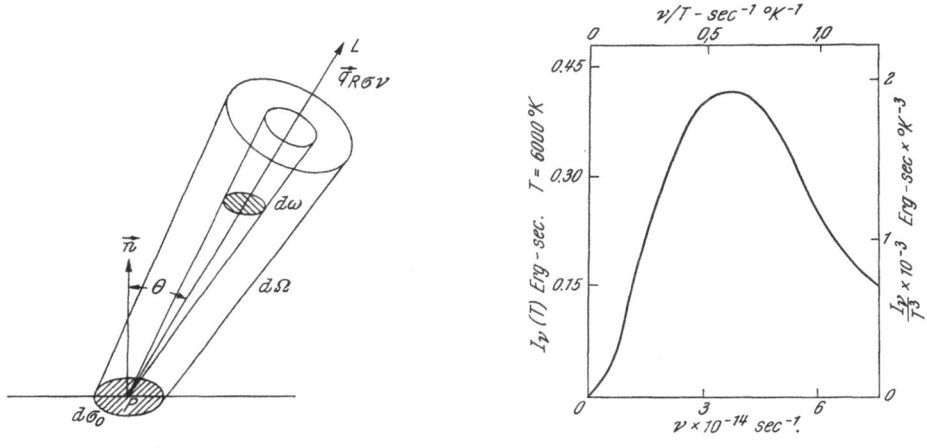

Fig. 2.1.   Fig. 2.2.

Fig. 2.1. Heat rays in a radiation field

Fig. 2.2. The specific intensity of a black body

As we shall see in chapter IV, one of the ideal radiators is known as a black body. The specific intensity of a black body is a function of temperature only [see equation (4.22)]. Fig. 2.2 gives the specific intensity of a black body against frequency at a temperature of 6,000 °K. We shall discuss this curve further in chapter IV. Now we shall only mention a few essential points. The specific intensity drops to zero at both the high frequency and the low frequency ends. There is a maximum in the intermediate frequency. The main source of radiation energy on the earth is the sun (4, 5). The solar spectra measured at the ground and corrected to the scattering and absorbing limit of the earth's upper atmosphere were found to be close to the spectrum curve of a black body at 6,000 °K over a large range of frequency or wave length. Hence the study of the black body radiation would be of great interest. Furthermore, we shall show in chapter V, the black body radiation is a good approximation for the optically thick medium under the assumption of local thermodynamic equilibrium.

2. The flux of radiation (6, 7). The net flux of radiation across $d\sigma_0$ per unit area and per unit time in the direction of L is the total amount of energy radiated over the whole spectrum in all directions, i.e.,

$$q_{R_\sigma} = \int \left(\frac{dE}{d\sigma_0 \, dt}\right) = \int I \cos\theta \, d\omega \qquad (2.5)$$

The direction L can be completely specified by the angular variables θ $(0 \leqslant \theta \leqslant \pi)$ and the azimuth angle ϕ $(0 \leqslant \phi \leqslant 2\pi)$. The elementary solid angle dω defined by the range $(\theta + \mathrm{d}\,\theta)$ and $(\phi + \mathrm{d}\,\phi)$ is

$$\mathrm{d}\,\omega = \sin\theta \; \mathrm{d}\theta \; \mathrm{d}\phi \qquad (2.6)$$

Hence equation (2.5) becomes

$$q_{R_3}(\theta, \phi, r, t) = \int\limits_0^\infty \int\limits_0^{2\pi} \int\limits_0^\pi I_\nu(\theta, \phi, r, t) \; \sin\theta \; \cos\theta \; \mathrm{d}\theta \; \mathrm{d}\phi \; \mathrm{d}\nu \qquad (2.7)$$

The flux of radiation is a vector quantity whose three components along the x-, y- and z-axis are respectively

$$q_{R_x} = \int I\,l\,\mathrm{d}\omega; \quad q_{R_y} = \int I\,m\,\mathrm{d}\omega; \quad q_{R_z} = \int I\,n\,\mathrm{d}\omega \qquad (2.8)$$

where l, m, and n be the direction cosine of the line of ray L, with respect to x-, y- and z-axis respectively.

In vector form, the resultant radiation flux is

$$\vec{q}_R = \vec{i}\,q_{R_x} + \vec{j}\,q_{R_y} + \vec{k}\,q_{R_z} \qquad (2.9)$$

where $\vec{i}$, $\vec{j}$ and $\vec{k}$ are respectively the unit vector along the x-, y- and z-axis.

The energy radiated out of a unit volume dx dy dz at the point P is

$$Q_R = \nabla \cdot \vec{q}_R = \frac{\partial q_{R_x}}{\partial x} + \frac{\partial q_{R_y}}{\partial y} + \frac{\partial q_{R_z}}{\partial z} \qquad (2.10)$$

The term Q_R should be added to the energy equation of gasdynamics if the radiative transfer is not negligible. This term is similar to the heat transfer by conduction in gasdynamics, Q_c. When the mean free path of the medium is large, Q_c should be expressed in the integral form while the mean free path of the medium is small, Q_c may be expressed in terms of thermal conductivity coefficient $\varkappa$ and the gradient of temperature, i.e., $Q_c = \nabla \cdot (\varkappa \nabla T)$. Similarly for finite mean free path of radiation [cf. equation (5.12)], the heat transfer by radiation Q_R or $\vec{q}_R$ should be expressed in the integral form as shown in equation (2.8) while for small mean free path of radiation, Q_R may be expressed in terms of radiation diffusion coefficient D_R and the gradient of temperature (9), i.e., $Q_R = \nabla \cdot (D_R \nabla E_R)$ where E_R is the radiation energy density given in equation (1.2).

In many of the engineering problems as well as astrophysical problems, we consider the radiative transfer with respect to a straight plane (Fig. 2.3). It is convenient to split the net radiative flux $q_{R\sigma}$ into two parts: one part $q_{R\sigma}^+$ represents the contribution coming from the side of the normal unit vector $\vec{n}$ and the other part $q_{R\sigma}^-$ represents the contribution from the opposite side, i.e.,

$$q_{R3}^+ = \int\limits_0^\infty \int\limits_0^{2\pi} \int\limits_0^{\pi/2} I_\nu(\theta, \phi \cdot r, t) \; \sin\theta \; \cos\theta \; \mathrm{d}\theta \; \mathrm{d}\phi \; \mathrm{d}\nu \qquad (2.7\,\mathrm{a})$$

$$q_{R3}^- = -\int\limits_0^\infty \int\limits_0^{2\pi} \int\limits_{\pi/2}^\pi I_\nu(\theta, \phi, r \cdot t) \; \sin\theta \; \cos\theta \; \mathrm{d}\theta \; \mathrm{d}\phi \; \mathrm{d}\nu \qquad (2.7\,\mathrm{b})$$

and the net radiative flux is

$$q_{R_3} = q_{R_3}^{+} - q_{R_3}^{-} \tag{2.7 c}$$

Customarily, we choose the normal $\vec{n}$ toward the wall (Fig. 2.3). Hence $q_{R_3}^{+}$ is usually referred to as the rays toward the wall while $q_{R_3}^{-}$ is referred to as the rays away from the wall. The net flux is the flux toward the wall.

3. Energy density of radiation, U_ν. The energy density of radiation U_ν is the amount of radiant energy per unit volume in the state frequency interval

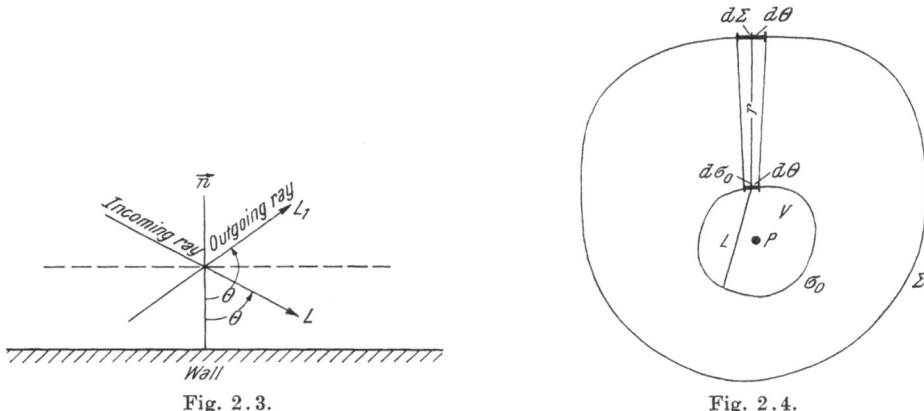

Fig. 2.3.　　　　　　　　Fig. 2.4.

Fig. 2.3. Incoming and outgoing rays with respect to a straight wall
Fig. 2.4. Radiative energy in a small volume bounded by a surface σ_0 at a point P in the medium

which is on course of transit in the neighborhood of the point considered. In Fig. 2.4 we consider an infinitesimal volume with surface area σ_0 around a point P. We assume that the surface σ_0 is convex everywhere. The surface σ_0 is surrounded by another convex area Σ. Both σ_0 and Σ are assumed to be small so that for first approximation, the specific intensity I_ν in the volume enclosed by Σ is constant. The radiation energy traversing the volume surrounded by σ_0 must have crossed some element of the surface Σ. We assume that the radiation ray which passes through $d\Sigma$ on the surface Σ will pass through $d\sigma_0$ on the surface σ_0. By equation (2.2) the energy flowing through $d\Sigma$ is then

$$dE_\nu = I_\nu \cos\Theta \, d\omega' \, d\nu \, dt \tag{2.11}$$

where Θ is the angle between the normal of area $d\Sigma$ and the ray r. The elementary solid angle $d\omega'$ is $d\sigma_0 \cos\theta/r^2$ and θ is the angle between the normal of $d\sigma_0$ and r and the elementary time dt should be considered the time through which the ray passes through the volume surrounded by σ_0, i.e., $dt = L/c$. Then equation (2.11) becomes

$$dE_\nu = I_\nu \, d\nu \, (d\sigma_0 \cos\theta \cdot L) \left(\frac{d\Sigma \cos\Theta}{r^2}\right) \frac{1}{c} = \frac{1}{c} I_\nu \, d\nu \, dV \, d\omega \tag{2.12}$$

where V is the volume surrounded by σ_0 and $d\omega = d\Sigma \cos\Theta/r^2$. Now we may define an energy density U_ν as follows:

$$V U_\nu d\nu = \frac{1}{c} \int I_\nu d\nu \, dV \, d\omega = \frac{V}{c} \int I_\nu \, d\omega \, d\nu$$

or   (2.13)

$$U_\nu = \frac{1}{c} \int I_\nu \, d\omega$$

The total energy density of radiation for the whole spectrum is

$$E_R = \int_0^\infty U_\nu \, d\nu = \frac{1}{c} \int I \, d\omega \qquad (2.14)$$

Thus E_R is the radiation energy per unit volume associated with the point P. It is similar to the internal energy of the gas. Hence E_R should be considered as a part of the total energy of the gas at the point considered in the energy balance of the gas flow.

4. The stress tensor of radiation. A quantum of energy $h\nu$ is associated with a momentum $\dfrac{h\nu}{c}$. Let us consider a surface whose normal is in the direction of x-axis. A heat ray of directional cosines l, m and n with respect to x-, y- and z-axis respectively passes through this surface. The total intensity of the ray is I. The rate of change of momentum associated to the energy is then

$$M_0 = \frac{1}{c} I l \, d\omega \qquad (2.15)$$

the normal component of the momentum M_0 transferred across the area is the normal stress on the surface $d\sigma_0$ in the x-direction, i.e.,

$$d\tau_{Rxx} = -\frac{1}{c} I l \, d\omega \, l \qquad (2.16)$$

The total normal radiation stress in the x-direction at the point P is the sum of the elementary normal stress over the complete sphere, i.e.,

$$\tau_{Rxx} = -\frac{1}{c} \int I l^2 \, d\omega \qquad (2.17)$$

Similarly, the stress tensor due to the radiation ray has the ijth component

$$\tau_{Rij} = -\frac{1}{c} \int I n_i n_j \, d\omega \qquad (2.18)$$

where n_i is the direction cosine of the heat ray with respect to the ith axis.

We may define a radiation pressure $\overline{p}_R$ such that

$$\overline{p}_R = -\frac{1}{3} (\tau_{Rxx} + \tau_{Ryy} + \tau_{Rzz}) = \frac{1}{3c} \int I \, d\omega = \frac{1}{3} E_R \qquad (2.19)$$

For isotropic radiation, we have

$$p_R = -\tau_{Rxx} = -\tau_{Ryy} = -\tau_{Rzz} = \frac{1}{3} E_R \qquad (2.20)$$

$$\tau_{Rij} = 0 \ \text{if} \ i \neq j$$

Both radiation pressure p_R and radiation energy density E_R are proportional to $1/c$. Hence they are of the same order of magnitude and are usually small in ordinary gasdynamics except in the case where the temperature is enormously high. If the radiation stress tensor is not negligible, we should add it in the equation of motion of radiation gasdynamics.

The body force due to the radiation stress τ_{Rij} is the divergence of the radiation stress tensor. Hence the ith component of the radiation body force is

$$F_{R_i} = \frac{\sigma \tau_{Rij}}{\partial x_j} \qquad (2.21)$$

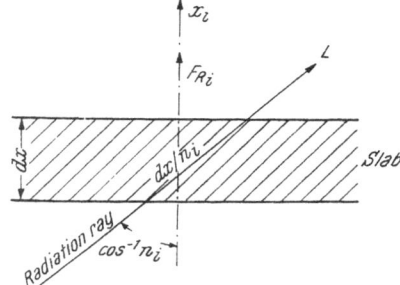

Fig. 2.5. A radiation ray passing through a slab of a medium

There the summation convention is used. In literature, some restricted expression for the body forces F_{Ri} has been used, such as the following formula (1, 10):

$$F_{Ri} = \frac{1}{c} \int_\nu \int_\omega k'_\nu \rho \, I_\nu (\theta, \phi, r, t) \, n_i \, \mathrm{d}\omega \, \mathrm{d}\nu \qquad (2.22)$$

where $k'_\nu \rho$ is the mass absorption coefficient allowing for the induced radiation which will be discussed in chapter III, and n_i, the directional cosine of the radiation ray with respect to the ith axis. Equation (2.22) may be obtained by considering a thin plane slab of thickness $\mathrm{d}x$ (Fig. 2.5), the amount of radiation energy absorbed by the slab in the direction L [cf. chapter III, eq. (3.1)] is

$$\mathrm{d}E_a = I_\nu \, n_i \, \rho \, k'_\nu \frac{\mathrm{d}x}{n_i} \, \mathrm{d}\omega \, \mathrm{d}\nu \qquad (2.23)$$

The force normal to each unit area of the slab is

$$\mathrm{d}\tau_{Ri} = \int_\nu \int_w \mathrm{d}E_a \, n_i/c = \frac{\mathrm{d}x}{c} \int_\nu \int_\omega k'_\nu \rho \, I_\nu \, n_i \, \mathrm{d}\omega \, \mathrm{d}\nu = \frac{\mathrm{d}x_i}{c} \, F_{Ri} \qquad (2.24)$$

The body force per unit volume is then

$$F_{Ri} = \frac{\mathrm{d}\tau_{Rii}}{\mathrm{d}x} = \frac{1}{c} \int_\nu \int_\omega \rho \, k'_\nu \, I_\nu \, n_i \, \mathrm{d}\omega \, \mathrm{d}\nu \qquad (2.25)$$

Equation (2.25) is rather a restricted case of equation (2.21). We shall use equation (2.21) in our equations of motion (cf. chapter V, § 4).

References

1. ALLER, L. H.: Astrophysics: The atmospheres of the sun and stars. Ronald Press Co., New York, second edition, 1963.
2. AMBARTSUMYAN, V. A.: Theoretical Astrophysics. Pergamon Press, New York, 1958.
3. CHANDRASEKHAR, S.: Stellar Structures. Dover Publications, In., New York, 1957.
4. GODSKE, C. L., T. BERGERON, J. BJERKNES, and R. C. BUNDGAARD: Dynamic Meteorology and weather forecasting. American Meteorological society and Carnegie Institution of Washington, 1957.
5. JOHNSON, J. C.: Physical Meteorology. John Willey & Sons, New York, 1954.
6. KOURGANOFF, V.: Methods in Transfer problems. Oxford Clarendon Press, 1952.
7. PAI, S. I.: Some considerations on radiation magnetogasdynamics. Proc. Non-linear Problem, Univ. of Wisconsin Press, pp. 47–67, 1963.
8. PLANCK, M.: The theory of heat radiation. Dover Publications, Inc., New York, 1959.
9. ROSSELAND, S.: Theoretical Astrophysics. Oxford Clarendon Press, 1936.
10. ZHIGULEV, V. N., YE. A. ROMISHEVSKII, and V. K. VERTUSHKIN: Role of Radiation in modern gasdynamics, AIAA Jour. vol. 1, No. 6, pp. 1473–1485, 1963.

Chapter III
Equation of Transfer of Radiation

1. Introduction. All of the terms of radiative transfer discussed in chapter II are expressed in terms of the specific intensity I_ν. Hence we have to find a relation to determine the specific intensity. The specific intensity is the result ·of the interaction between radiation and matter which should be studied by the quantum mechanics from the microscopic point of view. However for practical flow problem, our main interest is from the macroscopic point of view. The exact results of microscopic analysis may be represented in the average by means of some coefficients which represent the average physical properties of the medium considered. For the radiative transfer problems, the physical properties of the medium can be expressed in terms of an absorption coefficient k_ν and an emission coefficient j_ν. In the macroscopic analysis, we assume that these coefficients are known functions of the state of the medium (5, 7, 9). The exact form of these functions can be determined either from the microscopic analysis or from experiment. We shall discuss the determination of these coefficients in chapter XI. In this chapter, we shall assume that the medium is homogeneous so that the properties of absorption and emission change continuously in the medium. Special attention should be made for the case of radiation field with discontinuous media. For instance, when a solid body is in a gaseous medium, the radiative properties change suddenly at the surface of the body from those of the gas. We shall discuss the interface problems in chapter VI.

2. Absorption coefficient (*1, 2*). k_ν. In Fig. 3.1, we consider a ray of radiation of specific intensity I_ν entering an elementary volume of a homogeneous medium of density ρ. The intensity of this ray at a short distance ds from the entering point becomes $I_\nu + dI_\nu$. The change of specific intensity dI_ν may be expressed in terms of an absorption coefficient k_ν by the following formula:

$$dI_\nu = -k_\nu \rho I_\nu ds \qquad (3.1)$$

Equation (3.1) can be integrated from an initial point $s = s_0$ and $I_\nu = I_\nu(s_0)$, i.e.,

$$I_\nu(s) = I_\nu(s_0) \exp\left(-\int_{s_0}^{s} \rho k_\nu ds\right) = I_\nu(s_0) \exp(-\tau_\nu) \qquad (3.2)$$

where $\tau_\nu = \int_{s_0}^{s} k_\nu \rho ds = $ optical depth of the layer $\qquad (3.3)$

The optical depth τ_ν of the layer, sometimes called the optical thickness of the layer, is a non-dimensional length which characterizes the absorption of the radiation in the medium. For a given physical depth $(s - s_0)$, the optical depth

τ_ν may be larger or smaller than the physical length depending on the value $(k_\nu \rho)$. It is often to introduce a mean free path of radiation $L_{R\nu}$ such that

$$L_{R\nu} = \frac{1}{k_\nu \rho} \tag{3.4}$$

The mean free path of radiation is similar to the mean free path in the kinetic theory of gases. The mean free path of radiation represents an average distance of collision of a photon and a molecule while the ordinary mean free path represents the average distance of collision between molecules. Both the mean free path of radiation $L_{R\nu}$ and the optical depth τ_ν are functions of the frequency

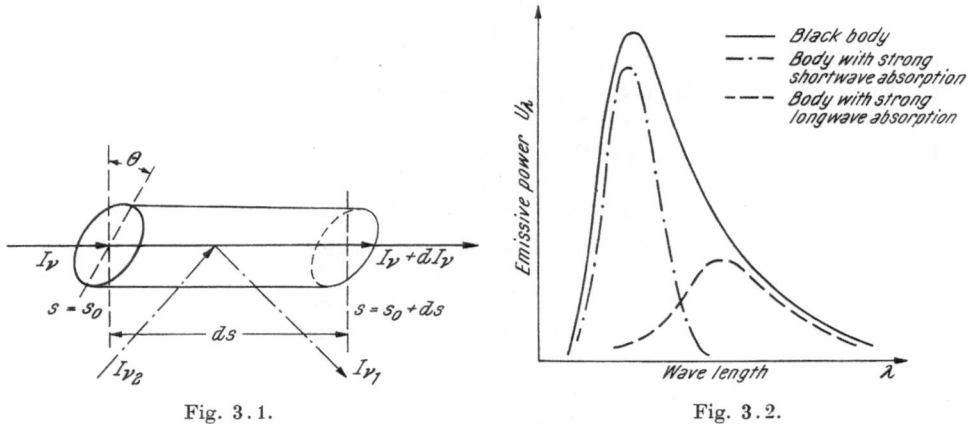

Fig. 3.1.   Fig. 3.2.

Fig. 3.1. Radiation rays in a medium

Fig. 3.2. Emissive power of three different media of different absorption at a given temperature

of radiation ν and the state variables such as temperature and density of the medium. Sometimes it is convenient to take the average value of $L_{R\nu}$ over the whole frequency range. We shall discuss such average values later because it depends on the situation whether the medium is optically thick or optically thin. When $L_{R\nu}$ is small in comparison with the physical dimension, the medium is said to be optically thick, on the other hand when $L_{R\nu}$ is large in comparison with the physical dimension, the medium is said to be optically thin.

For a black body, the radiation intensity or the emissive power depends only on the temperature T and the frequency ν or wave length $\lambda = c/\nu$ (see Fig. 2.2 and also chapter IV). For all other media, the radiation intensity or emissive power depends not only on temperature and frequency but also on the properties of the medium, i.e., its absorption coefficient k_ν. Since k_ν is a function of both temperature and frequency, it is not possible to draw a single curve for the emissive power of any real medium at all temperatures (cf. chapter IV, § 4), as in case of the black body radiation. However, the black body radiation represents the maximum emissive power at a given temperature T. Fig. 3.2 shows a few typical cases of actual medium. For those bodies with strong short wave absorption, the emissive power is closer to the black body values at short wave length, while for those with strong long wave absorption the emissive power is closer to the black body value at large wave length (3, 4).

The absorption coefficient k_ν consists of two parts: one is the true absorption coefficient $k_{\nu t}$ which represents the fact that the radiation energy of frequency ν has been transformed into other form of energy or radiation of other frequency, and the other is the scattering coefficients $k_{\nu s}$ which represents the fact that the energy lost from the incident ray will reappear as scattered radiation in other direction such as the ray $I_{\nu 1}$ in Fig. 3.1. Hence in general, we may write:

$$k_\nu = k_{\nu t} + k_{\nu s} \qquad (3.5)$$

Because $k_{\nu t}$ and $k_{\nu s}$ depend on different physical phenomena, we have to discuss them separately.

(i) True absorption coefficient $k_{\nu t}$. The true absorption coefficient represents the absorption of photon by atoms or molecules exposed in the radiation field. The true absorption coefficient may be expressed in terms of Einstein coefficient B_{mn} which represents the probability p_a of an atom in a state "m" exposed to radiation of frequency ν_{mn} absorbing a quantum $h\,\nu_{mn}$ in time $d\,t$, i.e.,

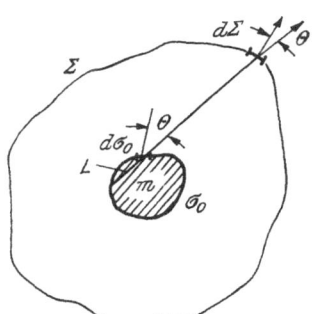

Fig. 3.3. Absorption of radiation energy

$$p_a = B_{mn}\,\mathrm{d}t \int I_{\nu mn}\,\mathrm{d}\omega \qquad (3.6)$$

Now we consider a small element of the medium of mass m, volume V and number density N_m for the atoms or molecules at state "m," the total energy absorbed by this element of the medium for the radiation of the frequency ν_{nm}

$$E_a = N_m\,V \cdot B_{mn}\,h\nu_{nm}\,\mathrm{d}\nu\,\mathrm{d}t \int I_{\nu nm}\,\mathrm{d}\omega \qquad (3.7)$$

The energy E_a of equation (3.7) may be also expressed in terms of the absorption coefficient. We may consider that the radiation absorbed by the element comes from a small closed surface Σ. The energy of radiation from an elementary surface $\mathrm{d}\,\Sigma$ on the surface Σ to an elementary area $\mathrm{d}\,\sigma_0$ of the element considered (Fig. 3.3) is

$$\mathrm{d}\,E_\nu = I_\nu \cos\theta\,\mathrm{d}\sigma_0 \cdot \frac{\cos\Theta\,\mathrm{d}\Sigma}{r^2}\,\mathrm{d}\nu\,\mathrm{d}t \qquad (3.8)$$

The energy absorbed by the element m from the ray of equation (3.8) is

$$\mathrm{d}\,E_a = (\mathrm{d}\,E_\nu) \cdot k_{\nu t}\,\rho\,L = I_\nu\,k_{\nu t}\,\mathrm{d}\,\omega\,\mathrm{d}m\,\mathrm{d}\nu\,\mathrm{d}t \qquad (3.9)$$

where $\mathrm{d}m = \rho\,L \cos\theta\,\mathrm{d}\sigma_0 =$ element of mass of the medium considered and L is the distance in m over which the ray travels.

The total energy absorbed by m is obtained by integrating the expression (3.9). Since we choose the element to be very small, we may assume the density and the specific intensity as constants. As a result, the integration of equation (3.9) gives

$$\mathrm{d}\,E_a = k_{\nu t}\,\rho\,V\,\mathrm{d}\nu\,\mathrm{d}t \int I_{\nu nm}\,\mathrm{d}\omega \qquad (3.10)$$

Comparing equations (3.7) and (3.10), we have the absorption coefficient for the initial state "m" is

$$k_{\nu t m} = \frac{N_m}{\rho} B_{mn} h\nu \qquad (3.11)$$

where we put $\nu = \nu_{nm}$. The total true absorption coefficient of a given medium which consists of molecules at various initial state "m" is the sum of $k_{\nu t m}$ over all initial states "m"

$$k_{\nu t} = \sum_m k_{\nu t m} \qquad (3.12)$$

where the difference of the energy level at the final state "n" after the absorption of the radiation energy from that at the initial state "m" is

$$E_n - E_m = h\nu \qquad (3.13)$$

In literature, we often take the product of the true absorption coefficient $k_{\nu t}$ and the density of the medium ρ and call it the linear absorption coefficient, because the dimension of $(1/\rho \, k_\nu)$ is a length [see eq. (3.4)], i.e.

$$K_{\nu t} = \rho \, k_{\nu t} = \sum_m N_m \sigma_m \qquad (3.14)$$

where

$$\sigma_m = B_{mn} h\nu \qquad (3.15)$$

is known as the cross section for absorption of photons of frequency ν by the molecules at state m. We may determine the cross section σ_m from the microscopic theory or experiments as we shall discuss in chapter XI. For macroscopic analysis, we may assume that the linear absorption coefficient K_ν is a known function of the frequency ν, and the state of the medium, i.e.,

$$K_\nu = K_\nu (\nu, T, \rho) \qquad (3.16)$$

(ii) Scattering coefficient $k_{\nu s}$. If we assume that the medium in radiation gasdynamics is a homogeneous medium in which the density varies in a smooth fashion from point to point, no scattering phenomenon will occur. However, in actual medium, small optical obstacles do exist. For instance, in the atmosphere, there are small smoke and dust particles. If the dimensions of these optical obstacles are of the same order of magnitude as the wave length of the radiation rays, scattering phenomena occur. If one considers the radiation processes in the atmosphere, scattering is very important in explaining many important phenomena such as the blue of the sky, the change in color of distant objects, radar echoes from storms, the polarization of sky light etc. In the present stage of radiation gasdynamics, the scattering phenomena may be considered as second order phenomena because we consider mainly homogeneous medium. In the following chapters, we shall neglect the scattering phenomena in radiation gasdynamics except otherwise specially mentioned. For the sake of future development of radiation gasdynamics, we shall briefly discuss the scattering phenomen in this chapter.

The energy of radiation scattered from an element of mass of cross section area $d\sigma_0$ and a length in the direction of the radiation ray ds for time dt is

$$dE_s = k_{\nu s}\, \rho\, ds \cdot I_\nu \cos\theta\, d\sigma_0\, d\nu\, d\omega = k_{\nu s}\, dm\, d\nu\, d\omega \qquad (3.17)$$

in all directions.

The total energy of radiation absorbed is then

$$dE_{ab} = dE_a + dE_s = (k_{\nu t} + k_{\nu s})\, ds \cdot \rho \cos\theta \cdot d\sigma_0\, d\nu\, d\omega\, dt \qquad (3.18)$$

Sometimes we would like to know the angular distribution of the scattered radiation which may be expressed in terms of phase function $p\,(\cos\Theta)$ such that

$$dE_{ab} \cdot p\,(\cos\Theta)\frac{d\omega'}{4\pi} = k_{\nu s}\, I_\nu\, p\,(\cos\Theta)\frac{d\omega'}{4\pi}\, dm\, d\nu\, d\omega \qquad (3.19)$$

gives the rate at which the energy of radiation is being scattered into an element of solid angle $d\omega'$ and in a direction inclined at an angle Θ to the direction of the incident ray of radiation on an element of mass dm. In order that equation (3.19) agrees with equation (3.18), we should normalize the phase function such that

$$p\,(\cos\Theta)\frac{d\omega'}{4\pi} = \frac{k_{\nu s}}{k_\nu} = \omega_0 \qquad (3.20)$$

where ω_0 represents the fraction of energy loss due to scattering and it is known as albedo for single scattering. For perfect scattering, we have $\omega_0 = 1$. In general $(1 - \omega_0)$ represents the fraction of loss due to true absorption.

For isotropic scattering, we have

$$p\,(\cos\Theta) = \omega_0 \qquad (3.21)$$

The well known Rayleigh's phase function is

$$p\,(\cos\Theta) = \frac{3}{4}\,(1 + \cos^2\Theta) \qquad (3.22)$$

In general, we may develop the phase function in terms of Legendre polynomials p_n such that

$$p\,(\cos\Theta) = \sum_n \bar{\omega}_n\, p_n\,(\cos\Theta) \qquad (3.23)$$

where ω_n are constants.

For exact treatment of scattering, the state of polarization of the radiation field is important. In other words, besides the intensity of the radiation rays, we have to specify the degree of polarization, the plane of polarization and the ellipticity of the radiation. Since we shall neglect the scattering phenomena in most of our problems, we shall not discuss the polarization of radiation in detail. For those readers who are interested in this subject, the special treatises on radiative transfer such as references number 2 and 7 at the end of this chapter should be referred to. For a first approximation as in radiation gasdynamics, the scattering phenomena may be expressed in terms of the scattering coefficient $k_{\nu s}$ and the phase function $p\,(\cos\Theta)$.

3. Emission coefficient j_ν. The radiant energy $\mathrm{d}E_e$ may be emitted by the medium which may be expressed in terms of emission coefficient as follows:

$$\mathrm{d}E_e = j_\nu\,\mathrm{d}m\,\mathrm{d}\omega\,\mathrm{d}t\,\mathrm{d}\nu \qquad (3.24)$$

where $\mathrm{d}E_e$ is the radiant energy emitted by the medium of mass $\mathrm{d}m$ in the solid angle $\mathrm{d}\omega$ in the time $\mathrm{d}t$ and in the frequency range ν and $\nu + \mathrm{d}\nu$. The emission coefficient j_ν is a function of the frequency as well as the state variables of the medium. In general, the emission of radiation is not uniform in all direction even though the element of mass is isotropic. As we shall see later, only in the case of an isotropic medium in an isotropic radiation field, the emission of radiation will be uniform in all directions.

The emission coefficient j_ν also consists of two parts: one is due to the creation of photons from the matter and the other is due to the contribution of photons from scattering from all other directions into the direction of radiation ray under consideration. For instance the radiation ray I_{ν_2} in Fig. 3.1 represents this phenomenon.

(i) Creation of photons. Emission of radiation takes place when certain atom changes from a higher state n to a lower state m and emits a quantum $h\,\nu_{nm}$, i.e.,

$$E_n - E_m = h\nu_{nm} = h\nu$$

where E_n and E_m are respectively the energy level of the states "n" and "m" and h is the Planck's constant, which is 6.62×10^{-27} ergs/sec. The actual process of emission has to be calculated by quantum mechanics. However for macroscopic treatment, the emission of radiation can be expressed by Einstein coefficients which consist of two different types of emission: One is known as spontaneous emission coefficient A_{nm} and the other is known as the induced emission coefficient B_{nm}.

The absence of external radiation field, the probability of an atom in the excited state "n" to emit a quanta of energy $h\nu_{nm}$ in the elementary solid angle $\mathrm{d}\omega$ and in the time interval $\mathrm{d}t$ in terms of Einstein coefficient A_{nm} is

$$A_{nm}\,\mathrm{d}\omega\,\mathrm{d}t$$

This is the spontaneous emission which is uniform in all directions.

The probability of the emission of a quanta $h\nu_{nm}$ is increased if the atom in the state n is exposed in a field of radiation of frequency ν_{nm}. The probability of this induced emission is proportional to the intensity of radiation and is as follows

$$B_{nm}I_{\nu_{nm}}\,\mathrm{d}\omega\,\mathrm{d}t$$

The total emission of energy by a single atom in the state n per unit time and in the solid angle $\mathrm{d}\omega$ is then

$$h\nu_{nm}\left(A_{nm} + B_{nm}I_{\nu_{nm}}\right)\mathrm{d}\omega$$

Now if the number density of atoms in state "n" is N_n, the total energy emitted in the solid angle $\mathrm{d}\omega$ per unit time is

$$\mathrm{d}\omega \cdot N_n h\nu_{nm}\left(A_{nm} + B_{nm}I_{\nu_{nm}}\right) = \rho\,j_{\nu_{nm}}\,\mathrm{d}\omega \qquad (3.25)$$

Hence the emission coefficient j_ν in terms of Einstein coefficients A_{nm} and B_{nm} is

$$j_\nu = \frac{N_n}{\rho} h\nu \, (A_{nm} + B_{nm} I_\nu) = j_{\nu_c} \qquad (3.26)$$

where we write ν for ν_{nm} without loss of generality. Equation (3.26) gives the emission coefficient due to creation of photons, i.e., j_{ν_c}.

(ii) Scattering. The scattering of a ray of radiation of energy from a direction (θ', ϕ') to the direction (θ, ϕ) is from equation (3.17) at a rate

$$k_{\nu_s} \, \mathrm{d}m \, \mathrm{d}\nu \, \mathrm{d}\omega \, p\,(\theta, \phi, \theta', \phi')\, I_\nu\,(\theta', \phi')\, \frac{\sin\theta' \, \mathrm{d}\theta' \, \mathrm{d}\phi'}{4\pi}$$

The integration of the above expression over all the direction (θ', ϕ') gives the emission coefficient due to scattering as follows:

$$j_{\nu_s} = k_{\nu_s} \frac{1}{4\pi} \int\limits_0^\pi \int\limits_0^{2\pi} p\,(\theta', \phi', \theta, \phi)\, I_\nu\,(\theta', \phi')\, \sin\theta' \, \mathrm{d}\theta' \, \mathrm{d}\phi' \qquad (3.27)$$

For scattering atmosphere,

$$j_\nu = j_{\nu_s} \qquad (3.28)$$

In general we have the contributions due to both scattering and creation and the total emission coefficient is then

$$j_\nu = j_{\nu_s} + j_{\nu_c} \qquad (3.29)$$

4. The equation of radiative transfer (6).
Now we are going to find an equation which governs the change of specific intensity of radiation in terms of the absorption coefficient k_ν and the emission coefficient j_ν. This equation is known as the equation of radiative transfer which is essentially a relation of conservation of radiative energy.

We consider a ray of radiation of intensity I_ν passing through an elementary cylinder of density ρ, base area $\mathrm{d}\sigma_0$ and length $\mathrm{d}s$. We assume that this ray is in the direction of the normal of the surface $(\cos\theta = 1)$. The radiant energy passing through the base into the cylinder is

$$\mathrm{d}E_i = I_\nu \, \mathrm{d}\omega \, \mathrm{d}\sigma_0 \, \mathrm{d}t \, \mathrm{d}\nu$$

The radiant energy coming out of the cylinder from the top where the specific intensity is $I_\nu + \mathrm{d}I_\nu$ is

$$\mathrm{d}E_0 = (I_\nu + \mathrm{d}I_\nu) \, \mathrm{d}\omega \, \mathrm{d}\sigma_0 \, \mathrm{d}t \, \mathrm{d}\nu$$

A part of the radiant energy is absorbed by the material in the cylinder and this energy absorbed is

$$\mathrm{d}E_a = -\, k_\nu \, I_\nu \, \rho \, \mathrm{d}s \cdot \mathrm{d}\omega \, \mathrm{d}\sigma_0 \, \mathrm{d}t \, \mathrm{d}\nu$$

Some of the radiant energy may be emitted by the material in the cylinder and this energy emitted is

$$\mathrm{d}E_e = j_\nu \, \rho \, \mathrm{d}\sigma_0 \, \mathrm{d}s \, \mathrm{d}\omega \, \mathrm{d}t \, \mathrm{d}\nu$$

There is always some radiant energy in the cylinder. Because of the above four process, the radiant energy in the cylinder will, in general, change with time. The total amount of radiant energy changed in the time interval dt is

$$dE_t = \frac{1}{c}\frac{\partial I_\nu}{\partial t}\,dt\,ds\,d\sigma_0\,d\omega\,d\nu$$

By conservation of the radiant energy, we have

$$dE_0 - dE_i = dE_e + dE_a - dE_t \tag{3.30}$$

or

$$\frac{\partial I_\nu}{\partial s} = -\frac{1}{c}\frac{\partial I_\nu}{\partial t} + \rho\,(j_\nu - k_\nu\,I_\nu) \tag{3.31}$$

Equation (3.31) is the equation of radiative transfer, which may also be written as follows:

$$\frac{1}{c}\frac{\partial I_\nu}{\partial t} + l\frac{\partial I_\nu}{\partial x} + m\frac{\partial I_\nu}{\partial y} + n\frac{\partial I_\nu}{\partial z} = \rho\,k_\nu\,(J_\nu - I_\nu) \tag{3.32}$$

where
$$J_\nu = \frac{j\nu}{k_\nu} = \text{source function of radiation} \tag{3.33}$$

and l, m and n are respectively the direction cosines with respect to the x-, y- and z-axis.

One of the main objects of radiation transfer problems is to determine the source function J_ν which depends on the process in the radiation problems.

If scattering is negligible as in most problems of radiation gasdynamics, the equation of radiative transfer (3.31) may be expressed in terms of Einstein coefficients A_{nm}, B_{nm} and B_{mn} as follows:

$$\frac{1}{c}\frac{\partial I_\nu}{\partial t} + \frac{\partial I_\nu}{\partial s} = N_n A_{nm} h\nu - N_m B_{mn} h\nu \left(1 - \frac{N_n B_{nm}}{N_m B_{mn}}\right) I_\nu \tag{3.34}$$

or

$$\frac{1}{c}\frac{\partial I_\nu}{\partial t} + \frac{\partial I_\nu}{\partial s} = \rho\,k_\nu\left(1 - \frac{N_n B_{nm}}{N_m B_{mn}}\right)\left[\frac{A_{nm}}{B_{mn}\left(\frac{N_m B_{mn}}{N_n B_{nm}} - 1\right)} - I_\nu\right] \tag{3.35}$$

We may define a reduced absorption coefficient $k_{\nu t}'$ which includes the effect of induced emission in the medium as follows:

$$k_{\nu t}' = k_{\nu t}\left(1 - \frac{N_n B_{nm}}{N_m B_{mn}}\right) \tag{3.36}$$

As we shall show in chapter IV, under thermodynamic equilibrium condition, the ratio of the Einstein coefficients occurred in equation (3.35) are as follows:

$$\frac{A_{nm}}{B_{nm}} = \frac{2h\nu^3}{c^2},\ \frac{N_m B_{mn}}{N_n B_{nm}} = \exp\,(h\nu/kT) \tag{3.37}$$

where k is the Boltzmann constant which is $1.379 \times 10^{-16}\,\text{ergs/}^\circ\text{K}$.

In the case where both scattering and creation and true absorption exist, equation (3.32) should be written as follows (8):

$$\frac{1}{c}\frac{\partial I_\nu}{\partial t} + \frac{\partial I_\nu}{\partial s} = \rho\, k_{\nu t}\,(J_{\nu t} - I_\nu) +$$

$$+ \rho\, k_s\left[-I_\nu + \frac{1}{4\pi}\int_0^\pi\int_0^{2\pi} p\,(\theta', \phi', \theta, \phi)\, I_\nu\,(\theta', \phi')\sin\theta'\,d\theta'\,d\phi'\right] \tag{3.38}$$

In general, the equation of radiative transfer (3.38) is a differentiointegral equation. When the scattering is neglected, it reduces to a differential equation. The term $J_{\nu t}$ is the true source function of radiation, i.e., $J_{\nu t} = j_{\nu t}/k_{\nu t}$.

5. A solution of the equation of radiative transfer. In all of the flow problems in radiation gasdynamics, we do not consider high frequency phenomena. Hence the time scale t is much larger than L/c where L is the typical dimension of the flow field. Since the distance s along the radiation ray considered should be of the same order of magnitude as L, we may neglect the unsteady term $\dfrac{1}{c}\dfrac{\partial I_\nu}{\partial t}$ in comparison with the spatial variation term $\dfrac{\partial I_\nu}{\partial s}$. In radiation gasdynamics, we shall always use this approximation and the equation and the equation of radiative transfer (3.32) becomes

$$\frac{\partial I_\nu}{\partial s} = \rho\, k_\nu\,(J_\nu - I_\nu) \tag{3.39}$$

Equation (3.39) is a first order total differential equation with respect to s. The formal solution of equation (3.39) from an initial point $s = s_0$ and $I_\nu = I_\nu(s_0)$ is

$$I_\nu(s) = I_\nu(s_0)\exp\left[-\tau_\nu(s, s_0)\right] + \int_{s_0}^s J_\nu(s')\exp\left[-\tau_\nu(s, s')\right]\rho\, k_\nu\,ds' \tag{3.40}$$

where τ_ν is defined by equation (3.3), i.e.,

$$\tau_\nu(s, s_0) = \int_{s_0}^s \rho\, k_\nu\,ds' \quad \text{and} \quad \tau_\nu(s, s') = \int_{s'}^s \rho\, k_\nu\,ds' \tag{3.41}$$

We may use the integral expression (3.40) instead of the differental equation (3.39) in the analysis of radiation gasdynamics.

References

1. CHANDRASEKHAR, S.: Stellar Structure. Dover Publications, Inc., New York, 1957.
2. CHANDRASEKHAR, S.: Radiative Transfer. Dover Publication, New York, 1960.
3. GODSE, C. L., T. BERGERON, J. BJERKNES, R. C. BUNDGAARD: Dynamic Meteorology and weather forecasting. American Meteorology Society and Carnegie Institution of Washington, 1957.

4. JOHNSON, J. C.: Physical Meteorology. John Wiley & Sons, New York, 1954.

5. KOURGANOFF, V.: Basic Methods in Transfer Problems. Oxford University Press, 1952.

6. PAI, S. I.: Some considerations of radiation magnetogasdynamics. Proc. Non-linear Problem, Univ. of Wisconsin Press, pp. 47–67, 1963.

7. PLANCK, M.: The theory of heat radiation. Dover Publications, Inc., New York, 1959.

8. VISKANTA, R., and R. J. GROSH: Heat Transfer in a Thermal Radiation Absorbing and Scattering Medium, Purdue Univ. M. E. Dept. Report.

9. ZHIGULEV, V. N., YE. A. ROMISHEVSKII, and V. K. VERTUSHKIN: Role of Radiation in Modern Gasdynamics. AIAA Jour. vol. 1, No. 6, June 1963, pp. 1473–1485.

Chapter IV
Radiative Equilibrium

1. Introduction. The simplest case of thermal radiation is the condition of radiative equilibrium at which the temperature of the medium must be a constant throughout the medium. Under the thermodynamic equilibrium, there is a definite relation between the absorption coefficient and the emission coefficient. We are going to discuss the laws of radiation under equilibrium condition in §§ 2 to 4. When there is a radiation field in the flow field, the thermodynamic relations change according to the relative strength of the radiation pressure and the gas pressure. We shall discuss these relations in §§ 4 and 5. For the case of complete thermodynamic equilibrium, the temperature of the whole region must be constant. In the flow field of a gas, the temperature is usually different at different points. Hence we usually will not have the thermodynamic equilibrium condition. However, since in ordinary flow problem, we usually may define a temperature at each point in the flow field, it is a good approximation to assume that the radiative equilibrium is reached locally according to the local temperature. In other words, we may assume that the emission of radiation is according to the local temperature as if the thermodynamic equilibrium condition is reached. Of course, we can not assume that the radiation absorbed is also according to the local temperature, because the radiation to be absorbed comes from other points in the flow field which has different temperature from the local temperature. We shall discuss the local thermodynamic equilibrium condition in § 7 which is very useful in the analysis of radiation gasdynamics.

2. Kirchhoff's law of radiation. In 1859, KIRCHHOFF found a theorem that under thermodynamic equilibrium, the ratio of the emission to absorption coefficients of a body depends only on the temperature of the body and not on its nature. In order to prove this theorem, we consider a system which is adiabatically inclosed and under thermodynamic equilibrium. First we consider a point far from the wall of the inclosure. Since the radiation fields is in equilibrium, the specific intensity I_v must be independent of position, direction and time. Equation (3.31) gives

$$I_v = j_v/k_v \qquad (4.1)$$

Equation (4.1) is one of the Kirchhoff's law of radiation which states that the under thermodynamic equilibrium the specific intensity at any frequency v is equal to the ratio of the corresponding emission coefficient and the absorption coefficient.

Now we consider a point which is arbitrarily close to the wall of the inclosure. Under radiative equilibrium condition, there is no net energy change at the point; hence the ray of radiation from the wall must be equal to that traveling in opposite direction. As a result, the state of radiation on the surface of the inclosure must be the same as that in the interior and that equation (4.1) holds true on the wall too.

The intensity I_ν does not depend on the nature of the medium. This theorem can be proved by the following scheme according to KIRCHHOFF and CHANDRASEKHAR (3). We consider an infinitesinal mass $d\,m = \rho\,ds\,d\sigma_0$, where ρ is the density of the mass, $d\,s$ is its length and $d\,\sigma_0$ is its base area. This mass is put inside a hollow spherical reflector and the whole system is inside an adiabatic inclosure. The whole inclosure is evacuated so that there is no absorption of radiation in the inclosure except those due to the small mass and the inner wall of the inclosure. There are two small openings on the spherical reflector of solid angle ω. The whole system is under thermodynamic equilibrium, hence the radiation energy emitted by the mass through these two openings is

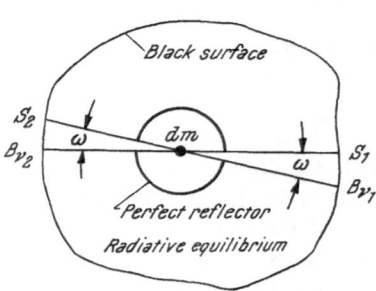

Fig. 4.1. Absorption and emission of radiation in equilibrium condition

$$E_e = j_\nu\,\rho\,\omega\,d\sigma_0\,ds\,d\nu\,dt \qquad (4.2)$$

This mass will absorb radiation energy which is emitted from the surfaces S_1 and S_2 (Fig. 4.1). Let the specific intensities at S_1 and S_2 be respectively $B_{\nu 1}$ and $B_{\nu 2}$. The total radiation energy absorbed is then

$$E_a = k_\nu\,\rho\,(B_{\nu 1} + B_{\nu 2})\,\omega\,d\sigma_0\,ds\,d\nu\,dt \qquad (4.2\,\mathrm{a})$$

Under radiative equilibrium, $E_e = E_a$. Hence we have

$$k_\nu\,(B_{\nu 1} + B_{\nu 2}) = 2\,j_\nu \qquad (4.3)$$

Since equation (4.3) holds if we deform the surface of the inclosure, the radiation intensity from the surface must be independent of the direction of the radiation ray. As a result we have

$$B_{\nu 1} = B_{\nu 2} = B_\nu \qquad (4.4)$$

If we change another material for this mass, equation (4.3) still holds if the thermodynamic equilibrium condition is maintained. Hence the function B_ν must be independent of the nature of the mass and is a function of temperature only. Finally we have the Kirchhoff's law of radiation:

Under thermodynamic equilibrium

$$I_\nu = B_\nu\,(T) = j_\nu/k_\nu \qquad (4.5)$$

We use the symbol B_ν because such a surface is usually referred as black surface. Hence the Kirchhoff's law of radiation is that under the thermodynamic equilibrium, the specific intensity is a function of the temperature only and not of the nature of the medium and the specific intensity is equal to the ratio of the emission coefficient to the absorption coefficient or equal to the specific intensity emitted by a black surface.

From equation (4.5) we see that if $k_\nu = 0$ and $I_\nu \neq \infty$, $j_\nu = 0$. Thus a medium does not emit any radiation of frequency ν which it does not absorb under the thermodynamic equilibrium condition. Thus in a medium which is transparent for a certain frequency of radiation, thermodynamic equilibrium can exist for any finite intensity of that frequency. KIRCHHOFF defined a black body which absorbs all the incident ray. Under thermodynamic equilibrium, the black body also emit radiation of all frequencies. According to Kirchhoff's law, the black body gives the greatest possible emission of radiation. The emission of all other kinds of bodies should be less than the black body radiation (4, 5). Good absorbers are also good emitters (6). A heated substance emits more strongly at the frequencies where the absorption coefficient is high than at frequencies where the absorption is low (see Fig. 3.2). Of course, these conclusions hold for thermal equilibrium condition only.

3. Wien's displacement law. Our next problem is to find the function B_ν as a function of frequency ν and temperature T. The function B_ν was found by PLANCK in 1900 by the quantum mechanics and is known as the Planck function of radiation. However, before we knew the exact expression of the Planck function of radiation B_ν, we already knew some general properties of this function, particularly about its integrated property such as the radiation energy density defined by equation (2.13).

WIEN found in 1893 the displacement law that the energy density U_ν is given by the following equation:

$$U_\nu = \nu^3 \, F \, (\nu/T) \tag{4.6}$$

where $F \, (\nu/T)$ is a function of (ν/T) only. The Wien displacement law can be derived by treating the radiation as a thermodynamic engine which can do work in virtue of the radiation pressure. On account of the Doppler effect, the motion of the engine causes a change of frequency and then a change of radiation energy density (2). It has been checked well experimentally by measuring the maximum intensity from an incandescent body maintained at constant temperature:

For radiative equilibrium condition, the specific intensity is independent of the direction and equation (2.13) gives

$$U_\nu = \frac{4\pi}{c} B_\nu \tag{4.7}$$

Hence the energy density U_ν and the specific intensity have the same spectral distribution. In many experimental data, we determine the energy density U_λ between the wave length interval λ and $\lambda + d\lambda$. The relation between U_λ and U_ν is

$$U_\lambda \, d\lambda = U_\nu \, d\nu \tag{4.8}$$

Since the velocity of propagation of the radiation is speed of light c, we have

$$c = \nu \, \lambda = \text{velocity of light} \tag{4.9}$$

From equations (4.8) and (4.9) we have then

$$U_\lambda = U_\nu \frac{d\nu}{d\lambda} = U_\nu \frac{\nu}{\lambda} = \frac{c^4}{\lambda^5} F\left(\frac{c}{\lambda T}\right) \tag{4.10}$$

In Fig. 4.2, we plot U_λ vs λ at a given temperature $T = 6,000$ °K. Because equation (4.10) may be written as

$$\frac{U_\lambda}{T^5} = G\left(\frac{c}{\lambda T}\right) \tag{4.11}$$

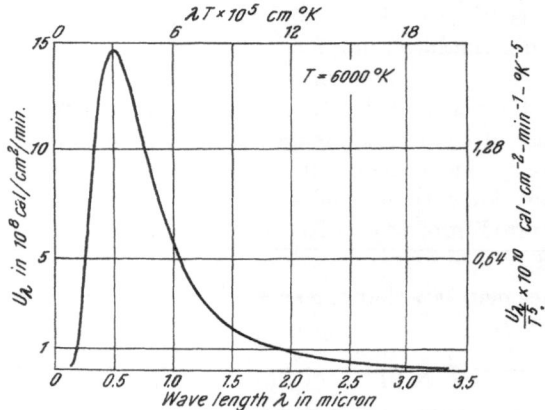

Fig. 4.2. The emissive power of a black body

the single curve of Fig. 4.2 is applicable to all temperature if we consider the ordinate as (U_λ/T^5) and the abscissa as (λT). The exact form of $F(c/\lambda T)$ or $G(c/\lambda T)$ can be expressed in terms of B_ν which will be discussed in the next section.

The reason that equation (4.6) or (4.10) is called displacement law is as follows:

The spectral intensity and the energy density tend to be zero at both very high frequency and very low frequency. There is a maximum energy density for a given temperature. This maximum energy density U_λ occurs when the wave length λ satisfies the relation:

$$\frac{d U_\lambda}{d\lambda} = \frac{c^4}{\lambda^5}\left[-\frac{5}{\lambda} F\left(\frac{c}{\lambda T}\right) - \frac{c}{\lambda^2 T} F'\left(\frac{c}{\lambda T}\right)\right] = 0$$

or

$$\tag{4.12}$$

$$\frac{c}{\lambda T} F'\left(\frac{c}{\lambda T}\right) + 5 F\left(\frac{c}{\lambda T}\right) = 0$$

where $F'(x) = dF/dx$.

Since equation (4.12) has only one variable $(c/\lambda T)$, the maximum energy density must occur at a constant value of $(c/\lambda T)$, i. e.,

$$\lambda_m T = \text{constant} = 0.290 \text{ cm-degree C.} \tag{4.13}$$

The constant is determined from the exact form of $F(c/\lambda T)$ as we shall show in the next section. Equation (4.13) shows that the wave length λ_m at which the maximum energy density occurs decreases with increase of temperature. In other words, the position of the maximum energy density is displaced as the temperature changes.

4. Planck's radiation law (7). The function $F\,(\nu/T)$ has to be determined from microscopic treatment. We have to consider the emission and absorption of quanta of energy. Since there is a large number of quanta, we have to use statistical mechanics. The classical theory of statistical mechanics failed to derive the correct expression of $F\,(\nu/T)$. It was PLANCK who in 1900 used the quantum theory of statistic mechanics to obtain the correct expression of $F\,(\nu/T)$.

Since the function $F\,(\nu/T)$ must be independent of the special mechanism PLANCK considered the radiation from a linear harmonic oscillator to derive the radiation function $F\,(\nu/T)$. Let $\bar{\varepsilon}$ be the mean value of the energy of the oscillator, whose frequency is ν, over a sufficient long time. The energy radiated by the oscillator per unit time is

$$\delta\varepsilon = \frac{2\,e^2}{3\,mc^3}\,(2\,\pi\nu)^2\,\bar{\varepsilon} \qquad (4.14)$$

where e is the absolute electric charge and m is the mass. The work done on the oscillator per unit time by the radiation field of energy density U_ν is

$$\delta W = \frac{\pi\,e^2}{3\,m}\,U_\nu \qquad (4.15)$$

Under equilibrium condition, $\delta\varepsilon = \delta W$, we have then

$$U_\nu = \frac{8\,\pi\,\nu^2}{c^3}\,\bar{\varepsilon} \qquad (4.16)$$

Our problem now is to find the mean energy $\bar{\varepsilon}$. From the kinetic theory of gases, the relative probability for an oscillator with energy ε in the state of thermodynamic equilibrium is, according to Boltzmann law (1), $\exp\,(-\,\varepsilon/k\,T)$. In the classical mechanics, the energy ε varies continuously. Hence the mean energy is

$$\bar{\varepsilon} = \frac{\displaystyle\int_0^\infty \varepsilon\,e^{-\varepsilon/k\,T}\,\mathrm{d}\varepsilon}{\displaystyle\int_0^\infty e^{-\varepsilon/k\,T}\,\mathrm{d}\varepsilon} = \frac{\mathrm{d}}{\mathrm{d}\,(1/k\,T)}\,\log\int_0^\infty e^{-\varepsilon/k\,T}\,\mathrm{d}\varepsilon = k\,T \qquad (4.17)$$

Substituting equation (4.17) into equation (4.16), we have the Rayleigh-Jeans radiation law:

$$U_\nu = \frac{8\,\pi\,\nu^2}{c^2}\,k\,T \qquad (4.18)$$

This law agrees with the Wien's displacement law and check well experimental results for low frequency. Since the energy density increases with ν^2, the total energy density E_R of equation (2.14) is infinite. We usually refer this fact as ultra-violet catastrophe. This shows that this radiation law is no good for very high frequency.

PLANCK introduced the concept of discrete finite quanta of energy ε_0 such that the energy ε must be an integral multiple of ε_0. Then the mean energy should be

$$\bar{\varepsilon} = \frac{\sum\limits_{n=0}^{\infty} n\,\varepsilon_0\,e^{-n\,\varepsilon_0/kT}}{\sum\limits_{n=0}^{\infty} e^{-n\,\varepsilon_0/kT}} = \frac{\varepsilon_0}{e^{\varepsilon_0/kT} - 1} \tag{4.19}$$

Substituting equation (4.19) into equation (4.16) we have the Planck's radiation law:

$$U_\nu = \frac{8\,\pi\,\nu^2}{c^3}\,\frac{\varepsilon_0}{e^{\varepsilon_0/kT} - 1} \tag{4.20}$$

In order that this formula should agree with the well established Wien's displacement law, we must have

$$\varepsilon_0 = h\,\nu \tag{4.21}$$

where

$$h = \text{Planck's constant} = 6.62 \times 10^{-27}\ \text{erg-sec.}$$

Under thermodynamic equilibrium, the specific intensity is the same in all direction, we have $U_\nu = 4\,\pi\,B_\nu/c$, and then we have the Planck's radiation law

$$B_\nu\,(\nu, T) = \frac{2\,h\,\nu^3}{c^2}\,\frac{1}{e^{h\,\nu/k\,T} - 1} \tag{4.22}$$

In terms of Einstein coefficients, from equation (3.35) we have

$$B_\nu\,(\nu, T) = \frac{A_{nm}}{B_{nm}}\,\frac{1}{\left(\dfrac{N_m\,B_{mn}}{N_n\,B_{nm}} - 1\right)} \tag{4.23}$$

Hence

$$\frac{A_{nm}}{B_{nm}} = \frac{2\,h\,\nu^3}{c^2}\,;\ \frac{N_m\,B_{mn}}{N_n\,B_{nm}} = \exp\,(h\,\nu/kT) \tag{4.24}$$

If we assume that the number density is according to Boltzmann distribution, i.e.,

$$N_i = \text{constant} \times g_i \exp\,(-E_i/kT) \tag{4.25}$$

equation (4.24) may be written as follows:

$$\frac{c^2}{2\,h\,\nu^3}\,g_n\,A_{nm} = g_n\,B_{nm} = g_m\,B_{mn} \tag{4.26}$$

where g_i is the degeneracy of the state i of the energy E_i and $h\,\nu = E_n - E_m$.

The integration of equation (4.22) with respect to the frequency gives

$$B\,(T) = \int\limits_0^{\infty} B_\nu\,(\nu, T)\,d\nu = \frac{2\,\pi^4\,k^4}{15\,c^2\,h^3}\,T^4 = \frac{\sigma}{\pi}\,T^4 \tag{4.27}$$

where σ is known as Stefan-Boltzmann constant with respect to radiative flux, i.e.,

$$\sigma = \frac{2\pi^5 k^4}{15 c^2 h^3} = 5.68 \times 10^{-5} \frac{\text{erg}}{\text{cm}^2 \, ^\circ K^4 \, \text{sec.}} \qquad (4.28)$$

The Stefan-Boltzmann constant σ is related to radiative flux over the half-plane, i.e., $q_{R_\sigma}^+$ or $q_{R_\sigma}^-$. For instance, equation (2.7a) for radiative equilibrium condition gives

$$q_{R_\sigma}^+ = \int\limits_0^\infty B_\nu \, d\nu \left(\int\limits_0^{2\pi} \int\limits_0^{\pi/2} \cos\theta \sin\theta \, d\theta \, d\phi \right) = \sigma T^4 = q_{R_\sigma}^- \qquad (4.29)$$

Of course, the total flux of radiation q_{R_σ} in the radiative equilibrium condition is zero.

In the radiative equilibrium condition, the total energy density of radiation for the whole spectrum is, from equation (2.14):

$$E_R = \frac{1}{c} \int\limits_0^\infty B_\nu \, d\nu \left(\int\limits_0^{2\pi} \int\limits_0^{\pi} \sin\theta \, d\theta \, d\phi \right) = \frac{4}{c} \sigma T^4 = a_R T^4 \qquad (4.30)$$

The constant a_R is also known as Stefan-Boltzmann constant which is related to the radiation energy density and radiation pressure and which is

$$a_R = \frac{4}{c} \sigma = \frac{8\pi^5 k^4}{15 c^3 h^3} = 7.67 \times 10^{-15} \, \text{erg. cm}^{-3} \, ^\circ K^{-4} \qquad (4.31)$$

It is interesting to see that at low frequency, the Planck's law of radiation agrees with the classical Rayleigh-Jeans law of radiation. When $h\nu/kT \ll 1$, we may develop the Planck's law of radiation in terms of powers of $h\nu/kT$ as follows:

$$U_\nu = \frac{8\pi h \nu^3}{c^2} \frac{1}{[1 + (h\nu/kT) + ----] - 1} = \frac{8\pi \nu^2}{c^2} kT + ---- \qquad (4.32)$$

If we take the first term of equation (4.32), we have the Rayleigh-Jeans law. However for short wave, $h\nu/kT \ll 1$, the asymptotic value of Planck's law differs greatly from the Rayleigh-Jeans law. The correct law for short wave is

$$U_\nu = \frac{8\pi h \nu^3}{c^3} \exp(-h\nu/kT) \qquad (4.33)$$

Hence we do not have the ultra-violet catastrophe in the correct law because the energy density tends to be zero as ν tends to infinity, or λ tends to be zero (see Fig. 4.2).

For a given temperature, the energy density has a maximum and the frequency or the wave length λ at this maximum point satisfies the following relation:

$$\frac{kT}{h\nu_{max}} = \frac{kT}{hc} \lambda_{max} = 0.2015 \qquad (4.34)$$

Equation (4.34) has been verified experimentally.

5. Stefan-Boltzmann Law of radiation. Before Planck's law was known Stefan-Boltzmann had derived some law of radiation from pure thermodynamic consideration. They considered an inclosure of volume V in which there is no matter but radiation. The radiation reaches the thermodynamic equilibrium condition. The first law of thermodynamics gives

$$dQ = T\,dS = dE + p_R\,dV = E_R\,dV + V\frac{dE_R}{dT}\,dT + \frac{1}{3}\,E_R\,dV$$

$$= V\frac{dE_R}{dT}\,dT + \frac{4}{3}\,E_R\,dV = T\left(\frac{\partial S}{\partial V}\right)dV + T\left(\frac{\partial S}{\partial T}\right)dT \qquad (4.35)$$

where $E = E_R V$ is the total internal energy and S is the entropy.

Equation (4.35) gives

$$\frac{\partial S}{\partial V} = \frac{4\,E_R}{3\,T}, \quad \frac{\partial S}{\partial T} = \frac{V\,dE_R}{T\,dT} \qquad (4.36)$$

From equation (4.36), since $\partial^2 S/\partial V\,\partial T = \partial^2 S/\partial T\,\partial V$, we have

$$\frac{dE_R}{dT} = 4\,\frac{E_R}{T}$$

or $\qquad\qquad\qquad\qquad\qquad\qquad\qquad\qquad\qquad\qquad\qquad\qquad (4.37)$

$$E_R = a_R \cdot T^4$$

where equation (4.37) is known as the Stefan-Boltzmann law of radiation. The constant a_R can not be determined from the thermodynamics. It should be determined experimentally or by exact theory as the Planck's law. In fact we have already shown the same relation from Planck's law in equation (4.30). The Stefan-Boltzmann law of radiation can also be derived from Wien's displacement law, i. e.,

$$E_R = \int\limits_0^\infty \nu^3\,F\,(\nu/T)\,d\nu = T^4\int\limits_0^\infty x^3\,F\,(x)\,dx = a_R\,T^4 \qquad (4.38)$$

6. Adiabatic changes in an inclosure containing matter and radiation (3). Before we study the interaction between radiation field and the flow field, it is useful to consider some simple thermodynamic relations between the interaction of matter and radiation in complete thermodynamic equilibrium.

We consider an inclosure in which there is an ideal gas and thermal radiation in thermodynamic equilibrium condition. The internal energy and the pressure consist of two parts: one is due to the gas and the other is due to the thermal radiation. Hence the internal energy per unit mass is

$$E = E_R/\rho + c_V T \qquad (4.39)$$

where ρ is the density and E_R given by equation (4.37) is the radiation energy per unit volume and C_V is the specific heat at constant volume. We assume that c_V is a constant for an ideal gas. The total pressure is

$$p = p_R + p_g = \frac{1}{3}\,E_R + \rho\,RT \qquad (4.40)$$

where R is the gas constant.

For adiabatic flow $dQ = 0$, the first law of thermodynamics gives

$$d E + p\, d\,(1/\rho) = (\partial E/\partial T)\, dT + (\partial E/\partial \rho)\, d\rho + p\, d\,(1/\rho) \qquad (4.41)$$

When we substitute the expressions E and p from equations (4.39) and (4.40) respectively into equation (4.41), we find the following relations between p, T and ρ:

$$\frac{dp}{p} - \Gamma_1 \frac{d\rho}{\rho} = 0 \qquad (4.42)$$

$$\frac{dp}{p} + \frac{\Gamma_2}{1 - \Gamma_2}\frac{dT}{T} = 0 \qquad (4.43)$$

$$\frac{dT}{T} - (\Gamma_3 - 1)\frac{d\rho}{\rho} = 0 \qquad (4.44)$$

where

$$\Gamma_1 = b + \frac{(4 - 3b)^2\,(\gamma - 1)}{b + 12\,(\gamma - 1)\,(1 - b)}$$

$$\Gamma_2 = \frac{(4 - 3b)\,\Gamma_1}{3\,(1 - b)\,\Gamma_1 + b} \qquad (4.45)$$

$$\Gamma_3 = 1 + \frac{\Gamma_1 - b}{4 - 3b}$$

$$b = p_g/p = \text{ratio of the gas pressure to the total pressure.}$$

When the radiation pressure is negligibly small, $b = 1$, then

$$\Gamma_1 = \Gamma_2 = \Gamma_3 = \gamma = c_p/c_v = \text{ratio of the specific heats of the gas.}$$

When the gas pressure is negligibly small, we have $b = 0$, then

$$\Gamma_1 = \Gamma_2 = \Gamma_3 = {}^4/_3$$

For finite value of b, the values of Γ_1, Γ_2 and Γ_3 are not equal. When the radiation effect dominants, the thermodynamic relations (4.42) to (4.44) behaves like an ideal gas with $\gamma = {}^4/_3$. However for finite value of b, the thermodynamic relations differ from those of ideal gas because the three values of Γ_1, Γ_2 and Γ_3 are not equal.

The specific heat at constant volume for the case of ideal gas and radiation is

$$c_{VR} = \left(\frac{\partial Q}{\partial T}\right)_{dV = 0} = \frac{c_v}{b}\,[b + 12\,(\gamma - 1)\,(1 - b)] \qquad (4.46)$$

The specific heat at constant pressure p is

$$c_{pR} = \left(\frac{\partial Q}{\partial T}\right)_{dp = 0} = \frac{c_v\,\Gamma_1}{b^2}\,[b + 12\,(\gamma - 1)\,(1 - b)] \qquad (4.47)$$

Hence

$$c_{pR}/c_{VR} = \Gamma_1/b \qquad (4.48)$$

7. Local thermodynamic equilibrium. In gasdynamics, the temperature of the gas is usually not constant throughout the whole field but at each point in the flow field we may define a local temperature T to describe the gas properties at that point. As a result, it is a good approximation for most radiation gasdynamics to assume that the gas is locally thermodynamic equilibrium where the spontaneous emission of radiation depends only on the local temperature T. However the induced emission of radiation and the absorption of radiation depends on the intensity of the radiation which in general differs from the black body radiation. We may find the expression of the equation of radiative transfer for the case of locally thermodynamic equilibrium as follows:

Substituting the expression of absorption coefficient $k_{\nu t}$ of equation (3.11) into the equation of radiative transfer (3.35)

$$\frac{1}{c}\frac{\partial I_\nu}{\partial t} + \frac{\partial I_\nu}{\partial s} = \rho\, k_{\nu t}{}'\, (B_\nu - I_\nu) \tag{4.49}$$

where

$$k'{}_{\nu t} = k_{\nu t}\,(1 - e^{-h\nu/kT}) = \text{reduced absorption coefficient} \tag{4.50}$$

The reduction of the absorption coefficient is due to the induced emission of radiation.

We shall use equation (4.50) for most of the flow problems except otherwise specified.

References

1. ALLER, L. H.: Astrophysics: The Atmosphere of the sun and stars. Ronald Press Co., New York, second edition, 1963.

2. BORN, M.: Atomic Physics. Hafner Publishing Company, Inc., New York, 1946.

3. CHANDRASEKHAR, S.: Stellar Structure. Dover Publications, Inc., 1957.

4. GODSKE, C. L., T. BERGERON, J. BJERKNES, R. C. BUNDGAARD: Dynamics meteorology and weather forecasting. American Meteorological Society and Carnegie Institution of Washington, 1957.

5. JOHNSON, J. C.: Physical Meteorology. John Wiley & Sons, New York, 1954.

6. JAKOB, M.: Heat Transfer. Vol. I, John Wiley & Sons, Inc., New York, 1955.

7. PLANCK, M.: The theory of heat radiation. Dover publications, Inc., New York, 1959.

Chapter V

Fundamental Equations of Radiation Gasdynamics

1. Introduction. In radiation gasdynamics, we study the interaction between the gasdynamic field and the radiation field. In the most general analysis, we should consider the distribution functions of a mixture of various types of particles of which photons may be considered as a special type of particles. In deriving the equations for these distribution functions, the relativistic character of photon must be considered. We shall discuss such an analysis in chapter X. In this general case, we may deal with both rarefied gasdynamics and continuum gasdynamics including the effects of radiation field. In order to bring some essential features of radiation effects, we shall consider only the case where the gas may be considered as a continuum in this chapter and the following four chapters. When the gas may be considered as a continuum, the gasdynamic variables are the pressure p, density ρ, temperature T, and three velocity components. In radiation gasdynamics, we have to add the specific intensity of radiation I_ν to these gasdynamic variables. Hence we need to find seven fundamental equations for these seven unknowns. These fundamental equations are

(i) Equation of state which connects the pressure, density, and temperature of the gas (§ 2).

(ii) Equation of continuity which expresses the conservation of mass of the medium (§ 3).

(iii) Equations of motion which are generally three in number and which express the conservation of momentum. In radiation gasdynamics, the radiation stresses should be included (§ 4).

(iv) Equation of energy which expresses the conservation of energy (§ 5).

(v) Equation of radiative transfer, which determines the specific intensity of radiation (§ 6).

It should be noticed that the above analysis may be easily generalized into the multi-fluid theory of gasdynamics in which the six gasdynamic equations for each species should replace the gasdynamic equations for the mixture as a whole.

In § 7, some general remarks on the fundamental equations of radiation gasdynamics will be discussed. Since the radiation terms in the fundamental equations have not been fully discussed in literature, we shall discuss them in some details in §§ 8 to 10. In § 8, the case of small mean free path of radiation will be discussed while in § 9, the case of finite mean free path of radiation will be discussed. Since in the general case the radiation terms are complicated integrals which is very difficult to be analyzed, some approximations have been used in literature in order to bring out the essential points of radiation effects. One of the approximations is the grey gas approximation which will be dis-

cussed in § 9 and the other is the one dimensional approximation which will
be discussed in § 10. In the one dimensional analysis, the radiation terms may
be expressed in terms of exponential integrals which will be discussed in § 11.

2. Equation of state. It is an empirical fact that there is a functional relation
between the pressure p, density ρ and the temperature T of a gas, i.e.,

$$p = \mathrm{f}\,(\rho,\,T) \tag{5.1}$$

For gas at very high temperature, it is a good approximation that the perfect
gas law holds. Hence equation (5.1) may be written as

$$p = \rho\,RT \tag{5.2}$$

where R is the gas constant and $R = k/m$ where k is the Boltzmann constant
and is 1.381×10^{-16} cm-dyne/°C and m is the mean mass of a particle of the gas.

3. Equation of continuity. The conservation of mass of the gas gives

$$\frac{\partial \rho}{\partial t} + \nabla\,(\rho\,\vec{q}) = \beta \simeq 0 \tag{5.3}$$

where $\vec{q} = \vec{i}u + \vec{j}v + \vec{k}w$ is the velocity vector of the flow and u, v, and w
are respectively the x-, y- and z-component of the velocity; $\vec{i},\vec{j}$, and $\vec{k}$ are respectively the x-, y- and z-unit vector; β is the change of mass due to radiation
of nuclear energy which is negligibly small in ordinary flow problems. We shall
neglect $\beta \cdot \nabla = \vec{i}\,(\partial/\partial x) + \vec{j}\,(\partial/\partial y) + \vec{k}\,(\partial/\partial z)$.

4. Equations of motion. The conservation of momentum gives the equation
of motion which is

$$\rho\,\frac{D\vec{q}}{Dt} = -\nabla p + \nabla \cdot \tau_s + \nabla \cdot \tau_R + \vec{F} \tag{5.4}$$

where $D/Dt = (\partial/\partial t) + \vec{q} \cdot \nabla$. The viscous stress tensor τ_s has the *ijth* component

$$\tau_s{}^{ij} = \mu\left(\frac{\partial u^i}{\partial x^j} + \frac{\partial u^j}{\partial x^i}\right) - \frac{2}{3}\,\mu\,(\nabla \cdot \vec{q})\,\delta^{ij} \tag{5.5}$$

$$\delta^{ij} = 0 \text{ if } i \neq j; \quad \delta^{ij} = 1, \text{ if } i = j$$

μ is the coefficient of viscosity.

The radiation stress tensor τ_R has the ij-component given by equation (2.18).
Hence equation (5.4) is an integro-differential equation.

$\vec{F}$ is the body force which may consist of the electromagnetic force $\vec{F}_e$ and
the gravitational force $\vec{F}_g = \rho\,\vec{g}$ where $\vec{g}$ is the gravitational acceleration. The
electromagnetic force $\vec{F}_e$ may be written as follows:

$$\vec{F}_e = \rho_e\,\vec{E} + \vec{J} \times \vec{B} \tag{5.6}$$

where ρ_e is the excess electric charge, $\vec{E}$ is the electric field intensity, $\vec{J}$ is the electric current density and $\vec{B} = \mu_e \vec{H}$ is the magnetic induction, μ_e is the magnetic permeability and $\vec{H}$ is the magnetic field intensity. At very high temperature, the gas will be ionized. For ionized gas, the electromagnetic field will produce an electromagnetic force in the gas. We shall discuss this in chapter VIII.

5. Equation of energy. The conservation of energy gives the equation of energy as follows:

$$\rho \frac{D\bar{e}_m}{Dt} = \nabla \cdot (p\,\vec{q}) + \nabla \cdot (\vec{q} \cdot \tau_s) + \nabla \cdot (\vec{q} \cdot \tau_R) + \nabla \cdot (\varkappa \nabla T) + \nabla \cdot \vec{q}_R + Q \quad (5.7)$$

where $\bar{e}_m = U_m + \frac{1}{2} q^2 + \phi + E_R/\rho$ = total energy of the gas per unit mass; U_m is the internal energy of the gas per unit mass, $\frac{1}{2} q^2$ is the kinetic energy of the gas per unit mass, ϕ is the potential energy of the gas per unit mass and E_R is the radiation energy density per unit volume which is given by equation (2.14).

The term on the left hand side of equation (5.7) is the rate of change of the total energy of the gas per unit volume. The first term on the right hand side of equation (5.7) is the work done by the pressure of the gas; the second term is the energy dissipation by the viscous stresses; the third term is the energy dissipation by the radiation stresses; the fourth term is the energy transfer by heat conduction with $\varkappa$ as the coefficient of heat conductivity; the fifth term is the heat transfer by radiation which is given by equations (2.8) and (2.10). The last term is the energy input by the electromagnetic fields and chemical reaction and other heat sources. The energy input by the electromagnetic field may be written as

$$Q_E = \vec{E} \cdot \vec{J} \qquad (5.8)$$

because the terms τ_R and $\vec{q}_R$ are generally in integral form, equation (5.7) is in general an integro-differential equation.

6. Equation of radiative transfer. The specific intensity of radiation is given by the equation of radiative transfer (3.31), i.e.,

$$\frac{1}{c} \frac{\partial I_\nu}{\partial t} + \frac{\partial I_\nu}{\partial s} = \rho\, j_\nu - \rho\, k_\nu\, I_\nu \qquad (5.9)$$

In general the emission coefficient is given by equation (3.29) which consists of the emission by scattering and the emission due to creation of photons. The emission due to creation of photons consists of the spontaneous emission and the induced emission. The absorption coefficient k_ν is given by equation (3.5) which consists of the true absorption coefficient and the scattering absorption coefficient.

In most of the problems of radiation gasdynamics, we may assume that the scattering of radiation is negligible. Hence we should consider only the true absorption coefficient and the true emission coefficient only. The determination of these two true absorption coefficient and emission coefficient depends on the radiation process in the nucleus of the gas molecules. These processes may vary from case to case, particularly when the radiation is not in the equilibrium state. No general statement can be made. However, for most of the gasdynamic problems in which we may define a local temperature, the condition of local thermodynamic equilibrium which has been discussed in chapter IV, §7 is a good approximation. For local thermodynamic equilibrium with negligible scattering, equation (5.9) becomes

$$\frac{1}{c}\frac{\partial I_\nu}{\partial t} + \frac{\partial I_\nu}{\partial s} = \rho \, k_\nu' \, (B_\nu - I_\nu) \tag{5.10}$$

For most of the flow problems, we do not consider very high frequency phenomena so that the unsteady term $(1/c)\,(\partial I_\nu/\partial t)$ is negligibly small in comparison with the spatial variation terms because the velocity of light is a very large quantity. Hence equation (5.10) becomes

$$l\frac{\partial I_\nu}{\partial x} + m\frac{\partial I_\nu}{\partial y} + n\frac{\partial I_\nu}{\partial z} = \rho \, k_\nu' \, (B_\nu - I_\nu) \tag{5.11}$$

In radiation gasdynamics, the absorption coefficient k_ν' may be considered as a known function of the temperature and the density of the gas.

7. General remarks on the fundamental equations. For radiation gasdynamics, we have to solve seven variables p, ρ, T, u, v, w and I_ν from the seven equations (5.2), (5.3), (5.4), (5.7), and (5.11) for given initial and boundary conditions. This system of equations is much more complicated than the corresponding fundamental equations of ordinary gasdynamics in which the radiation effects are negligible. Since it is not possible to solve the fundamental equations of ordinary gasdynamics in general, we can not solve the fundamental equations of radiation gasdynamics too. In order to bring out the essential features of radiation effects, we have to make reasonable approximations so that the fundamental equations may be simplified into a form which can be analyzed. The following are some of the approximations which will be discussed in detail in the following chapters when we treat various flow problems:

(i) All the well known approximations of gasdynamics may be used. For instance, we may consider

(a) The inviscid and non-heat-conducting flow. Outside the boundary layer or transition regions, the viscosity and heat conductivity may be neglected when the Reynolds number of the flow is high. In this case, we may add the radiation terms to the equation of inviscid and non-heat-conducting fluid; and

(b) Boundary layer flow. In the boundary layer region, the well known Prandtl boundary layer approximations may be applied.

(ii) Some approximations about the equation of radiative transfer may be made. Equation (5.11) shows that the dimension of $\rho \, k_\nu'$ is one over a length. Hence we may define a radiation mean free path $L_{R\nu}$ such that

$$L_{R\nu} = \frac{1}{\rho \, k_\nu'} \tag{5.12}$$

This mean free path of radiation represents the mean distance over which a photon travels before it is absorbed by the molecule of the gas. In a way it is similar to the mean free path in the kinetic theory of gases which represents the mean distance between collisions. When the mean free path of radiation is small, we say that the gas is optically thick. For optically thick case, the expressions of the radiation terms may be reduced to very simple forms which will be discussed in section 8. When the mean free path of radiation is large, we say that the gas is optically thin. When the gas is optically very thin, the radiation terms may be also reduced into simple forms which will be discussed in section 9.

In the intermediate case, we have to use the integral expression for the radiation terms. However in many of the flow problems, the following two approximations may be used:

(a) One dimensional approximation. In many of the flow problems, the variation of the state variables in a given direction is much larger than those in the other directions. We may consider the specific intensity I_ν as a function of this spatial coordinate only. By this assumption, the integral expression of radiation terms may be simplified as we shall discuss in section 9.

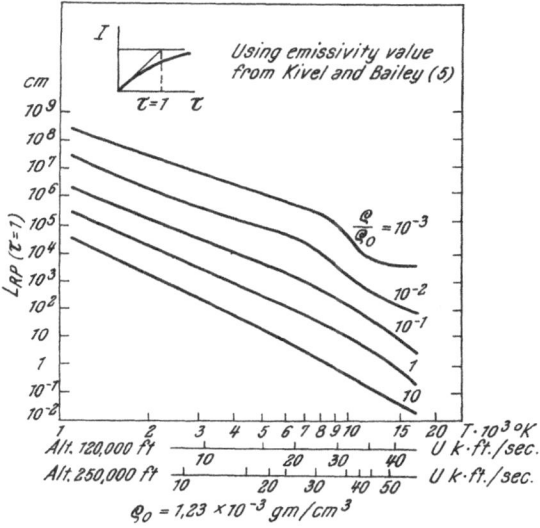

Fig. 5.1. Mean free path of radiation $L_{Rp} = 1/(\rho K_p)$ of air. (Fig. 3 of reference 3 by R. and M. GOULARD, Courtesy of Pergamon Press)

(b) The gray gas approximation. The absorption coefficient k_ν' is a function of the frequency ν in general. If the absorption coefficient k_ν' is independent of the frequency, we say that the gas is gray. By gray gas approximation, we assume that the absorption coefficient is independent of the frequency. In other words, we may use some mean value of the absorption coefficient which is independent of the frequency for the actual absorption coefficient. We shall also discuss the gray gas approximation in section 9.

The above radiation gasdynamic equation may be generalized into the fundamental equations of radiation magnetogasdynamics. We need only to add the electromagnetic variables $\vec{E}$, $\vec{H}$, $\vec{J}$ and ρ_e into our analysis. We have to add the electromagnetic equations too which are the Maxwell equations for the electromagnetic fields, the generalized Ohm's law for the electric current density and the conservation law of electrical charge. Of course, the well known magnetogasdynamic approximations may be used here. As far as the electromagnetic equations are concerned, they are not affected by the radiation terms. We shall write these equations in chapter VIII.

8. Case of small mean free path of radiation. Fig. 5.1 shows some mean free path of radiation at various temperature and density for air. Near the temperature of $10{,}000°\,\mathrm{K}$ and density $10\,\rho_0$, the mean free path of radiation is of the order of a few millimeters which is much smaller than the dimension of the flow field for practical problems. For small mean free path of radiation, we may express the solution of equation (5.11) for the specific intensity I_ν in term of power series of $L_{R\nu}$. The solution of equation (5.11) it then.

$$I_\nu = B_\nu - L_{R\nu}\left(l\frac{\partial B_\nu}{\partial x} + m\frac{\partial B_\nu}{\partial y} + n\frac{\partial B_\nu}{\partial z}\right) + 0\,(L^2{}_{R\nu}) \qquad (5.13)$$

For optically thick case, we may use the first two terms of equation (5.13) for the specific intensity of radiation I_ν and the corresponding radiation terms can be evaluated as follows:

(a) Radiation energy density: Equation (2.13) gives

$$U_\nu = \frac{1}{c}\int I_\nu\,d\omega = \frac{4\,\pi}{c}B_\nu \qquad (5.14)$$

and equation (5.14) gives

$$E_R = \int\limits_0^\infty U_\nu\,d\nu = \frac{4\,\pi}{c}\int\limits_0^\infty B_\nu\,d\nu = \frac{8\,\pi^5\,k^4}{15\,c^3\,h^3}\,T^4 = a_R\,T^4 \qquad (5.15)$$

where $a_R = 7.67\times 10^{-15}\,\mathrm{erg\text{-}cm}^{-3} - {}^0K^{-4}$ is known as the Stefan-Boltzmann constant. Equation (5.15) has been used to estimate the radiation energy density in table I of chapter I.

(b) Radiation stresses. Equations (2.18) and (5.13) give the following expressions for radiation stresses when the terms of second or higher order are neglected:

$$-\tau_{Rxx} = -\tau_{Ryy} = -\tau_{Rzz} = p_R = \frac{1}{3}\,a_R\,T^4$$
$$\tau_R{}^{ij} = 0 \quad \text{when } i \neq j \qquad (5.16)$$

For optically thick case, we have only the radiation pressure p_R.

(c) Radiation flux. Equation (2.8) gives

$$q_{Rx} = \frac{4\,\pi}{3}\int\limits_0^\infty L_{R\nu}\frac{\partial B_\nu}{\partial x}\,d\nu = \frac{4\pi}{3}\frac{\partial T}{\partial x}\int\limits_0^\infty L_{R\nu}\frac{\partial B_\nu}{\partial T}\,d\nu = \frac{4}{3K_R\rho}\frac{\partial B}{\partial x} = D_R\frac{\partial E_R}{\partial x} \qquad (5.17)$$

and similar expressions for q_{Ry} and q_{Rz}. The expression K_R is known as the Rosseland mean absorption coefficient which is defined by the relation:

$$\frac{1}{K_R}\int\limits_0^\infty\frac{\partial B_\nu}{\partial T}\,d\nu = \int\limits_0^\infty\frac{1}{k_\nu{}'}\frac{\partial B_\nu}{\partial T}\,d\nu \qquad (5.18)$$

The average mean free path of radiation in the sense of Rosseland mean is

$$L_R = \frac{1}{\rho\,K_R} = \text{average mean free path of radiation according} \qquad (5.19)$$

to Rosseland mean absorption coefficient.

Finally the Rosseland diffusion coefficient of radiation is

$$D_R = \frac{L_R c}{3} = \frac{c}{3 K_R \rho} \tag{5.20}$$

The radiation heat transfer flux is then

$$\vec{q}_R = - D_R \nabla E_R = - \varkappa_R \nabla T \tag{5.21}$$

where the effective coefficient of heat conductivity by radiation is

$$\varkappa_R = 4 D_R a_R T^3 \tag{5.22}$$

It is interesting to notice that all the radiation terms (5.15), (5.16) and (5.21) are reduced to simple expression instead of the integral forms in the general case. For optically thick case, the fundamental equations of radiation gasdynamics are partial differential equations instead of integro-differential equations. These fundamental equations for radiation gasdynamics of optically thick case are:

$$p = \rho R T \tag{5.23 a}$$

$$\frac{\partial \rho}{\partial t} + \nabla \cdot (\rho \vec{q}) = 0 \tag{5.23 b}$$

$$\rho \frac{D \vec{q}}{D t} = - \nabla (p + p_R) + \nabla \cdot \tau + \vec{F} \tag{5.23 c}$$

$$\rho \frac{D \vec{e}_m}{D t} = - \nabla \cdot [\vec{q} (p + p_R)] + \nabla \cdot (\vec{q} \cdot \tau) + \nabla \cdot [(\varkappa + \varkappa_R) \nabla T] + Q \tag{5.23 d}$$

where p_R, E_R and K_R are given by equations (5.16), (5.15) and (5.22) respectively.

9. Case of finite mean free path of radiation. If the mean free path of radiation $L_{R\nu}$ is not small, we have to use more terms in the series development of $L_{R\nu}$ (5.13). But since the question of convergence of such a series for large $L_{R\nu}$ is not known and it is difficult to determine how many terms should be used in the series (5.13), we should not use the series form of solution (5.13) and should solve exactly the radiative transfer equation with the boundary conditions. Equation (5.11) may be considered as a first order differential equation of the distance along a ray of radiation s where

$$ds = l \, dx + m \, dy + n \, dz \tag{5.24}$$

We define the optical thickness between two points s and s_1 along a ray of radiation as follows:

$$\tau_\nu (s, s_1) = \int_{s_1}^{s} k_{\nu}' \rho \, ds' \tag{5.25}$$

where s' is the s-coordinate for integration.

Equation (5.11) can be integrated and we have

$$I_\nu(s) = I_\nu(s_0) \exp\left[-\tau_\nu(s, s_0)\right] + \int_{s_0}^{s} B_\nu(s_1) \left\{\exp\left[-\tau_\nu(s, s_1)\right]\right\} \rho\, k_\nu'\, \mathrm{d}s_1 \quad (5.26)$$

where s_0 is certain initial point on the ray where the specific intensity $I_\nu(s_0)$ is known. The determination of $I_\nu(s_0)$ in general depends on the boundary condition of our problem and we shall discuss them in chapter VI. If we assume that $s_0 = -\infty$, the specific intensity due to the first term is zero because all the radiation will be absorbed by the medium. The equation (5.27) becomes

$$I_\nu(s) = \int_{-\infty}^{s} B_\nu(s') \left\{\exp\left[-\tau_\nu(s, s')\right]\right\} \rho\, k_\nu'\, \mathrm{d}s' \quad (5.27)$$

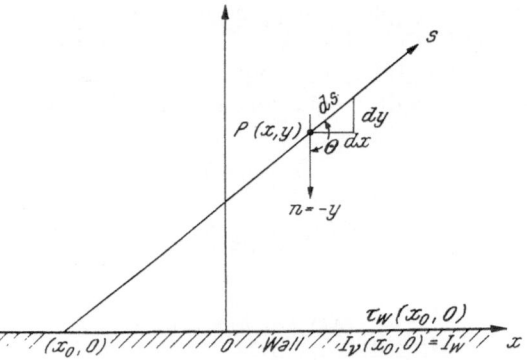

Fig. 5.2. Two dimensional radiation field over a straight wall

After we obtain the expressions of the specific intensity I_ν from equation (5.26) and (5.27), we may substitute these expressions into the integral expressions of radiation heat transfer, radiation stresses and radiation energy density, i.e., equation (2.8) etc. The evaluation of equations (2.8) etc. depends on the geometry of the configuration of the flow field. In order to illustrate such an evaluation, we consider a two dimensional radiation field over a straight plate as shown in Fig. 5.2. Let s be the distance measured along a heat ray. The increment $\mathrm{d}s$ is then

$$\mathrm{d}s = \sin\theta \cdot \mathrm{d}x - \cos\theta \cdot \mathrm{d}y \quad (5.28)$$

where θ is the angle between the ray s and the minus-y axis which is the normal toward the wall. Along each ray $\theta =$ constant. Hence we have for the two dimensional problem:

$$\mathrm{d}s = -\sec\theta \cdot \mathrm{d}y = -m\, \mathrm{d}y \quad (5.29)$$

where $m = \sec\theta$. Hence we may consider the specific intensity I_ν as a function of y and θ in our problem instead of x and y coordinates. The radiative transfer equation (5.11) may be written as follows:

$$m \frac{\partial I_\nu}{\partial \tau_\nu} = I_\nu - B_\nu \quad (5.30)$$

where we define the optical thickness τ_ν by the relation

$$d\tau_\nu = \rho \, k_\nu' \, dy \qquad (5.31)$$

$$\tau_\nu = \int_0^\cdot \rho \, k_\nu' \, dy \qquad (5.31\,a)$$

The general solution of equation (5.30) cf. (5.26) may be written as:

$$I_\nu(\tau_\nu, \theta) = \left\{ I_{\nu 0}(\tau_\nu^*, \theta) - \int_{\tau_\nu^*}^{\tau_\nu} m\, B(t, \theta) \exp(-mt)\, dt \right\} \exp(m\,\tau_\nu) \qquad (5.32)$$

where $\tau_\nu{}^*$ is any arbitrary value of τ_ν to be determined by the boundary condition and $I_{\nu 0}(\tau_\nu^*, \theta)$ is the arbitrary function of θ from the integration of equation (5.30) and to be also determined by the boundary condition. In order to determine the value $\tau_\nu{}^*$ and $I_{\nu 0}(\tau_\nu^*, \theta)$, it is convenient for our problem of Fig. 5.2 to divide the radiation rays into two groups: one is the group of rays directed toward the wall and the others are those directed away from the wall:

For rays directed toward the wall, we have $0 \leqq \theta \leqq \frac{1}{2}\pi$ and $1 \leqq m \leqq \infty$ and since the specific intensity at infinity will not affect the value at (τ_ν, θ), equation (5.32) becomes:

$$I_\nu(\tau_\nu, \theta) = \int_{\tau_\nu}^\infty B_\nu(t, \theta) \exp[-m(t - \tau_\nu)]\, m\, dt \qquad (5.33)$$

where τ_ν is the optical thickness at the point of the flow field at which we evaluate the radiation terms and t is the dummy variable of integration for the optical thickness.

For rays directed away from the wall, we have $\frac{1}{2}\pi \leqq \theta \leqq \pi$ and $-\infty \leqq m \leqq -1$ and since we may assume that the specific intensity on the wall ($\tau_\nu = 0$) is a known function $I_{\nu 0}(0, \theta)$ which is determined by the boundary condition of radiation rays as we shall discuss in chapter VI, equation (5.32) becomes

$$I_\nu(\tau_\nu, \theta) = I_{\nu 0}(0, \theta) \exp(m\,\tau_\nu) - \int_0^{\tau_\nu} B_\nu(t, \theta) \exp[-m(t - \tau_\nu)]\, m\, dt \qquad (5.34)$$

Substituting the expressions (5.33) and (5.34) into the radiation expressions (2.8) etc., we have the final expressions of radiation terms used in our fundamental equations. Here we consider only the two dimensional problem and all variables are independent of the azimuth angle ϕ. Hence the integration of the radiation terms with respect to ϕ can be easily carried out. For instance, the heat flux in the y-direction is

$$q_{Ry} = 2\pi \int_0^\infty \int_0^\pi I_\nu \cos\theta \, \sin\theta \, d\theta \, d\nu = 2\pi \int_0^\infty \left\{ \int_0^{\pi/2} \int_{\tau_\nu}^\infty B_\nu(t, \theta) \exp[-m(t - \tau_\nu)]\sin\theta \, d\theta \right.$$

$$\left. + \int_{\pi/2}^\pi \left[I_{\nu 0}(0, \theta)\cos\theta \exp(m\,\tau_\nu) - \int_0^{\tau_\nu} B_\nu(t, \theta)\exp[-m(t - \tau_\nu)]dt \right]\sin\theta \, d\theta \right\} d\nu \qquad (5.35)$$

It should be noticed that in general not only $I_{v0}(0, \theta)$ and $B_v(t, \theta)$ are functions of θ, the optical thickness τ_v and t are function of θ too because both ρ and k_v' are functions of θ. In the general case, the expression (5.35) can not be further simplified.

Equation (5.35) may be simplified by some approximations. The following two approximations have been frequently used:

(i) Gray gas approximation. If the absorption coefficient of a gas is independent of the frequency, the gas is known as a gray gas. By gray gas approximation, we may use some average absorption coefficient over the frequency range considered so that in the expression of specific intensity (5.33) and (5.34) we may assume that the optical thickness τ_v is independent of the frequency v. Furthermore we may also assume that the wall is a gray surface so that the emissivity and the reflectivity of the wall will be independent of the frequency v too. (The emissivity and the reflectivity of a surface will be discussed in chapter VI.) As a result of this gray gas approximation, the integration of equation (5.35) or similar radiation terms with respect to frequency v can be immediately carried out. We may use a Planck mean free path of radiation L_p to determine the average value of the mean free path of radiation in the gray gas approximation, i.e.,

$$\frac{1}{L_{RP}} = \rho\, K_P = \frac{\int\limits_0^\infty \rho\, k_v'\, B_v\, \mathrm{d}v}{\int\limits_0^\infty B_v\, \mathrm{d}v} = \frac{\int\limits_0^\infty \rho\, k_v'\, B_v\, \mathrm{d}v}{B(T)} \tag{5.36}$$

where $B(T) = (\sigma/\pi)\, T^4$ is given by equation (4.27). K_p is known as the Planck mean absorption coefficient of radiation. By the help of equation (5.36) and assuming that $\tau_v = \tau$ and t are independent of the frequency v, the integration of equation (5.35) with respect to v can be carried out and we have y-wise radiative flux in the gray gas approximation as follows:

$$q_{Ry} = 2 \left\{ \int\limits_0^{\pi/2} \int\limits_\tau^\infty \sigma\, T^4(t, \theta)\, \exp\left[-m(t-\tau)\right] \mathrm{d}t \sin\theta\, \mathrm{d}\theta \right. +$$
$$\left. + \int\limits_{\pi/2}^\pi \left[B_0(0, \theta) \cos\theta \exp(m\tau) - \int\limits_0^\tau \sigma\, T^4(t, \theta) \exp\left[-m(t-\tau)\right] \mathrm{d}t \right] \sin\theta\, \mathrm{d}\theta \right\} \tag{5.37}$$

where $\qquad B_0(0, \theta) = \int\limits_0^\infty I_{v0}(0, \theta)\, \mathrm{d}v$ and $\tau = \int\limits_0^y \rho\, K_p\, \mathrm{d}y$

It should be noticed that in Fig. 5.1, the mean free path of radiation is essentially based on Planck mean value. Since for a general two dimensional problem, the temperature $T(\tau, \theta)$ is a function of both the optical thickness τ or y and the angular coordinate θ, we can not simplify equation (5.37) any more.

(ii) One dimensional approximation. Equation (5.37) may be further simplified if the temperature T of the flow field is essentially one dimensional such that the variation of temperature in the y-direction is much larger than that in the x-direction. As a result, we may assume that the temperature is a function of y or τ only and independent of θ. Under this approximation, the integration with respect to θ in equations (5.35) and (5.37) can be expressed in terms of exponential integrals:

$$\varepsilon_n(t) = \int\limits_0^\infty m^{-n} \exp(-mt)\,dm = \int\limits_0^1 z^{n-2} \exp(-t/z)\,dz \tag{5.38}$$

where $m = 1/z$. Since the one dimensional approximation is the only case which has been extensively studied in radiation gasdynamics as well as in astrophysical problems, we shall discuss it in detail in the next section.

10. One dimensional radiative transfer. In order to include various cases of one dimensional problem, we consider the radiative transfer between two parallel plates (Fig. 5.3). We shall assume that the specific intensity of black radiation is a function of y or τ_ν only, but independent of the angle θ. We divide the specific intensity into two groups: one is directed upward I_ν^- for which the angle θ lies between $\frac{1}{2}\pi$ and π and the other is directed downward I_ν^+ for which the angle θ lies between 0 and $\frac{1}{2}\pi$. The plates

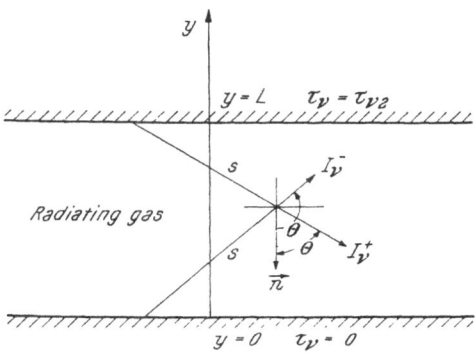

Fig. 5.3. Radiative transfer between two parallel plates

are situated respectively at $y = 0$ and $y = L$. Their optical thickness are respectively $\tau_\nu = 0$ and $\tau_\nu = \tau_{\nu2}$. The general expression of specific intensity of radiation (5.32) gives the following expressions for I_ν^+ and I_ν^-:

$$I_\nu^+ = \int\limits_{\tau_\nu}^{\tau_{\nu_2}} B_\nu(t) \exp[-m(t-\tau_\nu)]\,m\,dt + I_\nu^+(\tau_{\nu_2}) \exp[m(\tau_\nu - \tau_{\nu_2})] \tag{5.39 a}$$

$$I_\nu^- = -\int\limits_0^{\tau_\nu} B_\nu(t) \exp[-m(t-\tau_\nu)]\,m\,dt + I_\nu^-(0) \exp(m\tau_\nu) \tag{5.39 b}$$

where $I_\nu^+(\tau_{\nu_2})$ is the specific intensity at the upper wall $\tau_\nu = \tau_{\nu_2}$ and $I_\nu^-(0)$ id the specific intensity at the lower wall $\tau_\nu = 0$.

The flux of radiative energy across the surface $d\sigma_0$ parallel to the plate in the frequency interval ν and $\nu + d\nu$ is

$$q_{R\nu} = \int I_\nu \cos\theta\,d\omega = q_{R\nu}^+ - q_{R\nu}^- \tag{5.40}$$

where the net flux $q_{R\nu}$ is divided into the contribution due to I_ν^+ and that due to I_ν^- such that

$$q_{R\nu}^+ = 2\pi \int\limits_0^{\pi/2} I_\nu^+ \cos\theta\,\sin\theta\,d\theta = 2\pi \int\limits_{\tau_\nu}^{\tau_{\nu_2}} B_\nu(t)\,\varepsilon_2(t-\tau_\nu)\,dt + 2\pi I_\nu^+(\tau_{\nu_2})\,\varepsilon_3(\tau_{\nu_2} - \tau_\nu)$$

$$\tag{5.41}$$

and

$$q_{Rv}^- = -2\pi \int_{\pi/2}^{\pi} I_v^- \cos\theta \sin\theta \, d\theta = 2\pi \int_0^{\tau_v} B_v(t)\, \varepsilon_2(\tau_v - t)\, dt + 2\pi\, I_v^-(0)\, \varepsilon_3(\tau_v) \quad (5.42)$$

The total radiation flux is the integration of the specific heat flux q_{Rv} for all the frequencies, i.e.,

$$q_R = \int_0^\infty q_{Rv}\, dv = \int_0^\infty (q_{Rv}^+ - q_{Rv}^-)\, dv$$

$$= 2\pi \int_0^\infty \int_{\tau_v}^{\tau_{v_2}} B_v(t)\, \varepsilon_2(t - \tau_v)\, dt\, dv + 2\int_0^\infty q_{Rv}^+(\tau_{v2})\, \varepsilon_3(\tau_{v2} - \tau_v)\, dv \quad (5.43)$$

$$- 2\pi \int_0^\infty \int_0^{\tau_{v_2}} B_v(t)\, \varepsilon_2(\tau_v - t)\, dt\, dv - 2\int_0^\infty q_{Rv}^-(0)\, \varepsilon_3(\tau_v)\, dv$$

where

$$q_{Rv}^+(\tau_{v_2}) = \pi\, I_v^+(\tau_{v_2}), \qquad q_{Rv}^-(0) = \pi\, I_v^-(0) \quad (5.44)$$

both $q_{Rv}^+(\tau_{v_2})$ and $q_{Rv}^-(0)$ depend on the boundary conditions of the walls which will be discussed in chapter VI.

Equation (5.43) cannot be simplified without further assumption. One way to simplify equation (5.43) is to use the gray gas approximation so that the optical thickness τ_v is independent of the frequency v, i.e., $\tau_v = \tau$ and $\tau_{v_2} = \tau_2$. Under the gray gas approximation, the exponential integrals $\varepsilon_n(t)$ are independent of the frequency v and we may carry out the integration with respect to v in equation (5.43) and obtain the following expression:

$$q_R = 2\pi \int_\tau^{\tau_{v_2}} B(t)\, \varepsilon_2(t - \tau)\, dt - 2\pi \int_0^\tau B(t)\, \varepsilon_2(\tau - t)\, dt +$$

$$+ 2\, q_R(\tau_2)\, \varepsilon_3(\tau_2 - \tau) - 2\, q_R(0)\, \varepsilon_3(\tau) \quad (5.45)$$

where $q_R(\tau_2)$ and $q_R(0)$ are the integrated value of $q_{Rv}^+(\tau_{v_2})$ and $q_{Rv}^-(0)$ over all the frequencies respectively. $B(t)$ is given by equation (4.27).

In the equation of energy (5.7), we need the divergence of the radiation flux q_R which is in the present case for a non-gray gas (5.43) is

$$\frac{\partial q_R}{\partial y} = \frac{\partial \tau_v}{\partial y}\frac{\partial q_R}{\partial \tau_v} = \int_0^\infty \frac{2}{L_{Rv}}\left\{\left[\int_{\tau_v}^{\tau_{v_2}} \pi\, B_v(t)\, \varepsilon_1(t - \tau_v)\, dt - \pi\, B_v(\tau_v)\right]\right.$$

$$+ q_{Rv}^+(\tau_{v_2})\, \varepsilon_2(\tau_{v_2} - \tau_v) + q_{Rv}^-(0)\, \varepsilon_2(\tau_v) \quad (5.46)$$

$$\left.+ \left[\int_0^{\tau_y} \pi\, B_v(t)\, \varepsilon_1(\tau_v - t)\, dt - \pi\, B_v(\tau_v)\right]\right\} dv$$

In general the specific intensity on the walls, $q_{Rv}^+(\tau_{v_2})$ and $q_{Rv}^-(0)$ may be functions of frequency v. For gray surface, both $q_{Rv}^+(\tau_2)$ and $q_{Rv}^-(0)$ are independent of the frequency ν. We shall discuss the surface conditions in chapter VI.

For a gray gas, equation (5.46) becomes

$$
\frac{\partial q_R}{\partial y} = \frac{\partial \tau}{\partial y} \frac{\partial q_R}{\partial \tau} = \frac{c a_R}{L_{RP}} \left\{ \frac{1}{2} \int_0^{\tau_2} T^4(t) \, \varepsilon_1(|t-\tau|) \, dt - T^4(\tau) \right\} + \\
\left\{ + 2 \, q_R(\tau_2) \, \varepsilon_2(\tau_2 - \tau) + 2 \, q_R(0) \, \varepsilon_2(\tau) \right\} \frac{1}{L_{RP}}
$$
(5.47)

where $q_R(\tau_2)$ and $q_R(0)$ are the integral values of $q_{R\nu}^+(\tau_{\nu_2})$ and $q_{R\nu}^-(0)$ with respect to the frequency ν respectively.

If there is only one plate, we obtain the corresponding expression of radiative flux by putting τ_{ν_2} or τ_2 equal to infinity and setting the terms of $q_{R\nu}^+(\tau_{\nu_2})$ or $q_R(\tau_2)$ to be zero.

For finite mean free path of radiation, the radiation terms are integral forms. In some engineering problems, these integrals may be simplified by further approximations as we shall discuss later in chapters VI, VIII and IX. If the mean free path of radiation is very large, we may have some very simple form for the radiative flux as we shall show in chapter VI, section 6.

11. The exponential integrals. In evaluating the expressions (5.43) to (5.47) we need to know the properties of the exponential integral (5.38), i.e.,

$$
\varepsilon_n(t) = \int_1^\infty m^{-n} \exp(-mt) \, dm = \int_0^1 z^{n-2} \exp(-t/z) \, dz
$$
(5.38)

where $m = 1/z = \sec\theta$.

We shall be concerned only with the positive integral values of n. The following relations of the exponential integrals are useful in radiation gasdynamics:

(i) The value of the exponential integral at $t = 0$.
Except for $\varepsilon_1(0)$, we have

$$
\varepsilon_n(0) = \int_1^\infty m^{-n} \, dm = \frac{1}{n-1}
$$
(5.48a)

for $n = 2, 3, 4$ etc. But

$$
\varepsilon_1(0) = +\infty
$$
(5.48b)

(ii) The differentials of $\varepsilon_n(t)$ with respect to t is

$$
\varepsilon_n'(t) = \frac{d}{dt}[\varepsilon_n(t)] = -\varepsilon_{n-1}(t)
$$
(5.49a)

for $n = 2, 3, 4$ etc. and

$$
\varepsilon_1'(t) = -[\exp(-t)]/t
$$
(5.49b)

(iii) Recurrence formula.

From equation (5.38), we have

$$n \, \varepsilon_{n+1}(t) = - \int_1^\infty \exp(-xt) \frac{\mathrm{d}}{\mathrm{d}x}(x^{-n}) \, \mathrm{d}x \qquad (5.50)$$

Integrating equation (5.50) by parts, we have

$$n \, \varepsilon_{n+1}(t) = \exp(-t) - t \, \varepsilon_n(t) \qquad (5.51)$$

for $n \geqslant 1$.

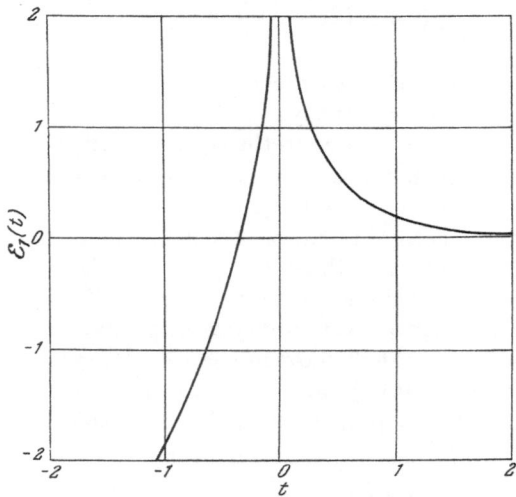

Fig. 5.4. The first exponential integral $\varepsilon_1(t)$

By means of the recurrence formula (5.51), all the exponential integrals with positive integral values of n can be expressed in terms of the first exponential $\varepsilon_1(t)$.

(iv) The first exponential integral

$$\varepsilon_1(t) = \int_1^\infty \exp(-mt) \frac{\mathrm{d}m}{m} = \int_t^\infty \exp(-m) \frac{\mathrm{d}m}{m} \qquad (5.38\,\mathrm{a})$$

has been tabulated in such book as the Table of functions (4). In reference 4, the notation $E_1(t)$ is used. The relation between $E_i(t)$ and $\varepsilon_1(t)$ is

$$\varepsilon_1(t) = - E_i(-t), \qquad (t > 0) \qquad (5.52)$$

where

$$E_i(t) = \int_{-\infty}^t \exp x \left(\frac{\mathrm{d}x}{x}\right), \qquad (-\infty < t < \infty) \qquad (5.53)$$

Equation (5.53) is integrated in the sense of the Cauthy principal value and valid for $t > 0$, i.e.,

$$E_i(t) = \lim_{\delta \to 0} \left(\int_\delta^\infty e^x \frac{\mathrm{d}x}{x} + \int_{-\infty}^{-\delta} e^x \frac{\mathrm{d}x}{x} \right), \qquad (t > 0) \qquad (5.54)$$

Fig. 5.4 shows the graph of the first exponential integral. At $t = -0.3725$, $\varepsilon_1(t) = 0$.

For all real values of t, we have the series form of $\varepsilon_1(t)$

$$\varepsilon_1(t) = -\gamma_0 - \log|t| + \sum_{n=1}^{\infty}(-1)^{n-1}\frac{t^n}{n \cdot n!} \tag{5.55}$$

where $\gamma_0 = 0.5772156--- =$ Euler-Mascheroni constant. Of course this series form is useful only for small values of t.

(v) The expansion of exponential integral as a power series of t.

From equation (5.50), we have the following expression:

$$(n-1)!\,\varepsilon_n(t) = (-t)^{n-1}\varepsilon_1(t) + e^{-t}\sum_{s=0}^{n-2}(n-s-2)!\,(-t)^s \tag{5.56}$$

Substituting equation (5.55) into equation (5.56) we have

$$\varepsilon_2(t) = 1 + t(\gamma_0 - 1 + \log t) - \frac{t^2}{2!1} + \frac{t^3}{3!2} - \frac{t^4}{4!3} + -- \tag{5.57}$$

$$\varepsilon_3(t) = \tfrac{1}{2} - t + \tfrac{1}{2}t^2\left(-\gamma_0 + \frac{3}{2} - \log t\right) + \frac{t^3}{3!1} - \frac{t^4}{4!2} + --- \tag{5.58}$$

The expressions (5.57) and (5.58) are very useful for small values of t. In the analysis of radiation gasdynamics, small value of t means optically thin case. We may approximate the exponential integrals by the first few terms of the series expansion of t. We shall use this approximation in studying the optically thin case.

(vi) Expression of the exponential integral for large value of t. For large value of t, we write $x = -1 + m$ and equation (5.37) gives

$$e^t\,\varepsilon_n(t) = \int_0^{\infty}\frac{e^{-tx}}{(1+x)^n}\,\mathrm{d}x = -\frac{1}{t}\int_0^{\infty}\frac{1}{(1+x)^n}\frac{\mathrm{d}e^{-tx}}{\mathrm{d}x}\,\mathrm{d}x$$

$$= \frac{1}{t} - \frac{n}{t}\int_0^{\infty}\frac{e^{-tx}\,\mathrm{d}x}{(1+x)^{n+1}} = \frac{1}{t}\left(1 - \frac{n}{t} + \frac{n(n+1)}{t^2} - :\dots\right) \tag{5.59}$$

For very large value of t, we have

$$\varepsilon_n(t) = e^{-t}/t \tag{5.60}$$

for all values of $n's$.

References

1. CHANDRASEKHAR, S.: An Introduction to the Study of Stellar Structure. Dover Publications, New York, 1957.

2. GOULARD, R.: Fundamental equations of radiation Gasdynamics. Purdue Univ. School of Aero. & Eng. Sci. report A & ES 62-4, 1962.

3. GOULARD, R. & M.: One dimensional energy transfer in radiant media. Int. Jour. Heat and Mass Transfer, vol. 1, pp. 81–91, 1960.

4. JAHNKE, E., and E. EMDE: Tables of Functions. Dover Publications, New York, 1945.

5. KIVEL, B., and K. BAILEY: Tables of Radiation from High temperature Air. AVCO Research Lab. Research Report 21, 1957.

6. KOURGANOFF, V.: Basic Methods in Transfer Problems. Oxford Press, 1952.

7. PAI, S. I.: Some considerations of radiation magnetogasdynamics. Proc. Non-linear Problem. University of Wisconsin Press, pp. 47–67, 1963.

8. PLANCK, M.: The Theory of Heat Radiation. Dover Publications, New York, 1959.

9. ROSSELAND, S.: Theoretical Astrophysics. Oxford Press, 1936.

10. SCALA, S. M., and D. H. SAMPSON: Heat Transfer in hypersonic flow with radiation and chemical reaction. Techn. Inform. Series R 63 D 46, Space Sciences Lab. General Electric Co., Phil. Pa., 1963. Also Supersonic Flow, Chemical Processes and Radiative Transfer, Pergamon Press, pp. 319–354, 1964

11. TELLEP, D. M., and D. K. EDWARDS: Radiant energy transfer in gaseous flows. Tech. report, Lockheed Missile & Space Div. LMSD-288139, 1960.

12. ZHIGULEV, V. N., YE. A. ROMISHEVSKII, and V. K. VERTUSHKIN: Role of Radiation in Modern Gasdynamics. AIAA Journal, vol. 1, No. 6, pp. 1473–1485, June 1963.

Chapter VI

Boundary Conditions of Radiation Gasdynamics

1. Introduction. For every particular problem of radiation gasdynamics, we have certain initial and boundary conditions. Our problem is to find solutions of the fundamental equations which satisfy these initial and boundary conditions. By initial conditions we mean the velocity distributions and the states of the gas as well as the specific intensity of radiation at certain initial time $t = 0$. Customarily in radiation gasdynamics, we do not give the spatial distribution of these initial conditions but we only require that the initial values be consistent with the boundary conditions for $t = 0$ and the fundamental equations. Hence we need to examine the boundary conditions only. In the case of dynamical system with a finite number of degrees of freedom, the motion is determined by the initial position and velocity. For a continuous medium, which has an infinite number of degrees of freedom, the motion is determined not only by the initial conditions but also by the boundary conditions, such as, for example, conditions on the velocity on the boundary of a domain considered at all the time $t \geqq 0$.

In radiation gasdynamics, we have to consider the boundary conditions of both the gasdynamic field and the radiation field. The boundary conditions of the gasdynamic field is well known and we shall discuss them briefly in § 2. The boundary conditions of the radiation field are not so well known to engineers and we shall discuss them in some detail. In many radiative transfer problems such as those in astrophysics, the effect of the boundary is usually negligibly small and simple approximation of the boundary effect is sufficient. However, in radiation gasdynamics of flow problem, the boundary effect is very important and we have to analyze it carefully. We discuss these boundary conditions in §§ 3 to 5. In § 6, we apply the results of section 5 to the case of the radiative transfer between two parallel plates while in § 7, to the case of emissivity of a constant gas layer which is used often in many engineering problems for the optically thin case. Finally in ordinary gasdynamics, if the mean free path is not small, we have the conditions of velocity slip and temperature jumps on the boundary. Similar situation occurs in radiation gasdynamics when the mean free path of radiation is not small. We shall discuss this case in § 8.

2. Boundary conditions of gasdynamic field (9). The boundary conditions of the gasdynamic field depend on the mean free path of the gas particles. When the mean free path is negligibly small, the gas may be considered as a continuum. The no-slip boundary condition is a good approximation. Under this condition, we have that across a surface separating a body and a fluid or two fluids, the velocity components, the stresses and the temperature are all continuous. In most of the problems discussed in this book, we consider the case that the gas may be considered as a continuum and these no-slip boundary conditions will be used.

When the mean free path of the gas is not small, even though the gas may still be considered as a continuum, there will be slip on the boundary. The velocity and the temperature of the gas at the wall may be different from those of the wall. Hence there will be a velocity slip and temperature jump across the boundary. This is known as slip flow in rarefied gasdynamics. We shall briefly discuss this case in chapter X. It is interesting to notice that there is a close analogy between the radiative transfer process and the molecular transport process which determines the boundary condition of the gasdynamics. In section 8 of this chapter we shall show that this temperature jump occurs when the mean free path of radiation is not negligibly small.

When the mean free path of the gas is larger than the dimension of the flow field, the gas cannot be considered as a continuum. We have the free molecular flow. The boundary condition depends on the smoothness of the surface. When the free molecules strike the wall, it may reflect diffusely or specularly. In general, part of the molecules will reflect diffusely and part of the molecules will reflect specularly. We shall discuss this case in chapter X. Again there is an analogy between the radiative transfer process and the molecular transport process when the mean free path of radiation is large. We shall discuss this case in sections 3 and 5.

Even though the no-slip boundary conditions are the correct boundary conditions for the gasdynamic field when the gas is considered as a continuum and when the mean free path of the gas is very small, such no-slip conditions may be relaxed under certain condition, particularly in the flow field far away from the boundary or transition region at very high Reynolds numbers. Under this condition, where the viscous effect is small, we may assume that the gas is inviscid. In the inviscid flow field, surface of discontinuity is allowable. The no-slip conditions should be relaxed. Two types of surface of discontinuity may occur in the inviscid flow field: the shock wave and the vortex surface. Across a shock wave, there is a discontinuity in the normal velocity but a continuity in the tangential velocity. On the other hand, across a vortex surface, there is a discontinuity in the tangential velocity but a continuity in the normal velocity. We shall discuss these conditions in detail when we treat special problems.

3. Boundary conditions of radiation field. The boundary condition of the radiation fields depends on the mean free path of radiation. When the mean free path of the radiation is very small, i.e., the optically thick case, all the radiation terms can be expressed in terms of the temperature, if we assume that the local thermodynamic equilibrium is attained as shown in chapter V, section 8. Under this condition, we do not need explicit the boundary condition of the specific intensity of radiation and we need only the boundary condition of temperature. Hence the consideration of gasdynamic boundary conditions is sufficient for this case of radiation gasdynamics.

When the mean free path of radiation is small but not negligibly small, we find that the fundamental equations of radiation gasdynamics obtained in chapter V, section 8 are sufficient to describe the flow field away from the boundary, but the no slip condition of temperature will not be satisfied. Hence in analogy to the slip flow of rarefied gasdynamics, we have a temperature jump in this case. We shall discuss this case in detail in section 8 of this chapter.

When the mean free path of radiation is not small, we have to consider the interaction of radiation at the interface of two media. When a ray of radiation strikes a surface, part of the radiative energy may penetrate into the second medium and part of the energy may reflect. Hence an incident ray may result a transmitted ray and a reflected ray. However radiation gasdynamics is a

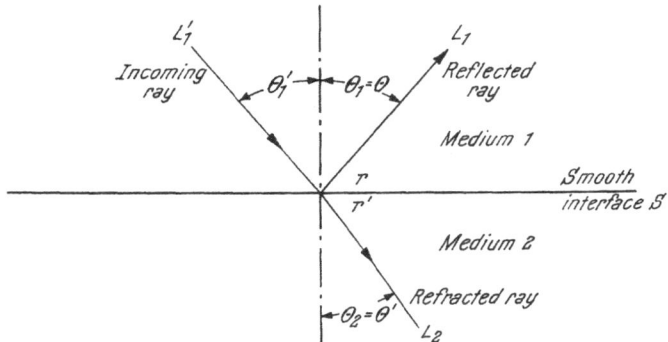

Fig. 6.1. Reflection and refraction of a ray on a smooth surface

macroscopic analysis. We do not study each ray individually but a large number of rays statistically. We have to use statistical average to consider the actual surface conditions in the engineering problems. In practice case we may classify the surface as "optically smooth" or "optically rough." For optically smooth surface the interaction of radiation and surface follows the well known Snell-Fresnel laws which will be discussed in section 4.

For most of the surfaces in engineering problem, there are not optically smooth. For optically rough surface, we have to use some phenomenological coefficients to describe the properties of the surface. We shall discuss the boundary conditions on such optically rough surface in section 5.

4. Smooth surface. We consider an optically smooth surface separating two media which have respectively the index of refraction n_1 and n_2 where

$$n_i = c/q_i \tag{6.1}$$

and c is the velocity of light in vacuum and q_i is the velocity of light in the ith medium (Fig. 6.1). When a ray of radiation L_1' from the medium 1 falls on the separating surface, it is divided into two rays: one is the reflected ray L_1 and the other is the transmitted ray L_2. The relations between the angles θ_1', θ_1 and θ_2 of these rays shown in Fig. 6.1 are given by the well known Snell-Fresnel laws as follows:

$$\sin \theta_1 = \sin \theta_1' \tag{6.2}$$

and

$$\frac{\sin\theta_1}{\sin\theta_2} = \frac{q_1}{q_2} = \frac{n_2}{n_1} \tag{6.3}$$

Let r be the coefficient of reflection of the surface S which is the fraction of radiation energy reflected by the surface. Then $(1-r)$ is the fraction of radiation energy transmitted into the second medium. In general, the coefficient of reflection r depends on the angle of incidence θ. We may use r' to denote the coefficient of reflection of a ray coming from the medium 2 and incident on the surface S. Let us consider the balance of radiation energy on an elementary surface $d\sigma_0$ of surface S. The radiative energy of the incident ray L_1' is

$$E_\nu = I_\nu \cos\theta \, d\sigma_0 \, dt \, d\nu \, d\omega \tag{6.4}$$

The radiative energy reflected by the surface L_1 is

$$E_{\nu r} = r \, I_\nu \cos\theta \, d\sigma_0 \, dt \, d\nu \, d\omega \tag{6.5}$$

where the elementary solid angle is $d\omega = \sin\theta \, d\theta \, d\phi$. The energy transmitted from the medium 2 to medium 1 is

$$E_\nu' = (1-r') \, I_\nu' \cos\theta' \, d\sigma_0 \, dt \, d\nu \, d\omega' \tag{6.6}$$

where $d\omega' = \sin\theta' \, d\theta' \, d\phi$. According to the law of refraction $\phi = \phi'$. Hence the relation $E_\nu = E_{\nu r} + E_\nu'$ gives

$$I_\nu \cos\theta \, d\omega \, (1-r) = (1-r') \, I_\nu' \cos\theta' \, d\omega' \tag{6.7}$$

By equation (6.3), we have

$$\cos\theta \, d\omega = \cos\theta' \, d\omega' \, (q^2/q'^2) \tag{6.8}$$

Substituting equation (6.8) into equation (6.7), we have

$$\frac{I_\nu}{I_\nu'} \frac{q^2}{q'^2} = \frac{1-r'}{1-r} \tag{6.9}$$

In general the coefficient of reflection depends on angle of incidence. For a special case, if $r = 0$ and then $r' = 0$. It shows that the lefthand side expression of equation (6.9) must be a constant unity and we have the general relations:

$$r = r' \tag{6.10}$$

and

$$q^2 I_\nu = q'^2 I_\nu' \tag{6.11}$$

The coefficient of reflection of a bounding surface are the same on both sides. For thermodynamic equilibrium condition, we have

$$q^2 I_\nu = c^2 B_\nu \tag{6.12}$$

The expression on the right-hand side is independent of the substance. Hence the expression on the left-hand side is also independent of the substance. Equation (6.12) gives that for thermodynamic equilibrium condition, the radiative specific intensity I_ν is

$$I_\nu = n^2 B_\nu \qquad (6.13)$$

In our previous analysis, we assume implicitly that for gas $n = 1$ and $I_\nu = B_\nu$ [cf. eq. (5.11)]. For a medium with index of refraction n, the radiative transfer equation for local thermodynamic equilibrium condition should be written as follows:

$$\frac{1}{\rho \, k_\nu'} \frac{\partial I_\nu}{\partial s} = n^2 B_\nu - I_\nu \qquad (6.14)$$

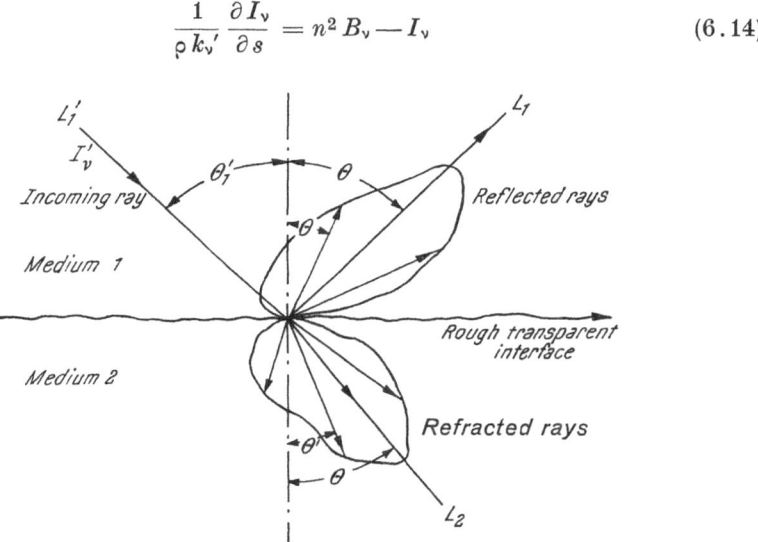

Fig. 6.2. Reflection and refraction of ray on a rough transparent surface

5. Rough surface. For smooth surface we need only to match the rays of radiation at the interface according to the Snell-Fresnel laws. However most of the interfaces between media are not "optically smooth." This means that for macroscopic point of view the incoming rays will not be refracted and reflected along two directions only but that both the refracted and reflected radiation will be distributed in many directions as shown in Fig. 6.2. We have to introduce special properties and distribution functions of the surface to describe this behavior. Fig. 6.2 represents a transparent rough surface. Another very common surface in engineering problems is opaque rough surface which absorbs all the radiative energy entering into it within a few molecular layers adjacent to the surface. Fig. 6.3 shows a sketch for such an opaque surface. We are going to discuss the properties of this opaque rough surface in detail and will briefly mention the results for transparent rough surface at the end of this section.

If we want to know the detail picture of the radiation field over the opaque surface, we have to know the directional absorptivity, reflectivity and emissivity. On the other hand for many problems of radiation gasdynamics, the overall absorption coefficient, reflection coefficient and emissivity coefficient of the surface will give sufficient data to solve our problem. The definitions of these coefficients are as follows:

(i) Directional absorptivity and reflectivity.

For an incoming ray of radiation L_1' of specific intensity I_ν', the radiative energy absorbed by the surface may be expressed in terms of a directional absorptivity $a_{\theta',\nu}$ of the surface which is in general a function of both θ' and ν. The radiative energy absorbed is then

$$\mathrm{d}E_{\nu,a} = a_{\theta',\nu}\, I_\nu'\, \cos\theta'\, \mathrm{d}\omega'\, \mathrm{d}\sigma_0\, \mathrm{d}\nu\, \mathrm{d}t \qquad (6.15)$$

The energy reflected by the surface may be expressed in terms of a directional

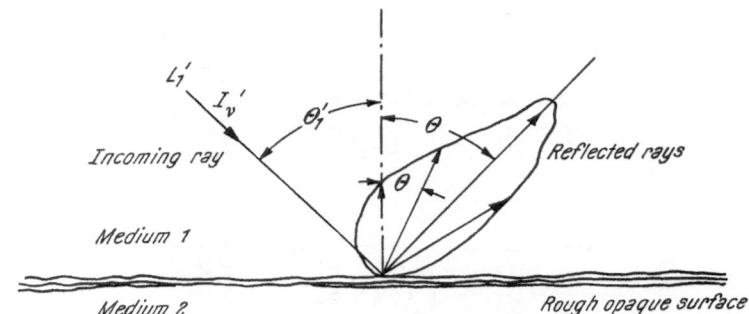

Fig. 6.3. Reflection of a ray on a rough opaque surface

reflectivity $r_{\theta',\nu}$ of the surface which is also a function of θ' and ν. The radiative energy reflected is then

$$\mathrm{d}E_{\nu,r} = r_{\theta',\nu}\, I_\nu'\, \cos\theta'\, \mathrm{d}\omega'\, \mathrm{d}\sigma_0\, \mathrm{d}\nu\, \mathrm{d}t \qquad (6.16)$$

For an opaque surface, the sum of the radiative energy absorbed and that reflected must be equal to the incoming energy $I_\nu'\, \cos\theta'\, \mathrm{d}\omega'\, \mathrm{d}\sigma_0\, \mathrm{d}\nu\, \mathrm{d}t$. We have

$$r_{\theta',\nu} + a_{\theta',\nu} = 1 \qquad (6.17)$$

The directional reflectivity $r_{\theta',\nu}$ gives the total amount of radiative energy reflected from an incoming ray of angle θ'. But the directional reflectivity does not give us how the reflected energy is distributed in terms of the reflected angle θ. This angular distribution of reflected energy of a given surface must be determined experimentally. Without any knowledge of the actual condition of the surface, we can only consider two extremely simple cases:

(a) Specular reflection in which $\theta = -\theta'$. In this case, the reflected rays follows the law of classical optics for a smooth surface and then

$$I_{\nu s} = r_{\theta',\nu s}\, I_\nu' \qquad (6.18)$$

where subscript s refers to specular reflection.

(b) Diffuse reflection in which the reflected energy is distributed uniformly in all direction of reflected angle θ. Let $I_{\nu d}$ represent the intensity of the reflected radiative energy due to an incoming ray of specific intensity I_ν', the total reflected energy may be written as

$$\mathrm{d}E_{\nu r} = r_{\theta',\nu d}\, I_\nu'\, \cos\theta'\, \mathrm{d}\omega'\, \mathrm{d}\sigma_0\, \mathrm{d}\nu\, \mathrm{d}t = \pi\, I_{\nu d}\, \mathrm{d}\omega'\, \mathrm{d}\sigma_0\, \mathrm{d}\nu\, \mathrm{d}t$$

or

$$I_{\nu d} = r_{\theta' \nu d} I_\nu' \frac{\cos \theta'}{\pi} \tag{6.19}$$

where $r_{\theta' \nu d}$ is the directional reflectivity in diffuse reflection.

In the actual case, it is most probable that the reflection pattern differs appreciably from both of these ideal cases. From experimental results, we may determine the true reflection distribution function $p_{\theta - \theta' \nu}$, analogous to the phase function of scattering of chapter III, § 2, such that

$$p_{\theta - \theta' \nu} = r_{\theta - \theta' \nu} / r_{\theta' \nu} \tag{6.20}$$

where $r_{\theta - \theta' \nu}$ is known as the reflection coefficient which represents the energy reflected from the incoming angle θ' to the reflected angle θ and which is defined by the following formula:

$$r_{\theta - \theta', \nu} = \frac{I_{\nu, \theta - \theta'}}{I_\nu'} \frac{\pi}{\cos \theta'} \tag{6.21}$$

where $I_{\nu, \theta - \theta'}$ is the specific intensity of the reflected ray in the direction θ from the incoming angle θ'. The true reflection distribution $p_{\theta - \theta' \nu}$ should be normalized by the relation

$$\int_{2\pi} p_{\theta \theta' \nu} \cos \theta \frac{d \omega}{\pi} = 1 \tag{6.22}$$

If the true reflection distribution is known, we may calculate the hemispherical reflectivity coefficient r_ν, usually called the reflection coefficient of the surface as follows:

$$r_\nu = \frac{\displaystyle\int_{2\pi} \cos \theta \, d\omega \int_{2\pi} r_{\theta' \nu} I_\nu' \, p_{\theta \theta' \nu} \frac{\cos \theta' \, d\omega'}{\pi}}{\displaystyle\int_{2\pi} I_\nu' \cos \theta' \, d\omega'} \tag{6.23}$$

In radiation gasdynamics, this reflection coefficient r_ν is used in the boundary conditions. Similarly, we may also define a hemispherical absorptivity coefficient a_ν, usually known as the absorption coefficient of the surface. By equation (6.17), we have

$$a_\nu + r_\nu = 1 \tag{6.24}$$

where both a_ν and r_ν are in general functions of frequency.

The above analysis may be easily extended into the case of transparent rough surface, if we introduce a hemispherical transparency coefficient tr_ν, usually known as transparency coefficient of the surface. Hence equation (6.24) should be replaced by the following equation:

$$a_\nu + r_\nu + tr_\nu = 1 \tag{6.25}$$

For opaque surface $tr_\nu = 0$.

(ii) Directional emissivity and radiosity.

As we have shown in chapter IV that the absorption of radiation is closely connected with emission, the same thing is true for the wall condition of a surface.

The energy emitted from the surface in the direction θ consists of the true emission and the energy due to reflection. The truly emitted radiative energy may be expressed in terms of the directional emissivity $e_{\theta v}$, i.e.,

$$\mathrm{d}E_{ve} = e_{\theta v} B_v \cos\theta \, \mathrm{d}\omega \, \mathrm{d}\sigma_0 \, \mathrm{d}v \, \mathrm{d}t = I_{ve} \cos\theta \, \mathrm{d}\omega \, \mathrm{d}\sigma_0 \, \mathrm{d}v \, \mathrm{d}t \qquad (6.26)$$

where B_v is the specific intensity of a black surface which is the maximum possible value of the specific intensity under the local thermodynamic equilibrium and I_{ve} is the specific intensity due to true emission. Equation (6.26) gives

$$I_{ve} = e_{\theta v} B_v \qquad (6.27)$$

The directional emissivity $e_{\theta v}$ is less or equal to unity and is in general a function of both the angle θ and the frequency v. Both theory and experiment show that for dielectric surface, the directional emissivity tends to be large and fairly independent of θ but for metallic surface, it is small and strongly dependent on the angle θ.

If the surface is in thermodynamic equilibrium with its environment, the radiative energy absorbed from the ray L is

$$\mathrm{d}E_{va} = a_{\theta v} B_v \cos\theta \, \mathrm{d}\omega \, \mathrm{d}\sigma_0 \, \mathrm{d}v \, \mathrm{d}t \qquad (6.28)$$

In equilibrium condition, the energy given by equation (6.28) must be equal to that of equation (6.26). Hence we have

$$a_\theta = e_\theta \qquad (6.29)$$

Equation (6.29) is the Kirchhoff's law for solid surface [cf. eq. (4.5)] which holds even when the medium is not in equilibrium with the body surface.

Similar to equation (6.23), we may define a hemispherical emissivity coefficient e_v, usually known as the coefficient of emission of the surface and from equation (6.29), we have

$$e_v = a_v \qquad (6.30)$$

The total radiative energy emitted per unit area per unit time and per unit frequency interval is known as Radiosity R_a which is defined as

$$R_a = \int_{2\pi} I_{\theta v T} \cos\theta \, \mathrm{d}\omega \qquad (6.31)$$

where $I_{\theta v T}$ is the specific intensity of radiation in the direction θ due to the total emission which consists of the true emission $I_{\theta ve}$ and the reflected radiation $I_{\theta v} i$, i.e.,

$$I_{\theta v T} = I_{\theta v e} + I_{\theta v} r = e_{\theta v} B_v + \int_{2\pi} r_{\theta' v} I_{v'} \, p_{\theta - \theta' v} \frac{\cos\theta' \, \mathrm{d}\omega'}{\pi} \qquad (6.32)$$

In summary, in radiation gasdynamics, for rough surface, we need to know three overall coefficients in determining the boundary conditions of specific intensity, i.e.,

(i) The hemispherical absorptivity coefficient a_v or simply called the absorption coefficient of the surface which is equal to the hemispherical emissivity coefficient e_v or simply called the emissivity coefficient of the surface.

(ii) Hemispherical reflectivity coefficient r_v or simply called the reflection coefficient of the surface, and

(iii) Hemispherical transparency coefficient tr_v or simply the transparency coefficient of the surface.

These three coefficients are connected by equation (6.25). But the value of each of these coefficients depends on the condition of the surface which should be determined experimentally.

There are a few cases which are of interest in the general discussion of radiation gasdynamics which are listed as follows:

(i) Rough opaque surfaces. For such surfaces, the coefficient of transparency tr_v is always zero and we have

$$a_v + r_v = 1 \tag{6.33}$$

(ii) Gray surface. For such surfaces, all the three coefficient a_v, r_v and tr_v are independent of the frequency v.

(iii) Black surface. For such surface, both the coefficients of reflectivity and of transparency are zero and then we have

$$a_v = e_v = 1 \tag{6.34}$$

The specific intensity of this kind of surface under the thermodynamic equilibrium is then

$$I_v(w) = n^2 B_v(T_w) \tag{6.35}$$

where T_w is the temperature of the surface or wall. Since for most engineering problems, the index of refraction n is very close to unity, we may take $n = 1$ and the specific intensity of a black wall is then

$$I_v(w) = B_v(T_w) \tag{6.35a}$$

For non-black but opaque surface, the specific intensity on the wall under thermo-dynamic equilibrium condition with $n = 1$ is

$$I_v(w) = e_v B_v(T_w) = (1 - r_v) B_v(T_w) \tag{6.36}$$

Equations (6.35) and (6.36) shall be used to solve various flow problems in radiation gasdynamics later.

6. Radiative transfer between two opaque parallel plates (5). Now we reconsider the problem discussed in chapter V, § 10, i.e., the radiative transfer between two parallel plates with some detailed considerations on the boundary values $I_v^-(0)$ and $I_v^+(\tau_2)$. We assume that gray gas approximation holds true here. The surfaces are opaque rough surfaces whose optical properties are given by the two coefficients e_w and r_w which are assumed to be independent of the frequency v. Hence the surface is also gray. These coefficients satisfy equations (6.24) and (6.30). The coefficients of emissivity for the upper and lower walls are respectively e_{wL} and e_{w0} and the corresponding coefficient of reflectivity are r_{wL} and r_{w0}. We consider first the radiative rays coming from the upper wall to the point P, i.e., $I_v^+(\tau_2)$, which is given by equation (5.39a), (see Fig. 6.4). The specific intensity $I_\nu{}^+ (\tau_2)$ is the cumulative effects of many radiative phenomena which consist of (i) the emission from the plate at point (2), (ii) the reflection of the ray (3), (iii) the reflection of the emitted ray at point (4), (iv) the double reflection of the ray (5) and so on. If we carry out the calculation of the attenuation of the various rays, the total radiative flux of equation (5.43) for the present case becomes:

$$
\begin{aligned}
q_R = q_R^+ - q_R^- = &\ 2\pi \int_\tau^{\tau_2} B(t)\, \varepsilon_2\,(t-\tau)\, \mathrm{d}t - 2\pi \int_0^\tau B(t)\, \varepsilon_2\,(\tau-t)\, \mathrm{d}t + \\
& + \frac{2}{1-4r_{wL}\, r_{w0}\, \varepsilon_3{}^2\,(\tau_2)} \Big\{ e_{wL}\, B(\tau_2)\, \varepsilon_3\,(\tau_2-\tau) - \varepsilon_{w0}\, B(0)\, \varepsilon_3\,(\tau) + \\
& + e_{w0}\, B(0)\, \varepsilon_3\,(\tau_2)\, 2\, r_{wL}\, \varepsilon_3\,(\tau_2-\tau) - e_{wL}\, B(\tau_2)\, \varepsilon_3\,(\tau_2)\, 2\, r_{w0}\, \varepsilon_3\,(\tau) + \\
& + 2 r_{wL}\, \varepsilon_3\,(\tau_2-\tau) \int_0^{\tau_2} B(t)\, \dot\varepsilon_2\,(t-\tau_2)\, \mathrm{d}t - 2 r_{w0}\, \varepsilon_3\,(\tau) \int_0^{\tau_2} B(t)\, \varepsilon_2\,(t)\, \mathrm{d}t + \\
& + 4 r_{w0}\, r_{wL}\, \varepsilon_3\,(\tau_2)\, \varepsilon_3\,(\tau_2-\tau) \int_0^{\tau_2} B(t)\, \varepsilon_2\,(t)\, \mathrm{d}t - \\
& - 4 r_{w0}\, r_{wL}\, \varepsilon_3\,(\tau_2)\, \varepsilon_3\,(\tau) \int_0^{\tau_2} B(t)\, \varepsilon_2\,(t-\tau_2)\, \mathrm{d}t \Big\}
\end{aligned}
\tag{6.37}
$$

Equation (6.37) is a general expression for the gray gas non-scattering medium. This expression can be simplified for special cases. Some of these simplied cases are listed below:

(i) Slab of very large optical thickness:

If the distance between the two plates is very large, we have $\tau_2 \to \infty$. Since both $\varepsilon_2(t)$ and $\varepsilon_3(t)$ tend to be zero as t tends to be infinite, we may neglect all the terms in equation (6.37) except the first two terms. Hence we have

$$
q_R = 2\pi \int_\tau^\infty B(t)\, \varepsilon_2\,(t-\tau)\, \mathrm{d}t - 2\pi \int_{-\infty}^\tau B(t)\, \varepsilon_2\,(\tau-t)\, \mathrm{d}t
\tag{6.38}
$$

This is the expression has been used often in the study of star photosphere or blast waves where the boundary effects are negligible.

(ii) Transparent medium between two gray parallel plates. For optically thin gas, the optical thickness is very small, i.e., $\tau \to 0$. Then we may use the approximation:

$$
\varepsilon_3\,(\tau) = \varepsilon_3\,(0) = \tfrac{1}{2}
\tag{6.39}
$$

Furthermore for opaque surface, we have $e_w = 1 - r_w$. Now all the integral terms in equation (6.37) can be neglected because $\tau \gtrsim \tau_2 \gtrsim 0$ and we have

$$
q_R = \sigma\,(T_2{}^4 - T_0{}^4)\, \frac{1}{(1/e_{w0}) + (1/e_{wL}) - 1}
\tag{6.40}
$$

This is an expression which is useful for rarefied gasdynamics, see chapter X.

(iii) Absorbing medium between two black plates at the same temperature T_w. For a black surface $r_w = 0$ and $e_w = 1$. Substituting these values of the coefficients of the surface into equation (6.37), we have

$$q_R = 2\,\pi \int_{\tau}^{\tau_2} B\,(t)\,\varepsilon_2\,(t - \tau)\,\mathrm{d}t - 2\,\pi \int_{0}^{\tau} B\,(t)\,\varepsilon_2\,(\tau - t)\,\mathrm{d}t + \tag{6.41}$$

$$+ 4\,\pi\,B\,(\tau_2)\,\varepsilon_3\,(\tau_2 - \tau) - 4\,\pi\,B\,(0)\,\varepsilon_3\,(\tau).$$

For optically thin gas, $\tau \ll 1$, then $\varepsilon_2 \cong 1$ and $\varepsilon_3 = \frac{1}{2} - t$. Equation (6.41) reduces to

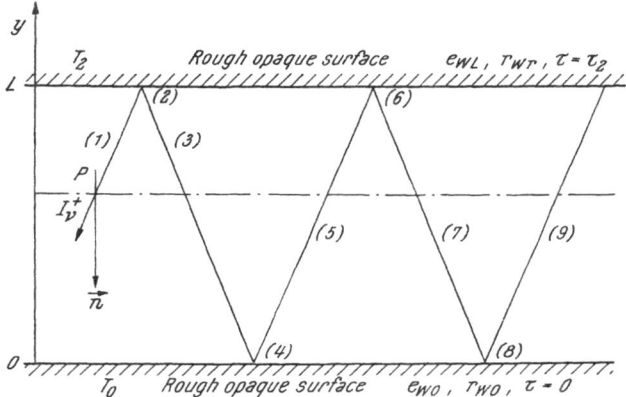

Fig. 6.4. Boundary conditions of radiative transfer between two parallel, rough, opaque and gray surface

$$q_R = 2 \int_{y}^{L} \sigma\,T^4\,\rho\,k_{\nu}'\,\mathrm{d}y - 2 \int_{0}^{y} \sigma\,T^4\,\rho\,k_{\nu}'\,\mathrm{d}y + \sigma\,(T_2{}^4 - T_0{}^4) - \tag{6.42}$$

$$- 2\,\sigma\,T_2{}^4\,\rho\,k_{\nu}'\,(L - y) + 2\,\sigma\,T_0{}^4\,\rho\,k_{\nu}'\,y$$

At the lower wall, $y = 0$, $\tau = 0$, equation (6.42) becomes (with $T_2 = T_0 = T_w$)

$$q_{Rw} = 2 \int_{0}^{L} \sigma\,T^4\,\rho\,k_{\nu}'\,\mathrm{d}y - 2\,\sigma\,T_w{}^4\,\rho\,k_{\nu}'\,L \tag{6.43}$$

The first term on the right-hand side is the emission of radiation in the gas and the second term may be considered as the emission of radiation from the wall. The net radiative flux to the wall is the difference of these two radiative flux.

7. Emissivity of a constant temperature gas layer (5). For many engineering problems, we would like to have a quick estimate of the radiative energy and to decide whether the effect of thermal radiation is large enough so that radiation gasdynamics is required to analyze the flow problem.   For instance, at hypersonic flow behind the detached shock of a blunt body (Fig. 6.5), the temperature of the gas behind the shock is very high and the radiative energy will not be negligible. In order to estimate the radiative energy approximately, we may make the following approximations:

(i) The shock layer may be replaced by a constant temperature gas layer of thickness L.

(ii) The detached shock may be considered as a transparent surface with $r_{wL} = 0$ and $e_{wL} = 0$.

(iii) The body may be considered as a black surface ($r_{w0} = 0$) with a low temperature so that $B(0) = 0$.

The radiative energy flux from the shock layer into the free stream is by equation (6.37):

$$q_R = 2\,\pi \int_0^{\tau_2} B(t)\,\varepsilon_2(t)\,\mathrm{d}t = \sigma\,T^4\,[1 - 2\,\varepsilon_3(\tau_2)] \qquad (6.44)$$

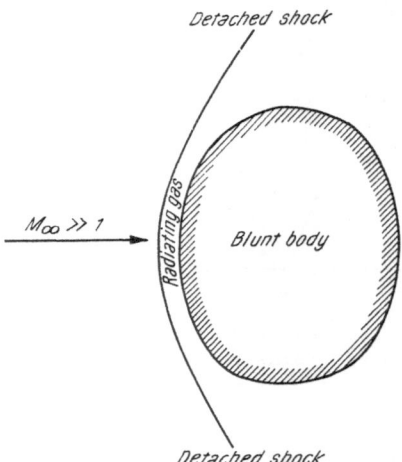

Fig. 6.5. Thermal radiation behind a detached shock in front a blunt body

If we further assume that the gas is optically thin, with $\varepsilon_3(t) = \frac{1}{2} - t$, equation (6.44) becomes

$$q_R = 2\,\rho\,k_{\nu}'\,L\,\sigma\,T^4 = e_g\,\sigma\,T^4 \quad (6.45)$$

where

$$e_g = 2\,\rho\,k_{\nu}'\,L = \text{emissivity}$$
$$\text{of the gas layer} \quad (6.46)$$

Hence the emissivity of the gas may be expressed in terms of the absorption coefficient. For instance, e_g/L has been tabulated in reference 7 for air.

It should be noticed that under the assumption of gray gas, we actually use certain mean absorption coefficient of the medium over the whole spectrum of frequency. In the optically thick case as we have shown in chapter V, § 8, the Rosseland mean absorption coefficient (5.18) is the proper mean value. In the optically thin case, as we have shown in chapter V, § 9, the Planck mean absorption coefficient (5.36) is the proper mean value. When we describe a special case as optically thick or optically thin, we compare the physical length with the mean free path of radiation. In other words, the optical thickness τ determines whether a special case is optically thick or optically thin. The value of the mean absorption coefficient or the corresponding mean free path of radiation L_R or L_p itself is not sufficient to determine whether a special case is optically thick or optically thin. In fact, for a given gas at a given temperature and density, the value of Planck mean absorption coefficient is usually larger than the Rosseland mean absorption coefficient. For instance, for high temperature air, Scala suggested that $K_p = 8.3\,K_R$. We shall discuss the variation of these two mean absorption coefficient in chapters VII and IX.

8. Radiation slip at finite mean free path of radiation. For finite mean free path of radiation, we have to solve the radiative transfer equation with the boundary conditions discussed in sections 3 to 6 and then to solve a system of integro-differential equations of radiation gasdynamics. The solution of this non-linear integro-differential equations system is very difficult to obtain. For small but finite mean free path of radiation, it is advisable to use a set of differential equations similar to those obtained in chapter V, § 8 for the case of very small mean free path of radiation. In chapter V, § 8, we assume that the mean free path of radiation is so small that it is not necessary to consider the boundary conditions of the radiation rays. However if the mean free path of radiation is not too small, we should consider both the fundamental equations and the boundary conditions and radiation slip will occur in such case.

· Let us reconsider the results of chapter V, § 10 by assuming that the upper wall is at infinity and the lower wall is an opaque surface of emissivity e_w and reflectivity r_w. We also assume that the gas is gray. Then equation (6.37) becomes:

$$q_R = 2\,\pi \int_\tau^\infty B\,(t)\,\varepsilon_2\,(t-\tau)\,\mathrm{d}t - 2\,\pi \int_0^\tau B\,(t)\,\varepsilon_2\,(\tau-t)\,\mathrm{d}t - 2\,e_w B\,(0)\,\varepsilon_3\,(\tau) -$$

$$\tag{6.47}$$

$$- 4\,\pi\,r_w\,\varepsilon_3\,(\tau) \int_0^\infty B\,(t)\,\varepsilon_2\,(t)\,\mathrm{d}t$$

We shall consider that the mean free path of radiation is small but not negligible. The exponential integral $\varepsilon_2\,(t-\tau)$ is negligibly small except in the vicinity $t \simeq \tau$. We thus may express the radiation function $B\,(t)$ in Taylor series of $(t-\tau)$ about the point $t=\tau$ in equation (6.47). First we consider a point in the medium far away from the wall so that $\varepsilon_3\,(\tau) = 0$. Equation (6.47) may be written as follows:

$$q_R = 2\,\pi \int_0^\infty \left[B\,(\tau) + \frac{(t-\tau)}{1!}\left(\frac{\partial B}{\partial t}\right)_\tau + \frac{(t-\tau)^2}{2!}\left(\frac{\partial^2 B}{\partial t^2}\right)_\tau + -- \right] \varepsilon_2\,(t-\tau)\,\mathrm{d}\,(t-\tau) -$$

$$- 2\,\pi \int_0^\infty \left[B\,(\tau) - \frac{(\tau-t)}{1!}\left(\frac{\partial B}{\partial t}\right)_\tau + \frac{(\tau-t)^2}{2!}\left(\frac{\partial^2 B}{\partial t^2}\right)_\tau --- \right] \varepsilon_2\,(\tau-t)\,\mathrm{d}\,(\tau-t)$$

$$= 2\,\pi \int_0^x \left[p\left(\frac{\partial B}{\partial t}\right)_\tau + \frac{p^3}{3!}\left(\frac{\partial^3 B}{\partial t^3}\right)_\tau + --- \right] \varepsilon_2\,(p)\,\mathrm{d}\,p \tag{6.48}$$

Now by the well known relation that

$$\frac{1}{m!} \int_0^\infty p^m\,\varepsilon_n\,(p)\,\mathrm{d}p = \frac{1}{m+n} \tag{6.49}$$

equation (6.48) gives

$$q_R = \frac{4}{3}\left(\frac{\partial B}{\partial t}\right)_\tau + 0\left[\left(\frac{\partial^3 B}{\partial t^3}\right)_\tau\right] = \frac{16}{3}\,\sigma\,T^3\,L_R\frac{\partial T}{\partial y} + 0\left[\left(\frac{\partial^3 B}{\partial t^3}\right)_\tau\right] \tag{6.50}$$

where L_R is the Rosseland mean free path of radiation defined by equation (5.19). If the second term of equation (6.50) is negligible in comparison with the first term, we have the same results obtained in chapter V, § 8 for the case of very small mean free path of radiation.

Now we consider the radiative flux at the wall, $y \gtrsim \tau \gtrsim 0$, i.e.,

$$q_R(0) = 2\pi e_w \left[\int_0^\infty B(t)\, \varepsilon_2(t)\, dt - B(0) \right] =$$

$$= \frac{8}{3}\, e_w\, \sigma\, T^3 L_R \frac{\partial T}{\partial y} + 0\left[\left(\frac{\partial^2 B}{\partial t^2} \right) \right] \tag{6.51}$$

where we assume that the wall is gray so that e_w is independent of the frequency ν. There is a jump of radiative flux at the wall, because the coefficient $8/3$ replaces $16/3$ if we take $e_w = 1$. Since this radiative flux jump is proportional to L_R, we may neglect this jump if the mean free path of radiation is negligibly small. Hence no-slip condition holds approximately when the mean free path of radiation is very small. On the other hand if the value L_R is not negligibly small, we should consider this jump of radiative flux by assuming that there is a temperature jump at the surface. Let T_0 be the temperature of the gas at the surface of the wall $y = 0$ and T_w be the temperature of the wall. For engineering purpose, we may assume a radiative film coefficient h_R such that the following relation holds:

$$h_R(T_0 - T_w) = (e_w - 2)\frac{8}{3}\, \sigma\, T_0{}^3 L_R \left(\frac{\partial T}{\partial y} \right)_0 \tag{6.52}$$

The radiative film coefficient h_R is analogous to the film coefficient in ordinary heat transfer problem and depends on the surface conditions. The value of h_R should be determined experimentally. As L_R tends to be zero, T_0 tends to be T_w.

References

1. BORN, M., and E. WOLF: Principle of Optics. Pergamon Press, New York.

2. CHANDRASEKHAR, S.: An Introduction to the Study of Stellar Structure. Dover Publications, New York, 1957.

3. GARDEN, R.: Calculation of temperature distribution in glass plates undergoing heat treatment. J. Am. Ceramic Soc. vol. 41, No. 6, pp. 200–209, 1958.

4. GOULARD, R.: Fundamental Equations of Radiation Gasdynamics. Purdue Univ. School of Aero. & Eng. Scien. report A & ES 62-4, 1962.

5. GOULARD, R. & M.: One dimensional energy transfer in radiant media. Int. J. Heat & Mass Transfer vol. 1, pp. 81–91, 1960.

6. JOHNSON, J. C.: Physical Meteorology. John Wiley & Sons, Inc., New York, 1954.

7. KIVEL, B., and K. BAILEY: Tables of radiation from high temperature air. AVCO Res. Lab. research report 21, 1957.

8. KOURGANOFF, V.: Basic methods in transfer problem. Oxford Univ. Press, 1952.

9. PAI, S. I.: Introduction of the theory of Compressible Flow. D. Van Nostrand Co., 1959.

10. PAI, S. I.: Some consideration of radiation magnetogasdynamics. Proc. Non-linear Problem. Univ. of Wisc. Press, pp. 47–67, 1963.

11. PENNER, S. S.: Quantitative molecular spectroscopy and gas emissivities. Addison Wesley, 1960.

12. PLANCK, M.: The Theory of Heat Radiation. Dover Publications, New York, 1959.

13. PROBSTEIN, R. F.: Radiation slip. M. I. T. Fluid Mechanics Lab. Dept. of Mech. Eng. report No. 63-2, 1963.

14. SCALA, S. M., and D. H. SAMPSON: Heat Transfer in hypersonic flow with radiation and chemical reaction. Tech. Inf. series R 630 SD 46, Space Sci. Lab. General Electric Co., Phil. Pa., 1963. Also Supersonic Flow, chemical processes and radiative transfer, Pergamon Press, pp. 319—354, 1964.

15. TELLEP, D. M., and D. K. EDWARDS: Radiant energy transfer in gaseous flows Tech. report Lockheed Missile & Space Div. LMSD-288139, 1960.

16. ROSSELAND, S.: Theoretical Astrophysics. Oxford Univ. Press, 1936.

Chapter VII

Similarity Parameters of Radiation Gasdynamics

1. Introduction. In general, the fundamental equations of radiation gas-dynamics is a system of non-linear integro-differential equations. It is extremely difficult to solve these equations for given boundary conditions. Even in the simple case of optically thick case that the integro-differential equations reduce to a set of non-linear differential equations (cf. chapter V, § 8), the fundamental equations are still much more complicated than those of Navier-Stokes equations of a compressible fluid without radiation effects. There is no general method of finding the solutions of these non-linear equations. In order to bring out the essential features of the flow problems of radiation gasdynamics it is desirable to find the important parameters which characterize the flow problems.

These parameters in non-dimensional form are known as similarity parameters because they show the relative effects of various forces and transfers in the flow field. From these parameters, one may have some guide how the flow with radiation effects differs from those without radiation effect. Thus we may divide the radiation flow field into various regions and proper approximations may be applied in these regions so that practically important flow problems can be solved.

These parameters are also useful in correlating the experimental results. In experimental investigations, it frequently happens that the test model is of a different size from the actual body, that the test fluid is in a different thermodynamic state from the actual fluid or that the test fluid is a different fluid than that existing in the actual conditions. If we want to compare the test results with the actual condition, we have to know the relation between the test condition and the actual situation. This relation depends mainly on some important parameters of the problem. If we know these parameters, it will be easy to correlate the experimental results.

There are two methods to find these important parameters: one is known as inspection analysis and the other is known as dimensional analysis, particularly by the well known π-theorem of dimensional analysis. In section 2, we first discuss briefly the dimensional analysis and the π-theorem and in section 3, the non-dimensional form of the fundamental equations of radiation gasdynamics will be discussed from the inspection analysis point of view. In the last two sections, the important parameters of radiation gasdynamics will be treated and many approximations based on the values of these parameters will be discussed.

2. Dimensional analysis and π-theorem. Every physical quantity has certain dimensions. But there are a few basic dimensions by which the dimensions of all the other physical quantities may be expressed. In radiation gasdynamics, there are four basic dimensions which are usually taken as the length L, the time t, the mass m and the temperature T. The dimensions of all other physical quantities of radiation gasdynamics can be expressed in terms of these four basic dimensions. For instance, the following are a few examples:

(i) Velocity U. The dimensions of a velocity are

$$U = \text{velocity} = \text{length/time} = L/t \tag{7.1}$$

(ii) Acceleration a_c. The dimensions of an acceleration are

$$a_c = \text{acceleration} = \text{velocity/time} = L/t^2 \tag{7.2}$$

(iii) Force F. The dimensions of a force are

$$F = \text{mass} \times \text{acceleration} = m\,L/t^2 \tag{7.3}$$

(iv) Pressure p. The dimensions of pressure p are

$$p = \text{pressure} = \text{force/area} = (m\,L/t^2)/L^2 = m/(L\,t^2) \tag{7.4}$$

(v) Stefan-Boltzmann constant a_R [cf. eq. (5.15)]. The dimensions of a_R may be determined by the following relation:

$$a_R = \text{pressure}/T^4 = m/(L\,t^2\,T^4) \tag{7.5}$$

(vi) The Rosseland diffusion coefficient of radiation D_R [cf. eq. (5.20)]. The dimensions of D_R are

$$D_R = \text{length} \times \text{velocity} = L^2/t \tag{7.6}$$

The basic principle of dimensional analysis is the homogeneity of dimensions in any equations of physical problem. In other words, the dimensions of every term in any equation must be the same. The basic theorem on which the application of dimensional analysis for finding the non-dimensional quantities characterizing the dynamic similarity of the flow field rests is known as the π-theorem of VISHY and BUCKINGHAM (2). We are not going to discuss the mathematical foundations of this famous theorem but to state theorem with some examples of its application.

The π-theorem may be stated as follows:

If there are n quantities, $Q_1, Q_2, ---, Q_n$, which are important in a physical problem and there are m independent basic units in this system, the complete equation for this problem may be written in functional form as

$$\phi\,(Q_1, Q_2, ---, Q_n) = 0 \tag{7.7}$$

and the final solution of the problem may be written in the form

$$f\,(\pi_1, \pi_2, ---, \pi_{n-m}) = 0 \tag{7.8}$$

where the non-dimensional parameters $\pi_1, \pi_2, ---, \pi_{n-m}$, are formed by those Q's. There are only $n - m$ independent non-dimensional parameters known as the π's but the choice of these non-dimensional parameters is arbitrary. Equation (7.8) may be also written as

$$\pi_1 = F\,(\pi_2, \pi_3, ---, \pi_{n-m}) \tag{7.9}$$

It should be noted that information as to the nature of this function $f(X)$ or $F(X)$ can naturally be obtained only through experiments or an actual theory of the physical phenomena in question.

The proof of this theorem may be found in standard textbook of dimensional analysis (2). We are going to treat a simple problem of radiation gasdynamics in order to illustrate the application of this π-theorem. We consider the flow field behind a detached shock in which the radiation phenomena are important but the viscosity and heat conductivity are negligible. The important physical quantities in this problem are the following ten quantities:

$Q_1 \;= L \;=$ typical length of the body in the flow

$Q_2 \;= U \;=$ typical velocity of the flow field

$Q_3 \;= \rho \;=$ typical density of the gas in the flow field

$Q_4 \;= T \;=$ typical temperature of the gas in the flow field

$Q_5 \;= p \;=$ typical pressure of the gas in the flow field

$Q_6 \;= R \;=$ gas constant which describe the properties of the gas

$Q_7 \;= t \;=$ typical time scale of the flow field

$Q_8 \;= L_R =$ mean free path of radiation which characterizes the heat transfer
     of radiation $= 1/(\rho\, K_R)$

$Q_9 \;= c \;=$ velocity of light which is the speed of the photons

$Q_{10} = a_R =$ Boltzmann-Stefan constant which characterizes the radiation
     stresses or radiation energy density

Since there are ten Q's and four basic units, we may form six non-dimensional parameters π_1, π_2, π_3, π_4, π_5, and π_6. We take the first four Q's, i.e., L, U, ρ and T as the fundamental units. Then we can determine the non-dimensional parameters by considering the dimensions of any one of the Q_5 to Q_{10} together with those of Q_1 to Q_4. For example, if we consider a_R, we have the following relations of dimensions:

	L	t	m	T
L	1	0	0	0
U	1	-1	0	0
ρ	-3	0	1	0
T	0	0	0	1
a_R	-1	-2	1	-4

We define a non-dimensional parameter from these five Q's, i.e.,

$$\pi_1 = a_R\, L^{b_1}\, U^{b_2}\, \rho^{b_3}\, T^{b_4} \tag{7.10}$$

where b_1, b_2, b_3 and b_4 are constants to be determined so that the parameter π_1 is non-dimensional, i.e.,

$$\begin{aligned}
1 &= b_1 + b_2 - 3\,b_3 \\
2 &= -\,b_2 \\
-1 &= b_3 \\
4 &= b_4
\end{aligned} \tag{7.11}$$

Hence we $b_1 = 0$, $b_2 = -\,2$, $b_3 = -\,1$ and $b_4 = 4$.

The parameter π_1 can then be written as

$$\pi_1 = \frac{\frac{1}{3} a_R T^4}{\frac{1}{2} \rho U^2} = \frac{\text{radiation pressure}}{\text{dynamic pressure}} \qquad (7.12)$$

As far as dimensions are concerned, equation (7.12) gives the ratio of the radiation pressure ($\frac{1}{3} a_R T^4$) and the dynamic pressure of the gas ($\frac{1}{2} \rho U^2$). The constant (1/3) and $\frac{1}{2}$ are used because the special physical significance of the radiation pressure and dynamics pressure.

Similarly the following non-dimensional parameters can be easily obtained:
(i) For pressure p, we have

$$\pi_2 = \frac{\frac{1}{2} \rho U^2}{p} = \frac{\text{dynamic pressure}}{\text{gas pressure}} \qquad (7.13)$$

(ii) For gas constant R, we have

$$\pi_3 = \frac{\rho R T}{p} = \text{compressibility factor} \qquad (7.14)$$

This parameter refers to the equation of state of the gas. For ideal perfect gas, it is unity. For real gas, it may be different from unity.
(iii) For time t, we have

$$\pi_4 = \frac{t U}{L} = R_t \qquad (7.15)$$

The parameter R_t represents the unsteadiness of the flow field.
(iv) For the mean free path of radiation L_R we have

$$\pi_5 = L_R / L = K_r \qquad (7.16)$$

This parameter is closely related to the optical thickness.
(v) For the velocity of light c, we have

$$\pi_6 = U / c = R_r \qquad (7.17)$$

This parameter R_r is known as the relativistic parameter.

There are only six independent parameters in the present problem. However we may form other non-dimensional parameters by combining several of these parameters. For instance, combining equations (7.12) and (7.13) we have

$$R_p = \pi_1 / \pi_2 = \frac{\frac{1}{3} a_R T^4}{p} = \frac{\text{radiation pressure}}{\text{gas pressure}} \qquad (7.18)$$

$$= \text{radiation pressure number}$$

The significance of these parameters will be discussed in section 4.

3. Non-dimensional equations of radiation gasdynamics. The non-dimensional parameters may be also obtained from the fundamental equations of radiation gasdynamics by transforming the equations into non-dimensional forms. We introduce the following non-dimensional quantities:

$$x_i^* = x_i/L; \quad \nabla^* = L\nabla; \quad t^* = t/t_0, \quad \vec{q}^* = \vec{q}/U; \quad \rho^* = \rho/\rho_0;$$

$$p^* = p/(\rho_0 \, U^2); \quad T^* = T/T_0; \quad \tau_s^* = \tau_s/(\mu_0 \, U L); \quad \tau_R^* = \tau_R/(a_R \, T_0^4); \quad (7.19)$$

$$\varkappa^* = \varkappa/\varkappa_0; \quad C_p^* = C_p/C_{p0}; \quad L_{Rv}^* = L_{Rv}/L_{R0}; \quad q_R^* = q_R/(\varkappa_0 \, T_0/L)$$

$$\bar{e}_m^* = \bar{e}_m/U^2, \quad I_v^* = I_v/I_{v0}, \quad B_v^* = B_v/I_{v0}$$

where * refers to the non-dimensional quantities and the subscript 0 refers to some references values. U and L are respectively the reference velocity and reference length of the flow field.

Substituting the expressions of equation (7.19) into the fundamental equations of radiation gasdynamics discussed in chapter V, § 2 to 6, we have the following non-dimensional equations:

(i) Equation of state (5.2)

$$\gamma \, M_0^2 \, p^* = \rho^* T^* \tag{7.20}$$

(ii) Equation of continuity (5.3)

$$\frac{1}{R_t} \frac{\partial \rho^*}{\partial t^*} + \nabla^* \cdot (\rho^* \vec{q}^*) = 0 \tag{7.21}$$

(iii) Equation of motion (5.4)

$$\rho^* \left[\frac{1}{R_t} \frac{\partial \vec{q}^*}{\partial t^*} + (\vec{q}^* \cdot \nabla^*) \, \vec{q}^* \right] = - \nabla^* p^* + \frac{1}{R_e} (\nabla^* \cdot \tau_s^*) + \frac{3 \, R_p}{\gamma \, M_0^2} (\nabla^* \cdot \tau_R^*) \tag{7.22}$$

where we neglect the body force $\vec{F}$ for simplicity.

(iv) Equation of energy (5.7)

$$\rho^* \left(\frac{D \, \bar{e}_m^*}{Dt^*} \right) = - \nabla^* \cdot (p^* \vec{q}^*) + \frac{1}{R_e} \nabla^* \cdot (\vec{q}^* \, \tau_s^*) + \frac{3 \, R_p}{\gamma \, M_0^2} \nabla^* \cdot (\vec{q}^* \cdot \tau_R^*) +$$

$$+ \frac{1}{P_r \, R_e \, (\gamma - 1) \, M_0^2} [\nabla^* \cdot (\varkappa^* \, \nabla^* \, T^*) + \nabla^* \cdot \vec{q}_R^*] \tag{7.23}$$

where we neglect the energy source for simplicity.

(v) Equation of radiative transfer (5.10)

$$\frac{R_r}{R_t} \frac{\partial I_v^*}{\partial t^*} + \frac{\partial I_v^*}{\partial s^*} = \frac{1}{L_{Rv}^*} (B_v^* - I_v^*) \tag{7.24}$$

From the non-dimensional equations (7.20) to (7.24), we obtain the following non-dimensional parameters:

(i) Ratio of the specific heats γ

$$\gamma = \frac{C_p}{C_v} = \frac{\text{specific heat at constant pressure}}{\text{specific heat at constant volume}} \tag{7.25}$$

(ii) The time parameter R_t

$$R_t = \frac{U\,t_0}{L} \qquad (7.26)$$

(iii) Mach number M_0

$$M_0 = \frac{U}{(\gamma\,RT_0)^{\frac{1}{2}}} \qquad (7.27)$$

(iv) Reynolds number R_e

$$R_e = \frac{U L\,\rho_0}{\mu_0} = \frac{U L}{\nu_g} \qquad (7.28)$$

where $\nu_g = \mu_0/\rho_0$ is the coefficient of kinematic viscosity.
(v) Prandtl number P_r

$$P_r = \frac{C_{p0}\,\mu_0}{\varkappa_0} \qquad (7.29)$$

(vi) Radiation pressure number R_p

$$R_p = \frac{a_R\,T_0^4}{3\,p_0} = \frac{\text{radiation pressure}}{\text{gas pressure}} \qquad (7.30)$$

(vii) Relativistic parameter

$$R_r = \frac{U}{c} \qquad (7.31)$$

(viii) Radiation Knudsen number

$$L_R{}^* = \frac{L_R}{L} = K_r \qquad (7.32)$$

Comparing equations (7.12) to (7.18) to equations (7.25) to (7.32), we see that many of the non-dimensional parameters are exactly the same. Hence we show that there are two ways to find these non-dimensional parameters, i.e., the dimensional analysis of section 2 and the inspection analysis of section 3. It should be noticed that these non-dimensional parameters are not unique. It is always possible to combine several non-dimensional parameters into a new one as we shall discuss in section 4.

4. Important parameters of radiation gasdynamics. Each of the non-dimensional parameters has a special significance which represents a special kind of dynamical similarity or kinematic similarity in which the ratio of two special physical quantities, such as forces, remains constant for the constant value of this non-dimensional parameter. It is interesting to know the physical significance of these parameters. In the analysis of the flow problems of radiation gasdynamics, proper approximations may be made in different range of the values of these parameters as we shall discuss in section 5.

The non-dimensional parameters of radiation gasdynamics may be divided into two groups: One group consists of the parameters of ordinary gasdynamics and the other group consists of special parameters which are due to essentially the effect of radiation. We are going to study these parameters according to such a division as follows:

(A) Non-dimensional parameters of ordinary gasdynamics.

The following are some of the most important non-dimensional parameters of ordinary gasdynamics:

(i) The time parameter R_t:

$$R_t = \frac{t_0}{(L/U)} \tag{7.26}$$

This parameter characterizes the time scale of the flow problem with respect to the flow velocity and dimension of the flow field. The quantity (L/U) may be considered as a characteristic time of the flow field. In the steady flow problem, R_t is infinite so that the unsteady terms $(\partial Q/\partial t)$ is zero. However for the convection terms, the value of R_t should be taken as unity. In general we may assume that the value of R_t in ordinary gasdynamics is of the order of unity. Only in the case of high frequency phenomena such as flutter problems or high frequency wave motion, the value of R_t may be very small.

(ii) The ratio of specific heats γ defined by equation (7.25).

This parameter is a measure of the relative complexity of the molecules of the gas, because the specific heat of a gas is closely related to its molecular structure. The exact expressions of specific heats in terms of state variables of a gas should be derived on the basis of statistical mechanics. However, a simple formula for γ may be derived from the simple kinetic theory of gases. If n_1 be the number of degrees of freedom of a molecule of a gas, we have

$$\gamma = \frac{n_1 + 2}{n_1} \tag{7.33}$$

For monatomic gas, $n_1 = 3$ and $\gamma = {}^5/_3$ and for diatomic gas with translational and rotational energies only, $n_1 = 5$ and $\gamma = {}^7/_5$.

(iii) Mach number M defined as follows:

$$M = \frac{U}{a} = \frac{\text{flow velocity}}{\text{sound speed}} \tag{7.27a}$$

where sound speed $a = (\gamma RT)^{1/2}$. From now on we drop the subscript 0 for simplicity. Without subscript o, we may also refer to the local value of a variable instead of the typical value of the variable. In many cases, it is interesting to notice the difference of the value of a non-dimensional parameter based on the local values of the variables from that based on the typical values of the variables. The Mach number is a measure of compressibility of the gas due to a high flow velocity. It is easy to show that the ratio of the variation of the density of a gas to the variation of velocity is, to a first approximation, proportional to the square of the Mach number of the flow. Hence for very small Mach number, the variation of density, i.e., the compressibility effect, due to the variation of velocity of the flow field is negligibly small and the gas may be considered as an incompressible fluid. For large Mach number, the effect of compressibility is important. When $M < 1$, the flow is said to be subsonic and when $M > 1$, the flow is supersonic. The flow field of a subsonic flow differs greatly from that of a supersonic flow. For instance, shock waves occur only in the supersonic flow field for the steady case but never in the corresponding subsonic case. When Mach number M is of the order of unity, the flow is considered as transonic while Mach number M is much larger than unity, the flow is hypersonic. Many new features of the flow field in the regions of transonic flow and hypersonic flow differ from those of the ordinary subsonic and supersonic flow. Hence Mach number is one of the most important parameters in gasdynamics as well as in radiation gasdynamics. Because of the variation of the local Mach number in the flow field, the flow may be subsonic in one part of the flow field and supersonic in the other part of the flow field with a transonic flow field between them. For instance, if we consider a body in a uniform stream, the reference Mach number may be determined based on the variables in the uniform stream, i.e., $M_0 = U/a_0$. The value of M_0 may be greater or smaller than unity. For a given value of M_0, the local Mach number $M = q/a$ may differ greatly from M_0.

As we shall show in chapter VIII, the sound speed of a gas will be increased by the effect of thermal radiation. With radiation effects, we have an effective sound speed and a corresponding effective Mach number which is the ratio of the flow velocity to the effective sound speed with radiation effect.

(iv) Reynolds number R_e defined as follows:

$$R_e = \frac{UL\rho}{\mu} = \frac{UL}{\nu_g} = \frac{\text{inertial force}}{\text{viscous force}} \tag{7.28 a}$$

This is the most important parameter for the fluid dynamics of a viscous fluid, which represents the ratio of the inertial force to the viscous force. When the Reynolds number is small, the viscous force is predominant and the effect of viscosity is important in the whole flow field. When the Reynolds number is large, the inertial force is predominant and the effect of viscosity is important only in a narrow region near a solid boundary or other restricted region which is known as boundary layer region or transition region. Outside these transition or boundary layer regions, the flow may be considered as inviscid. If the Reynolds number is enormously large, the flow becomes turbulent.

(v) Prandtl number defined as follows:

$$P_r = \frac{C_p \mu}{\varkappa} = \frac{\nu_g}{(\varkappa/C_p \rho)} = \frac{\text{kinematic viscosity}}{\text{thermal diffusivity}} \tag{7.29 a}$$

The value of the kinematic viscosity $\nu_g = \mu/\rho$ shows the effect of viscosity of a gas or a fluid. If other things are the same, the smaller the value of kinematic viscosity is, the narrow the region which is affected by viscosity will be. This region is generally known as the boundary layer of the velocity field when the kinematic viscosity is very small. The thickness of the boundary layer of the velocity field of a laminar flow is proportional to the square root of the kinematic viscosity $\sqrt{\nu_g}$. Thus the kinematic viscosity ν_g shows the momentum diffusivity due to the viscous effect. The value of $(\varkappa/C_p \, \rho)$ shows the thermal diffusivity due to the heat conductivity. The smaller the value of the thermal diffusivity $(\varkappa/C_p \, \rho)$ is, the narrower will be the region which is affected by the heat conductivity and which is known as the thermal boundary layer when the thermal diffusivity $(\varkappa/C_p \, \rho)$ is very small. Thus the thickness of the thermal boundary layer is proportional to the square root of the thermal diffusivity $(\varkappa/C_p \, \rho)^{1/2}$. The Prandtl number P_r shows the relative importance of the heat conductivity and the viscosity of a fluid. As a first approximation, the Prandtl number gives the square of the ratio of the thickness of the velocity boundary layer to that of the thermal boundary layer. The Prandtl number depends only on the physical properties of the fluid but not on the flow conditions such as the velocity and the size of the flow field.

In radiation gasdynmaics, the thermal diffusivity of a gas depends on both the heat conduction and thermal radiation. In certain cases we shall discuss in chapter IX, we may define an effective thermal conductivity which includes both the heat conduction and thermal radiation effects. We may then define an effective Prandtl number in radiation gasdynamics. It is evident that the thermal radiation will increase the effective thermal conductivity and decrease the effective Prandtl number. As a result, the effect of thermal radiation will increase the thickness of the thermal boundary layer. We shall discuss this result in detail in chapter IX.

(vi) Schmidt number S_c defined as follows:

$$Sc = \frac{\nu_g}{\mathfrak{D}} \tag{7.34}$$

where $\mathfrak{D}$ is the coefficient of diffusion of the fluid. In many problems of gas-dynamics at high temperature, we have to consider the gas as a mixture of several species. The diffusion phenomena between these species are very important to analyze the flow field of the mixture. These diffusion phenomena may be characterized by various diffusion coefficients of various species in the mixture or by the corresponding Schmidt numbers of these species.

(vii) Nusselt number Nu. This non-dimensional number differs from the above non-dimensional parameters in the fact that it is a non-dimensional form of heat transfer per unit area and per unit time through a surface. Hence it expresses a resultant effect of a flow field instead of special condition of the flow field expressed by a parameter.

Let $q_c \, (x)$ be the quantity of heat transferred through unit area per unit time along a surface. We may define a coefficient of heat transfer $h_c \, (x)$, which is sometimes called a film coefficient, as follows:

$$h_c \, (x) = \frac{q_c \, (x)}{T_w - T_{wi}} \tag{7.35}$$

where both $q_c \, (x)$ and $h_c \, (x)$ are in general a function of the position x along the surface. T_w is the temperature of the surface or solid wall and T_{wi} is the temperature of the solid surface in the case of no heat transfer. Some authors use a reference temperature T_f of the fluid such as the free stream temperature instead of T_{wi} in equation (7.35) to define the coefficient of heat transfer. The Nusselt number is defined as

$$N_u = \frac{h_c(x)\, L}{\varkappa} \qquad (7.36)$$

The Nusselt number may be considered as the total heat transfer to the conductive heat transfer at the wall. In radiation gasdynamics, the heat transfer $q_c(x)$ depends on both heat conduction and thermal radiation. But the Nusselt number way be used as an expression for the total heat transfer.

(B) Non-dimensional parameters of radiation gasdynamics.

There are a few parameters which are due essentially to the effects of thermal radiation and which are given below:

(viii) Relativistic parameter R_r defined as follows:

$$R_r = \frac{U}{c} = \frac{\text{flow velocity}}{\text{velocity of light}} \qquad (7.17)$$

For ordinary gas particles, the relativistic parameter is usually very small. Hence the relativistic effect on the flow of gas particles is negligible. However the thermal radiation effect is due to the motion of photons which move with the speed of light c. As a result, the relativistic parameter plays an important role in the heat transfer by thermal radiation.

(ix) Knudsen number of radiation K_r defined as follows:

$$K_r = L_R{}^* = \frac{L_R}{L} = \frac{\text{mean free path of radiation}}{\text{characteristic length}} \qquad (7.16\,\text{a})$$

This number characterizes the distance travelled by photons before they are absorbed by the molecules in the gas. When Knudsen number of radiation is very small, we say that the medium is optically thick. When the Knudsen number of radiation is very large, we say that the medium is optically thin.

(x) Radiation pressure number R_p defined as follows:

$$R_p = \frac{a_R\, T^4}{3\,p} = \frac{\text{radiation pressure}}{\text{gas pressure}} \qquad (7.18\,\text{a})$$

This number in ordinary gasdynamics is usually very small. Hence in ordinary gasdynamics, the effects of radiation are negligible. However, for the condition of very high temperature and low pressure, this number will be large. When R_p is not negligibly small, we have to consider the effects of radiation pressure and radiation energy density. Even when R_p is small, the radiation heat flux may not be negligible as we shall see in next section. Fig. 7.1 shows the values of R_p at various pressures and temperature. Ordinarily, R_p is very small except at very high temperature over 40,000 °K and very low pressures.

5. Some further remarks for the non-dimensional parameters. The ten non-dimensional parameters are the most important parameters in radiation gasdynamics. But they are not unique. We may obtain other important parameters from various combinations of these ten parameters. These new parameters may be more appropriate for special problems than those described in last section. Some of these new parameters are given below:

(i) Peclet number P_e defined as follows:

$$P_e = \frac{UL}{(\varkappa/C_p\,\rho)} = P_r \cdot R_e \qquad (7.37)$$

The Peclet number plays the same role in the thermal boundary layer as the Reynolds number in the velocity boundary layer. In equation (7.23), we see

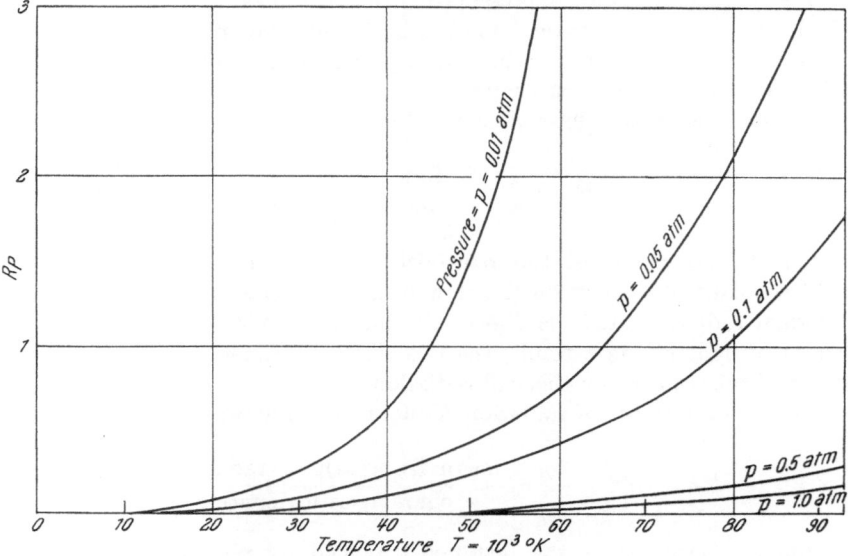

Fig. 7.1. Radiation pressure number as a function of temperature and pressure

that the Peclet number occurs in the heat transfer terms while the Reynolds number occurs in the momentum diffusive terms.

(ii) Stanton number S_t defined as follows:

$$S_t = \frac{h(x)}{UC_p\rho} = \frac{N_u}{P_e} \qquad (7.38)$$

Stanton number is another way to express the heat transfer rate which may be considered as the total heat transfer to the typical convective heat transfer.

(iii) Knudsen number K_f defined as follows:

$$K_f = \frac{\text{mean free path of a gas}}{\text{characteristic length}} = \frac{L_f}{L} = 1.255\,\sqrt{\gamma}\,\frac{M}{R_e} \qquad (7.39)$$

where the mean free path of a gas L_f may be taken as follows:

$$L_f = 1.255\,\sqrt{\gamma}\,\frac{v_g}{a} \qquad (7.40)$$

In most of the flow problem discussed in this book, we consider that the Knudsen number is very small so that the gas may be considered as a continuum. If the

Knudsen number is not small, the discrete character of the gas must be taken into account. For large Knudsen number, we have the slip flows, free molecule flows and other special transition flows. We shall discuss some of these cases in chapter X. The Knudsen number of equation (7.39) is similar to the Knudsen number of radiation of equation (7.16a). There are many similarities of the gasdynamic terms and the radiative transfer terms according to the values of these two Knudsen numbers. For small Knudsen numbers, the radiative transfer terms as well as the gasdynamic transfer terms such as viscous terms, heat conduction terms may be expressed in terms of gradients of some local variables. When Knudsen numbers are large, the radiative as well as the gasdynamic terms should be expressed in terms of integral forms or other complicated form such as some types of differential equations. We shall discuss them in the following chapters.

(iv) Radiative flux number defined as follows:

$$R_F = \frac{\text{radiative heat flux}}{\text{heat conduction flux}} \tag{7.41}$$

Since the exact expression of radiative heat flux depends on the Knudsen number of radiation, the expression of the radiative flux number also depends on the Knudsen number of radiation. For optically thick case, i.e., small value of Knudsen number of radiation K_r, the radiative heat flux is given by equation (5.21). The corresponding radiative flux number may be written as follows:

$$R_{F1} = \frac{\varkappa_R}{\varkappa} = \frac{4 D_R a_R T^3}{\varkappa} = 4\left(\frac{\gamma - 1}{\gamma}\right) P_r R_e L_R^* R_p / R_r \tag{7.42}$$

For the case of optically thin gas, the radiative heat flux is given by equation (5.37) or (6.45). The corresponding radiative flux number is then

$$R_{F2} = \frac{c a_R T^4 L^2}{L_R \varkappa T} = 3\left(\frac{\gamma - 1}{\gamma}\right) P_r R_e R_p / (L_R^* R_r) \tag{7.43}$$

The main difference between R_{F1} and R_{F2} lies in the parameter Knudsen number of radiation L_R^*. For a very small mean free path of radiation, $L_R^* \ll 1$, the radiative flux number R_{F1} is proportional to the Knudsen number of radiation L_R^* while for large Knudsen number of radiation, $L_R^* \gg 1$, the radiative flux number R_{F2} is inversely proportional to L_R^*. Hence for both very large L_R^* and very small L_R^*, the radiative heat flux tends to be zero.

We may also define the radiative flux number as the ratio of radiative heat flux to the convective heat flux, i.e.

$$R_F' = \frac{\text{radiative heat flux}}{\text{convective heat flux}} = R_F \frac{T}{C_p \rho U T L} = \frac{R_F}{P_r R_e} \tag{7.44}$$

Comparing equation (7.44) to equation (7.42) or (7.43), we have for the optically thick case:

$$R_{F1}' = \frac{\varkappa_R T}{C_p \rho U T L} = 4\left(\frac{\gamma - 1}{\gamma}\right) L_R^* R_p / R_r \tag{7.45}$$

and for the optically thin case:

$$R'_{F2} = \frac{c \, a_R \, T^4 \, L^2}{L_R \, C_p \, \rho \, U \, T \, L} = 3 \left(\frac{\gamma - 1}{\gamma} \right) R_p / (L_R^* \, R_r) \qquad (7.46)$$

It is interesting to notice that all the radiative flux numbers are directly proportional to the radiation pressure number R_p and inversely proportional to the relativistic parameter R_r. Usually both R_p and R_r are small. Hence the radiative flux number may not be small. Particularly in many engineering problems, as shown in tables I and II of chapter I, the radiative pressure number R_p is so small that the radiation pressure and the radiation energy density are negligible but the radiative flux number is not small and hence we should include the radiative heat flux in the basic equations of gasdynamics.

In the radiative flux number, the mean free path of radiation L_R or the mean absorption coefficient $\rho \, K_R = 1/L_R$ is the only new physical property of the medium which determines the radiative heat transfer. This property represents a new diffusive phenomenon in the medium in addition to the momentum diffusivity by viscosity and thermal diffusivity by thermal conductivity. Many new and interesting phenomena may be introduced by this radiative diffusivity as we shall discuss later. The value of the mean free path of radiation L_R should be determined either experimentally or from a microscopic treatment which will be discussed in chapter XI.

In general the mean free path of radiation $L_{R\nu}$ is a function of the frequency ν and the state variables such as temperature T and density ρ. In engineering problems, it is advisable to use some simple expressions of the average value of the mean free path of radiation over the whole spectrum of frequency. The following simplifications have been extensively used:

(i) For the variation of mean free path of radiation with frequency, some average values over the whole spectrum have been used. Two of the most popular values of the average mean free path of radiation are:

(a) Rosseland mean absorption coefficient K_R defined as follows:

$$K_R = \frac{1}{\rho \, L_R} = \frac{\displaystyle\int_0^\infty \frac{\partial B_\nu}{\partial T} \, d\nu}{\displaystyle\int_0^\infty \frac{1}{k_\nu'} \frac{\partial B_\nu}{\partial T} \, d\nu} \qquad (7.47)$$

As we have shown in chapter V equation (5.18), this average value is the appropriate value for the optically thick case.

(b) Planck mean absorption coefficient Kp

$$K_p = \frac{1}{\rho \, L_{Rp}} = \frac{\displaystyle\int_0^\infty k_\nu' \, B_\nu \, d\nu}{B(T)} \qquad (7.48)$$

As we have shown in chapter V equation (5.36), this average value is the appropriate value for the optically thin case or in the intermediate range where the integral expression is used for radiation terms.

It should be noticed that the value of the mean free path of radiation itself does not determine whether the medium is optically thick or thin. We have to compare the mean free path of radiation to the representative length L which characterizes the flow field investigated in order to determine whether the medium may be regarded as optically thin or thick. In other words, the optical thickness defined by equation (3.3) or similar expression determines the optical property of the medium. In fact for a given state of a medium, the Planck mean free path of radiation L_{Rp} is usually smaller than the Rosseland mean free path of radiation L_R. SCALA and SAMPSON (5) have recommended the formula $L_R = 8 \cdot 3\, L_{Rp}$ as a first approximation for high temperature air up to a temperature of 20,000 °K.

(ii) After we have taken the average of the mean free path of radiation over the whole spectrum, we should find some simple formula for the variation of the average mean free path of radiation with temperature and density or pressure. Since the absorption coefficient increases with the density, the mean free path of radiation decreases with the increase of density (cf. chapter XI: Figs. 11.12 or 11.14). The variation of the mean free path of radiation with temperature is very complicated. At low temperature, say $T < 10,000$ °K, the mean free path of radiation decreases with increase of temperature while at high temperature, say $T > 100,000$ °K, the mean free path increases with the temperature and there is a minimum in the intermediate temperature range depending on the density and the property of the medium. If the temperature range in the flow field is not too large, a power law for the average value of the mean free path of radiation may give a good approximation of the variation of the mean free path of radiation with the state variables, i.e.,

$$\frac{L_R}{L_{R0}} = \frac{L_{Rp}}{L_{Rp0}} = \left(\frac{T_0}{T}\right)^{m_1}\left(\frac{\rho_0}{\rho}\right)^{m_2} \tag{7.49}$$

where subscript 0 refers to the values at some reference conditions. The powers m_1 and m_2 should be so chosen that the formula (7.49) gives the best fit with the opacity data over the temperature range considered. The value of m_2 lies between 1 and 2 while the value of m_1 may be positive at low temperature range and negative at high temperature range. For air, it was found that in the temperature range of 7,000 °K to 12,000 °K, the following values may be used (cf. Fig. 5.1):

$$m_1 = 4.4, \; m_2 = 1, \; \rho_0 = 1.23 \text{ gr/cm}^3, \; T_0 = 10,000 \text{ °K}$$

$$L_{R\,p0} = 0.5 \text{ meter}$$

For higher temperature in the neighborhood of 20,000 °K, $m_1 = 2.5$ and $m_2 = 1$ will give better results. At very high temperature in some astrophysical problem $m_1 = -\,^7/_2$ and $m_2 = 2$ have been used.

If we consider a large range of temperature including the minimum value of the mean free path of radiation, the following formula may be used:

$$\frac{L_{Rp0}}{L_{Rp}} = \frac{L_{R0}}{L_R} = \frac{L_{Rp\,10}}{L_{Rp\,1}} + \frac{L_{Rp\,20}}{L_{Rp\,2}} \tag{7.50}$$

where $L_{R\,p1}$ is the mean free path of radiation given by the power law (7.49) which is good for low temperature range, i.e., m_1 is positive and L_{Rp_2} is the corresponding value for the high temperature range, i.e., m_1 is negative.

Some authors express the average mean free path of radiation as a function of temperature and pressure. For instance Scala and Sampson (5) give the following formulae for the Rosseland mean free of radiation for air:

(i) For linear representation:

$$\rho\, K_R = \frac{1}{L_R} = 4.86 \times 10^{-7}\, p^{1.31} \exp (4.56 \times 10^{-4}\, T) \tag{7.51}$$

and (ii) for quadratic representation:

$$\rho\, K_R = \frac{1}{L_R} = 4.52 \times 10^{-7}\, p^{1.31} \exp (5.18 \times 10^{-4}\, T - 7.13 \times 10^{-9}\, T^2) \tag{7.52}$$

where the mean free path of radiation L_R is in centimeters, the pressure p is in atmospheres and the temperature T is in °K. Equation (7.51) corresponds to equation (7.49) with $m_1 = 4.4$ while equation (7.52) corresponds to equation (7.50).

Even though we have list many important parameters, these parameters are not of the same importance in a specific problem. In certain problems, we may omit some of these parameters. For instance, outside the boundary layer region, the influence of Reynolds number is negligible. Under the conditions of Tables I and II of chapter I, for the temperature of the order of 10,000 °K, the radiation pressure number R_p is negligibly small while the radiative flux number is of the order of unity or higher. Under this condition, the radiation stresses and radiation energy density are negligible but the radiative heat flux is not negligible.

Under different range of the values of these parameters, we may simplify the fundamental equations of radiation gasdynamics accordingly as we shall show in the next two chapters.

In the above discussions, we consider essentially the parameters in the cases where only the radiation effects and ordinary gasdynamic effects are important. Under the conditions of very high temperature, some of the other physical phenomena may also be important in the flow problems. For instance, at very high temperature, the gas will be ionized. Then the electromagnetic fields play important roles in the flow field. Besides those parameters discussed above, we have to consider also those parameters due to the electromagnetic variables such as electrical field strength, magnetic field strength and electrical conductivity. We shall discuss some of these parameters in next two chapters.

References

1. Armstrong, B. H., J. Sokoloff, R. W. Nicholls, D. H. Holland, and R. E. Meyerott: Radiative properties of high temperature air. Jour. Qual. Spec. Rad. Transfer. vol. 1, pp. 143–162, Pergamon Press Ltd., 1961.

2. Bridyeman, P. W.: Dimensional Analysis. Yale Univ. Press, second edition, 1931.

3. Goulard, R.: Similarity parameters in radiation gasdynamics. Purdue Univ. School of Aero. & Eng. Sci. report A & ES 62-8, 1962.

4. KIVEL, B., and K. BAILEY: Tables of radiation from high temperature air. AVCO Research report 21, Dec. 1957.

5. SCALA, S. M., and D. H. SAMPSON: Heat transfer in hypersonic flow with radiation and chemical reaction. Tech. Inf. ser. R 63-SD 46, Space Sci. Lab. General Electric Co., Phil. Pa., 1963.

6. TRAUGOTT, S. C.: Shock structure in a radiating heat conducting and viscous gas. Res. report RR 57, Martin Co., Baltimore, Md., 1964.

7. ZHIGULEV, V. N., YE. A. ROMISHEVSKII, and V. K. VERTUSHKIN: Role of radiation in modern gasdynamics. AIAA Jour. vol. 1, No. 6, pp. 1473–1485, 1963.

Chapter VIII

Waves and Shock Waves in Radiation Gasdynamics

1. Introduction. In order to consider the influence of thermal radiation on the flow field, we study the wave motion of infinitesimal amplitude in a radiating gas. Such wave motion will bring out many essential features of radiation gasdynamics. In section 2, we discuss the wave motion of infinitesimal amplitude in an optically thick medium at very high temperature so that the gas is in a state of plasma. We consider the wave motion with the influence of both thermal radiation and electromagnetic forces. In section 3, we consider the wave motion in a radiating gas of finite mean free path of radiation. This case has not been thoroughly investigated as the case of optically thick medium. However some essential features will be discussed for this general case.

Since the radiation effects are most important in extremely high speed flow in which shock waves usually occur, the shock wave phenomena are the most important phenomena in radiation gasdynamics. In the next four sections, we deal with the shock waves in radiation gasdynamics. First the well known Rankine-Hugoniot relations will be extended in radiation gasdynamics. Then the effect on the shock wave structure by thermal radiation will be studied. Finally we consider the flow behind shock waves in radiation gasdynamics.

2. Wave of small amplitude in an optically thick medium. We consider a gas which is in a very high temperature state so that it is ionized and radiating. We assume that the gas is of sufficient opacity that the radiation can be considered as being trapped in it and as in equilibrium with the plasma. Hence the fundamental equations of radiation gasdynamics of small mean free path of radiation discussed in chapter V, § 8 hold true for our case. Since the gas is ionized, electromagnetic forces should be considered. These electromagnetic forces are determined by electromagnetic fields which are governed by the Maxwell's equations of electromagnetic fields. In addition to the radiation gasdynamic variables: $\vec{q}$, p, ρ, and T, we have also the electromagnetic variables: electrical field strength $\vec{E}$, magnetic field strength $\vec{H}$, electric current density $\vec{J}$ and electrical excess charge ρ_e. These electromagnetic variables are governed by the following equations (9).

Maxwell equations for the electromagnetic fields are

$$\nabla \times \vec{H} = \vec{J} + \frac{\partial \epsilon \vec{E}}{\partial t} \tag{8.1}$$

$$\nabla \times \vec{E} = -\frac{\partial \mu_e \vec{H}}{\partial t} \tag{8.2}$$

where μ_e is the magnetic permeability and ϵ is the inductive capacity.

The generalized Ohm's law is

$$\vec{J} = \sigma_e (\vec{E} + \vec{q} \times \vec{B}) + \rho_e \vec{q} \tag{8.3}$$

where σ_e is the electric conductivity of the gas and $\vec{B} = \mu_e \vec{H}$ is the magnetic induction.

The conservation of the electrical charge gives

$$\frac{\partial \rho_e}{\partial t} + \nabla \cdot \vec{J} = 0 \tag{8.4}$$

The effects of electromagnetic variables on the equations of radiation gasdynamics (5.23) are:

(i) The electromagnetic force $\vec{F}_e$ in equation (5.23 c), i.e.,

$$\vec{F}_e = \rho_e \vec{E} + \vec{J} \times \vec{B} \tag{8.5}$$

and (ii) The electromagnetic energy flux Q in equation (5.23 d), i.e.,

$$Q = \vec{J} \cdot \vec{E} \tag{8.6}$$

Now we assume that originally the plasma is at rest with a pressure p_0, temperature T_0 and density ρ_0 and that it is subjected to an externally applied uniform magnetic field $\vec{H}_0 = \vec{i} H_x + \vec{j} H_y + \vec{k} 0$ where $\vec{i}, \vec{j}$, and $\vec{k}$ are respectively the x-, y-, and z-wise unit vectors; and H_x and H_y are constants. There is no electric current, nor excess electric charge, nor externally applied electric field. The plasma is perturbed by a small disturbance so that in the resultant disturbed motion we have:

$$u = u(x, t), \; v = v(x, t), \; w = w(x, t),$$
$$p = p_0 + p'(x, t), \; T = T_0 + T'(x, t),$$
$$\rho = \rho_0 + \rho'(x, t), \; \vec{E} = \vec{E}(x, t), \tag{8.7}$$
$$\vec{H} = \vec{H}_0 + \vec{h}(x, t), \; \vec{J} = \vec{J}(x, t), \; \rho_e = \rho_e(x, t)$$

where u, v and w are respectively the perturbed x-, y- and z-velocity components; p', ρ' and T' are respectively the perturbed pressure, density and temperature; and $\vec{E}, \vec{h}, \vec{J}$ and ρ_e are respectively the perturbed electric field strength, magnetic field strength, electric current density and the excess electric charge. We assume that all the perturbed quantities are small so that the second or higher order terms in these quantities are negligible. For simplicity, we assume that the perturbed quantities are functions of one space coordinate x and time t only. Thus we discuss only wave propagation in the direction of x-axis.

Substituting equation (8.7) into the fundamental equations of radiation magnetogasdynamics, i.e., equations (5.23) and (8.1) to (8.6) and neglecting the higher order terms of perturbed quantities, we have the following linear equations for the wave motion in radiation magnetogasdynamics:

(i) Maxwell equations for the electromagnetic fields (8.1) and (8.2)

$$\frac{\partial\, \mu_e h_x}{\partial\, t} = 0 \tag{8.8a}$$

$$\frac{\partial\, \mu_e h_y}{\partial\, t} = \frac{\partial\, E_z}{\partial\, x} \tag{8.8b}$$

$$\frac{\partial\, \mu_e h_z}{\partial\, t} = -\frac{\partial\, E_y}{\partial\, x} \tag{8.8c}$$

$$J_x + \frac{\partial\, \epsilon E_x}{\partial\, t} = 0 \tag{8.8d}$$

$$J_y + \frac{\partial\, \epsilon E_y}{\partial\, t} = -\frac{\partial\, h_z}{\partial\, x} \tag{8.8e}$$

$$J_z + \frac{\partial\, \epsilon E_z}{\partial\, t} = \frac{\partial\, h_y}{\partial\, x} \tag{8.8f}$$

where the subscript x, y or z refers to the corresponding component of a vector.

(ii) The generalized Ohm's law (8.3)

$$J_x = \sigma_e\, (E_x - \mu_e\, w\, H_y) \tag{8.8g}$$

$$J_y = \sigma_e\, (E_y + \mu_e\, w\, H_x) \tag{8.8h}$$

$$J_z = \sigma_e\, [E_z + \mu_e\, (u\, H_y - v\, H_x)] \tag{8.8i}$$

(iii) The equation of conservation of electric charge (8.4)

$$\frac{\partial\, \rho_e}{\partial\, t} + \frac{\partial\, J_x}{\partial\, x} = 0 \tag{8.8j}$$

(iv) The equation of state (5.23a)

$$\frac{p'}{p_0} = \frac{\rho'}{\rho_0} + \frac{T'}{T_0} \tag{8.8k}$$

(v) The equation of continuity (5.23b)

$$\frac{\partial\rho'}{\partial t} + \rho_0\, \frac{\partial\, u}{\partial\, x} = 0 \tag{8.8l}$$

(vi) The equations of motion (5.23c) and (8.5)

$$\rho_0\, \frac{\partial\, u}{\partial\, t} = -\frac{\partial\, p'}{\partial\, x} + 4\, R\, R_p\, \rho_0\, \frac{\partial T'}{\partial\, x} + \frac{4}{3}\, \mu\, \frac{\partial^2\, u}{\partial\, x^2} - \mu_e\, J_z\, H_y \tag{8.8m}$$

$$\rho_0\, \frac{\partial\, v}{\partial\, t} = \mu\, \frac{\partial^2\, v}{\partial\, x^2} + \mu_e\, J_z\, H_x \tag{8.8n}$$

$$\rho_0\, \frac{\partial\, w}{\partial\, t} = \mu\, \frac{\partial^2\, w}{\partial\, x^2} + \mu_e\, (J_x\, H_y - J_y\, H_x) \tag{8.8o}$$

where R_p is the radiation pressure number based on the undisturbed state variables, i.e., $R_p = a_R\, T_0{}^3/(3\, R\, \rho_0)$. The coefficient of viscosity μ corresponds to the temperature T_0.

(vii) The energy equation (5.23d) and (8.6):

$$C_p{}^* \, \rho_0 \, \frac{\partial T'}{\partial t} - 4 R R_p T_0 \frac{\partial \rho'}{\partial t} = \frac{\partial p'}{\partial t} + K^* \frac{\partial^2 T'}{\partial x^2} \tag{8.8p}$$

where

$$C_p{}^* = C_p + 12 \, R \, R_p = \text{effective specific heat at constant pressure}$$
$$\text{including the radiation effect} \tag{8.9}$$

$$K^* = \varkappa + 12 \, R \, R_p \, \rho_0 \, D_R = \varkappa + \varkappa_R = \text{effective coefficient of}$$
$$\text{heat conductivity including the radiation effect} \tag{8.10}$$

The second term on the left-hand side of equation (8.8p) is due to the variation of the energy density of radiation per unit mass E_R/ρ and the work done by the radiation pressure.

Examining the linearized equations (8.8), we see that the quantity h_x is independent of all the other quantities and is given by equation (8.8a). Since the divergence of $\vec{h}$ is zero, h_x must be a constant which may be put equal to zero.

The rest of the perturbed quantities may be divided into two groups: One group consists of the variables w, h_z, J_x, J_y, E_x, E_y and ρ_e which are governed by equations (8.8c), (8.8d), (8.8e), (8.8g), (8.8h), (8.8j), and (8.8o). This group characterizes a transverse wave, since it deals with the velocity component w and the magnetic field component h_z which are perpendicular to the externally applied magnetic field H_0.

The second group consists of the variables u, v, p', T', ρ', h_y, J_z and E_z, which are governed by the rest equations of (8.8). This group characterizes a longitudinal wave in which ordinary sound wave is a special case.

We are looking for periodic solutions in which all the perturbed quantities are proportional to

$$\exp\left[i \, (\omega \, t - \lambda \, x)\right] = \exp\left[- i \, \lambda \, (x - V t)\right] \tag{8.11}$$

where ω is a given angular frequency, λ is the wave number which is equal to $2 \, \pi/(\text{wave length})$, V is the velocity of wave proportion and $i = \sqrt{-1}$. Substituting these variables into equations (8.8), we obtain one determinantal equation for each group of the above quantities. The eigen values of these determinantal equations give the different modes of wave propagation through the plasma.

(i) Transverse waves.

The determinantal equation for the transverse waves gives the following relation:

$$\left(i \, \omega - \nu_H \frac{\omega^2}{c^2}\right)\left[(i \, \omega + \nu_g \lambda^2)\left(i \, \omega + \nu_H \lambda^2 - \nu_H \frac{\omega^2}{c^2}\right) + V_x{}^2 \left(\lambda^2 - \frac{\omega^2}{c^2}\right)\right] -$$
$$- \frac{\omega^2}{c^2} V_y{}^2 \left(i \, \omega + \nu_H \lambda^2 - \nu_H \frac{\omega^2}{c^2}\right) = 0 \tag{8.12}$$

where

$$V_x = H_x \, (\mu_e/\rho_0)^{\frac{1}{2}} = x\text{-component of the Alfven's wave}$$

$$V_y = H_y \, (\mu_e/\rho_0)^{\frac{1}{2}} = y\text{-component of the Alfven's wave} \qquad (8.13)$$

$$\nu_g = \mu/\rho_0; \quad \nu_H = \frac{1}{\mu_e \, \sigma_e} = \text{magnetic diffusivity}$$

Equation (8.12) does not contain the radiation parameters. Hence the transverse wave is independent of the radiation field and it is simply the regular transverse wave of magnetofluid dynamics which is independent of the compressibility properties of the plasma. Since equation (8.12) is a quadratic equation in λ^2, there are two different modes of transverse waves. These transverse waves are the interaction of an electromagnetic wave and a viscous wave through the action of the externally applied magnetic field. These two basic waves can be obtained, if the external magnetic field is zero, i.e., $V_x = V_y = 0$. Then equation (8.12) gives the following two roots:

(a) A damped electromagnetic wave in a conducting medium:

$$\lambda_1{}^2 = \frac{\omega^2}{c^2} - i \, \frac{\omega}{\nu_H} \qquad (8.14)$$

For non-conducting medium $\nu_H = \infty$, this wave propagates at the speed of light c without damping.

(b) A damped wave in a viscous fluid:

$$\lambda_2{}^2 = - \, i \, \frac{\omega}{\nu_g} \qquad (8.15)$$

These two basic waves (8.14) and (8.15) are coupled through the magnetic field H_0. For an ideal plasma, i.e., $\nu = 0$ and $\nu_H = 0$, inviscid and infinitely electrically conducting fluid, equation (8.12) gives

$$V = \frac{\omega}{\lambda} = V_x \left(1 + \frac{V_x{}^2 + V_y{}^2}{c^2} \right)^{-\frac{1}{2}} \qquad (8.16)$$

This may be called the modified Alfven's wave. For magnetogasdynamic waves, $(V_x{}^2 + V_y{}^2)$ is much smaller than c^2, equation (8.16) becomes

$$V = \frac{\omega}{\lambda} = V_x \qquad (8.17)$$

This is the well known Alfven's wave which shows that the disturbance in an inviscid and infinitely electrically conducting fluid of density ρ_0 propagates as a wave in the direction of H_x with the Alfven's wave speed V_x.

It is interesting to notice that the first bracket of equation (8.12) gives a special value of ω, i.e.,

$$\omega = i \, \frac{c^2}{\nu_H} = i \, \frac{\sigma_e}{\varepsilon} \qquad (8.18)$$

This is the well known damped electromagnetic wave in ordinary electrodynamics.

(ii) Longitudinal waves.

The dispersion relation of the longitudinal waves has the following general form:

$$
\left[K^* \left(\frac{1}{\rho_0} + \frac{4}{3} i\, \omega\, \frac{\nu_g}{p_0} \right) \lambda^4 - \left\{ \frac{\omega^2 K^*}{p_0} + \frac{4}{3} \frac{\nu_g\, \omega^2}{T_0\,(\gamma-1)} (1 + 12\,(\gamma-1)\,R_p) - \right. \right.
$$

(8.19)

$$
\left. \left. - i\,\omega\,(C_p + 20\,R\,R_p + 16\,R\,R_p{}^2) \right\} \lambda^2 - \frac{i\,\omega^3}{T_0\,(\gamma-1)} (1 + 12\,(\gamma-1)\,R_p) \right]
$$

$$
\left[\left(\lambda^2\,\nu_H + i\,\omega - \nu_H\,\frac{\omega^2}{c^2} \right) (i\,\omega + \nu_g\,\lambda^2) + V_x{}^2 \left(\lambda^2 - \frac{\omega^2}{c^2} \right) \right] - \left(\lambda^2 - \frac{\omega^2}{c^2} \right) (i\,\omega + \nu_g\,\lambda^2)
$$

$$
\left[\frac{\omega^2}{T_0\,(\gamma-1)} (1 + 12\,(\gamma-1)\,R_p) - \frac{i\,\omega\,K^*\,\lambda^2}{p_0} \right] V_y{}^2 = 0
$$

Equation (8.19) is a quartic equation in λ^2. Hence there are four different modes of longitudinal waves in the present cases. These waves are the modified heat wave and modified magnetogasdynamic waves.

The heat wave depends on the effective coefficient of heat conductivity K^*. When K^* vanishes, there will be no heat wave. Even in a non-heat conducting and inviscid fluid, the radiative heat transfer will introduce a heat wave in the medium in a similar manner as the heat conductivity introduces the heat wave in ordinary non-radiating fluid.

Even though equation (8.19) is rather complicated, the meaning of various terms is very clear. The first square bracket characterizes sound waves in a viscous, heat conducting and radiating fluid. If there is no transverse magnetic field, i.e., $V_y = 0$, these sound waves will not interact with the electromagnetic variables. Hence the first square bracket gives the ordinary sound waves in a viscous, heat-conducting and radiating medium. Since this bracket gives a quadratic equation in λ^2, there are two modes of sound waves: One corresponds the ordinary sound wave and the other corresponds the heat wave. There are two types of radiation effects on these sound waves: One is due to the radiation pressure and radiation energy density and the other is due to radiative heat flux.

Let us consider the radiation pressure and radiation energy density first which are characterized by the radiation pressure number R_p. For an inviscid and non-heat conducting fluid without radiative heat flux, i.e., $\nu_g = 0$ and $K^* = 0$, the first square bracket of equation (8.19) equating to zero gives the sound wave in a radiating gas with the speed of propagation C_R given by the relation:

$$
C_R{}^2 = \frac{\omega^2}{\lambda^2} = a_0{}^2 \cdot \frac{1 + 20\left(\dfrac{\gamma-1}{\gamma}\right) R_p + 16\left(\dfrac{\gamma-1}{\gamma}\right) R_p{}^2}{1 + 12\,(\gamma-1)\,R_p}
$$

(8.20)

where $a_0 = (\gamma\,R\,T_0)^{1/2}$ is the ordinary sound speed without radiation effects, while C_R is the radiation sound speed which includes the effects of both the radiation pressure and the radiation energy density. For a given temperature T_0, the radiation sound speed increases with the radiation pressure number R_p

(see Fig. 8.1). At very high value of R_p, C_R increases as the square root of R_p. This is the mode that corresponds to the ordinary sound speed.

If the radiative heat flux is different from zero, K^* will be different from zero. It will introduce two effects on the sound waves: One is to introduce a damping on the above sound wave and modifies its speed and the other is to introduce a heat wave. In order to show the effect of radiative heat flux, we consider a radiating, inviscid and non-heat conducting gas with a small amount of radiative heat flux so that K^* is a very small quantity. Now the first bracket of equation (8.19), i.e.,

$$\frac{K^*}{p_0}\lambda^4 - \left\{\frac{\omega^2 K^*}{p_0} - i\,\omega\,(C_p + 20\,R\,R_p + 16\,R\,R_p{}^2)\right\}\lambda^2 -$$

$$-\frac{i\,\omega^3}{T_0\,(\gamma-1)}\,[1 + 12\,(\gamma-1)\,R_p] = 0 \tag{8.21}$$

gives the following two modes as a first approximation:

(a) Radiation sound wave:

$$\lambda_1 = \pm\,\frac{\omega}{C_R}\,[1 - i\,\omega\,f\,(R_p)\,D_R] \tag{8.22}$$

and (b) Radiation heat wave:

$$\lambda_2 = \pm\,\frac{\omega}{C_R}\left[\frac{g\,(R_p)}{\omega\,D_R}\right]^{\frac{1}{2}}\,(-1 + i)$$

where

$$f\,(R_p) = \left(\frac{\gamma}{a_0{}^2} - \frac{1}{C_R{}^2}\right)\frac{a_0{}^2\,6\,(\gamma-1)\,R_p}{C_R{}^2\,\gamma\,[1 + 12\,(\gamma-1)\,R_p]} \tag{8.23}$$

and

$$g\,(R_p) = \frac{C_R{}^4\,\gamma\,[1 + 12\,(\gamma-1)\,R_p]}{24\,a_0{}^4\,(\gamma-1)\,R_p}$$

It is evident that the first mode λ_1 of equation (8.22) is the radiation sound wave whose speed of propagation is C_R given by equation (8.20). The radiative heat flux introduce the damping term in this sound wave which is indicated by the imaginary part of λ_1. The second mode λ_2 of equation (8.23) is the heat wave which is propagated at a speed:

$$V_2 = \frac{\omega}{\lambda_{2R}} = C_R\left(\frac{\omega\,D_R}{g}\right)^{\frac{1}{2}} \tag{8.24}$$

For small radiative heat flux, V_2 is very small. V_2 is proportional to C_R and square root of ω. This heat wave is a damped wave with the damping factor given by the imaginary part of λ_2. This damping factor is inversely proportional to the square root of D_R. As D_R tends to zero, the damping of the sound wave due to the radiative heat flux tends to be zero while the damping of the heat wave tends to be infinity so that the heat wave will disappear. On the other hand for a large value of D_R, the sound wave will be highly damped and the heat wave will be only slightly damped.

As the radiation pressure number R_p decreases, the radiation sound speed C_R tends to the value of the ordinary sound speed a_0 and the damping of the sound wave tends to zero. Thus as R_p tends to zero, the radiation sound speed tends to the ordinary sound speed. As R_p tends to zero, the damping of heat wave tends to be infinity and the heat wave will disappear. As R_p tends to infinity, the damping of the sound wave tends to zero because $f(R_p)$ tends to zero. The sound wave will be undamped at very large value of R_p. As R_p tends to infinity, the damping of the heat wave tends to infinity and the heat wave will again disappear.

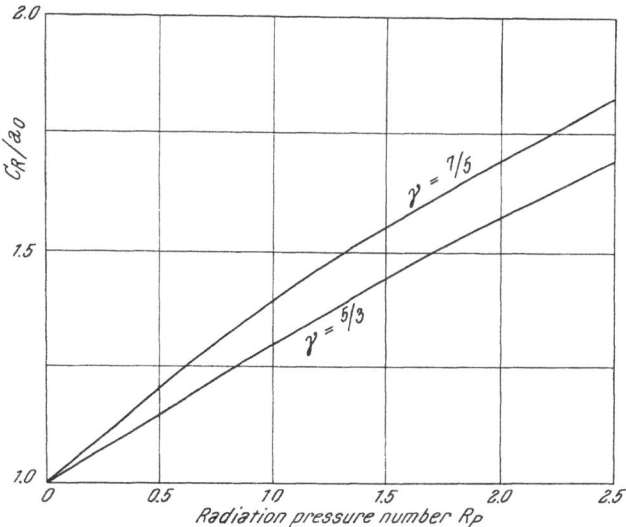

Fig. 8.1. Radiation sound speed as a function of radiation pressure number R_p

Another interesting result is that the damping of the sound wave is proportional to the square of the frequency ω while the damping of the heat wave is proportional to the square root of the frequency ω. Thus at very high frequency, the sound wave will damped out more than the heat wave.

The second square bracket of equation (8.19) is the transverse wave of equation (8.12) without the transverse magnetic field, i.e., $V_y = 0$ and the frequency is different from that given by equation (8.18). The last term of equation (8.19) gives the effect on the longitudinal wave due to transverse magnetic field. When V_y is different from zero, we have four waves: one modified heat wave and three modified magnetogasdynamic waves. This modified heat wave depends essentially on the effective heat conductivity K^*. When K^* is zero, there will be no heat wave and we have only three magnetogasdynamic waves. First we examine the longitudinal waves in an ideal plasma, i.e., inviscid, non-heat conducting and non-radiative transferring and infinitely electrically conducting gas with $K^* = \nu_g = \nu_H = 0$. Equation (8.19) becomes

$$\left[\lambda^2 \left(C_p + 20\,R\,R_p + 16\,R\,R_p{}^2 \right) - \frac{\omega^2}{T_0\,(\gamma - 1)}\,\left(1 + 12\,(\gamma - 1)\,R_p \right) \right]$$

(8.25)

$$\left[\left(\lambda^2 - \frac{\omega^2}{c^2} \right) V_x{}^2 - \omega^2 \right] - \left(\lambda^2 - \frac{\omega^2}{c^2} \right) V_y{}^2\,\frac{\omega^2}{T_0\,(\gamma - 1)}\,\left[1 + 12\,(\gamma - 1)\,R_p \right] = 0$$

Now equation (8.25) is a quadratic equation of λ^2. Hence it gives two modes of longitudinal waves in an ideal plasma. When the effective heat conductivity K^* is zero, the heat wave disappears. On the other hand when both the viscosity and magnetic diffusivity are zero, we lose another mode of the longitudinal wave which is essentially the interaction of the transverse wave with the thermal effects.

The two roots of equation (8.25) give the so-called fast and slow waves in radiation electromagnetogasdynamics which are the interaction of the sound wave with the Alfven's wave. If there is no transverse magnetic field, i.e., $V_y = 0$, equation (8.25) gives the following two roots:

$$V_1{}^2 = \frac{\omega^2}{\lambda_1{}^2} = C_R{}^2 \tag{8.26a}$$

$$V_2{}^2 = \frac{\omega^2}{\lambda_2{}^2} = V_x{}^2 \left(1 + \frac{V_x{}^2}{c^2}\right)^{-1} \tag{8.26b}$$

The first mode (8.26a) is simply the radiation sound wave and the second mode is the modified Alfven's wave. If V_y is different from zero, we also have two roots which are given by the following equation:

$$\left(C_R{}^2 - \frac{\omega^2}{\lambda^2}\right)\left[V_x{}^2 - \frac{\omega^2}{\lambda^2}\left(1 + \frac{V_x{}^2}{c^2}\right)\right] - \frac{\omega^2}{\lambda^2}V_y{}^2\left(1 + \frac{\omega^2}{c^2\lambda^2}\right) = 0 \tag{8.27}$$

Equation (8.27) has the same form of the relation between the fast and slow waves of magnetogasdynamics except that the radiation sound speed C_R replaces the ordinary sound speed a_0. In magnetogasdynamics, both V_x and V_y are negligibly small in comparison with the speed of light c. Let the roots of equation (8.27) be

$$V_{\text{fast}} = \frac{\omega}{\lambda_1} \quad \text{and} \quad V_{\text{slow}} = \frac{\omega}{\lambda_2}$$

We have the inequalities:

$$V_{\text{slow}} \lessgtr C_R \lessgtr V_{\text{fast}} \tag{8.28a}$$

$$V_{\text{slow}} \lessgtr V_x \lessgtr V_{\text{fast}} \tag{8.28b}$$

In an ideal plasma, the radiation effect on the longitudinal wave is restricted entirely to the value of the sound speed of the plasma. The interaction of the radiation sound speed C_R with the Alfven's wave speeds V_x and V_y is the same as that of the ordinary sound speed with the Alfven's wave speed when the radiation effects are negligible.

When any one of the transport coefficients is different from zero, i.e., $\nu_g \neq 0$, $\nu_H \neq 0$ or $K^* \neq 0$, it will introduce damping on the fast and slow waves and also introduce a new damped wave which will be a modified heat wave if K^* is different from zero and which will be a transverse longitudinal viscous or electromagnetic wave if ν_g or ν_H or both are different from zero. In order to show this damping effect, we now consider a simple case that $\nu_g = 0$, $\nu_H = 0$ and $K^* \neq 0$ but small. Now the damping terms of the waves as well as their speeds or propagation depends on the acute angle θ between the planar externally applied magnetic field and the direction of wave propagation. If $\theta = 0$, i.e., $V_y = 0$ and $V_x \neq 0$, there is no interaction of radiation gasdynamic wave (8.21) and the electromagnetic waves. From equation (8.19), we see that the radiation gasdynamic waves (8.21) and the magnetogasdynamic transverse wave (8.12) are not coupled. If θ is different from zero, we have three damped waves which are the interaction of the heat wave, fast wave and slow wave. We are of special interest to see the interaction of the heat wave and magnetogasdynamic wave. To show the essential feature of this interaction, we consider the case $\theta = 90°$, i.e., $V_y \neq 0$ and $V_x = 0$. In this case, the slow wave vanishes and we have only the fast wave. Equation (8.19) reduces again a quadratic equation which gives two modes of waves which are the results of the interaction of heat wave and the fast magnetogasdynamic wave. If we assume K^* or D_R to be small so that the square or higher order terms of K^* may be neglected. Then the two roots of λ are as follows:

$$\lambda_1 = \pm \frac{\omega}{C_R} \frac{A_2}{A_3}\left[1 - i\omega \frac{A_2{}^2}{A_3{}^2} f(R_p) D_R\right] \tag{8.29 a}$$

$$\lambda_2 = \pm \frac{\omega}{C_R} \frac{A_3}{A_1}\left[g(R_p)/D_R\right]^{\frac{1}{2}} (-1 + i) \tag{8.29 b}$$

where

$$A_1{}^2 = i + \frac{\gamma V_y{}^2}{a_0{}^2}; \quad A_2{}^2 = 1 + \frac{V_y{}^2}{c^2}; \quad A_3{}^2 = 1 + \frac{V_y{}^2}{C_R{}^2}$$

the functions $f(R_p)$ and $g(R_p)$ are given in equations (8.22) and (8.23) respectively.

When $V_y = 0$, equations (8.29 a) and (8.29 b) reduce respectively to equations (8.22) and (8.23). Hence the root λ_1 represents a modified damped fast magnetogasdynamic wave and the root λ_2 represents a modified heat wave.

The effect of the magnetic field increases the speed of propagation of the radiation sound speed (8.22) by a factor A_3/A_2 and decreases the damping factor by a factor $A_2{}^3/A_3{}^3$. The resultant speed of propagation is the same as that by equation (8.27) with $V_x = 0$, i.e.,

$$V_{\text{fast}} = \frac{\omega}{\lambda_{1R}} = (C_R{}^2 + V_y{}^2)^{\frac{1}{2}}\left(1 + \frac{V_y{}^2}{c^2}\right)^{-\frac{1}{2}} \tag{8.30}$$

The variation of the damping terms with ω and D_R for the cases with and without V_y are exactly the same. When D_R tends to zero, the heat wave disappears and only the fast wave survives.

The effects of the magnetic field on the heat wave are that it changes the speed of propagation by a factor A_1/A_3 and the damping term by a factor A_3/A_1. It is interesting to notice that in the factor A_1 the isothermal sound speed occurs instead of the adiabatic sound speed, i.e., $a_T = a_0/\gamma$.

As V_y increases from zero to infinity, the speed of propagation for the modified fast wave varies monotonously from radiation sound speed C_{R2} to the speed of ligth c and the corresponding damping factor from $-(\omega^2/C_R) f(R_p) D_R$ to $-(\omega^2/c)(C_R{}^2/c^2) f(R_p) D_R$. Similarly for the heat wave, the speed of propagation varies from $C_R [\omega\, D_R/g\,(R_p)]^{1/2}$ to $(C_R^2/a_T)\,[\omega\, D_R/g\,(R_p)]^{1/2}$ and the damping factor varies from $-\,(1/C_R)\,[\omega\, g\,(R_p)/D_R]^{1/2}$ to $-\,(a_T/C_R^2)\,[\omega\, g\,(R_p/D_R)]^{1/2}$.

3. Wave of small amplitudes in a radiating gas of finite mean free path of radiation. For a radiating gas of finite mean free path of radiation, we should use the general equation of radiation gasdynamics discussed in chapter V, sections 2 to 6 to derive the dispersion relations similar to equations (8.12) and (8.19) instead of the equations (5.23). Such dispersion relations have not been studied in a general manner as we discussed in the last section including the electromagnetic effects. However it has been studied by V. A. PROKOF'EV (*12*) for the case without electromagnetic field similar to our equation (8.21).

PROKOF'EV's analysis (*12*) is more general than our equation (8.21) by the following three accounts:

(i) He considered a general equation of state (5.1) with the internal energy U_m as a function of p, ρ and T.

(ii) He assumed two coefficients of viscosity μ_1 and μ while we assumed that $\mu_1 = -\,2\,\mu/3$.

(iii) He considered the specific intensity of radiation and the absorption coefficient of radiation as functions of frequency $\nu = \omega/2\,\pi$ while in our analysis the Rosseland mean value was used. Hence the most important new feature of PROKOF'EV's analysis is that the dispersion relation depends greatly on the frequency of the waves.

PROKOF'EV found the following dispersion relation instead of equation (8.21):

$$[\gamma + (1 + i\,\gamma\, X)\, m^2]\,[i + Z\, X^{(0)})\, m + 6\, i\, \beta\, R_p\, X^{(1)} - \theta\, m^3] +$$
$$i\, m^2\,(\gamma - 1 + b_1)\,(m - 12\, R_p\, X^{(2)}/h_2) = 0 \tag{8.31}$$

where γ is the ratio of specific heats and R_p is the ratio of radiation pressure to gas pressure in the undisturbed stream and the relations of the other symbols with our previous notations are as follows:

$$
\begin{aligned}
m &= i\,\lambda\, a_0/\omega, & h_2 &= (\partial\, \ln p/\partial\, \ln T)_0\\
X &= (\mu_1 + 2\,\mu)_0\,\omega/\gamma\, p_0\, h_1, & h_3 &- (\partial\, U_m/\partial\, \rho)_0\\
\theta &= \varkappa_0\, T_0\,\omega/\gamma\, p_0\, h_1\, h_4, & h_4 &= C_{v0}\, T_0 & (8.32)\\
Z &= 12\,\beta\, R_p\, c/a_0, & \beta &= p_0/\rho_0\, h_4\\
h_1 &= (\partial\, \ln p/\partial\, \ln \rho)_0, & b_1 &= 4\,(h_2/h_1)\, R_p
\end{aligned}
$$

where the subscript 0 refers to the values in the undisturbed state.

The frequency-dependent terms $X^{(0)}$, $X^{(1)}$ and $X^{(2)}$ are given by the following integrals:

$$X^{(0)} \int_0^\infty \frac{\partial B_\nu}{\partial T}\, d\nu = \int_0^\infty \frac{\partial B_\nu}{\partial T}\, w_\nu \left(1 - \frac{1}{2\,q_\nu}\, \ln \frac{1 + q_\nu}{1 - q_\nu}\right) d\nu$$

$$X^{(1)} \int_0^\infty \frac{\partial B_\nu}{\partial T}\, d\nu = \int_0^\infty \frac{\partial B_\nu}{\partial T}\, w_\nu\, \ln \frac{1 + q_\nu}{1 - q_\nu}\, d\nu \tag{8.33}$$

$$X^{(2)} \int_0^\infty \frac{\partial B_\nu}{\partial T}\, d\nu = \int_0^\infty \frac{\partial B_\nu}{\partial T}\, \frac{w_\nu}{q_\nu} \left(1 - \frac{1}{2\,q_\nu}\, \ln \frac{1 + q_\nu}{1 - q_\nu}\right) d\nu$$

where
$$q_\nu = \frac{m}{w_\nu + i\,(a_0/c)}, \quad w_\nu = \rho\,k'_{\nu 0}\,a_0/\omega$$

Hence equation (8.31) is a very complicated equation for m. However without the radiation effects, i.e., the terms of (8.33) all vanish, equation (8.31) reduces to a quadratic equation of m^2 which gives the heat wave and sound wave of the classical results. It is very difficult to solve the dispersion equation (8.31). PROKOF'EV made no attempt to solve it exactly nor to give an analytic theory of its solution. He discussed in detail some limiting cases about the two main modes of waves, i.e., the radiation sound wave [cf. eq. (8.22)] and the radiation heat wave [cf. eq. (8.23)] at various frequency ranges. The following conclusions have been drawn:

The non-dimensional parameters: X which is the reciprocal of the generalized Reynolds number, θ which is the reciprocal of the generalized Peclet number, Z which is a generalized radiation flux number, R_p and (a_0/c), are all small over an extensive range of frequencies for liquid and gases under normal conditions for not too high temperatures. For instance, for air at zero degree centigrade and 760 mm Hg pressure, we have

$$X = 1.60 \times 10^{-10}\,\omega, \quad \theta = 2.26 \times 10^{-10}\,\omega, \quad Z = 6.07 \times 10^{-6}$$
$$R_p = 1.38 \times 10^{-12}, \quad (a_0/c) = 1.105 \times 10^{-6} \tag{8.34}$$

For small values of the above parameters (8.34), small amplitude waves are almost adiabatic sound waves. The damping factor is a monotonically increasing function of frequency and is controlled by radiation transfer at low frequencies; by viscosity, heat conductivity and radiative transfer at medium frequencies and by viscosity and heat conductivity alone at very high frequencies, because the relative role of radiative transfer tends to (zero the radiative transfer rises asymptotically with frequency to a frequency-independent value). The damping factor per wave length due to radiative heat transfer has only a single maximum at medium frequencies and tends to monotonically to zero at both very high and very low frequencies.

At high temperatures (Z of the order of unity), with small values of X and θ, the speed of sound wave is nearly adiabatic value a_0 for very high and very low frequencies and it approaches but not attains the isothermal value a_T at medium frequencies.

At very high temperatures, Z can be large while X, θ, R_p and (a_0/c) are still small. In this case, the speed of propagation falls monotonically from the adiabatic value at low frequencies to isothermal value at medium frequencies and the transition from one speed to the other occurs in the region at which $Z/\rho\,K_R = 0(1)$ where K_R is the mean value of absorption coefficient according to Rosseland rule (5.18). Higher frequencies, up to about the limit $Z\,\rho\,K_{p1} = 0(1)$ where K_{p1} is the modified mean value of Planck absorption coefficient, i.e., [cf. eq. (5.36)]

$$K_{p1}\int_0^\infty \frac{\partial B_\nu}{\partial T}\,d\nu = \int_0^\infty k_{\nu}{}' \frac{\partial B_\nu}{\partial T}\,d\nu \tag{8.35}$$

have isothermal sound speed and this limit specifies the transition from isothermal sound speed back to adiabatic value at higher frequencies. At very high frequencies, the speed of wave propagation will again be adiabatic value, when X and θ remain small. As we have seen in the last section, for large value of X and θ which corresponds to large values of μ and K^*, the speed of wave propagation of both waves will differ considerably from the adiabatic value a_0.

When both Z and R_p are not small while X and θ are still small, the main feature is that the radiation sound speed C_R replaces the adiabatic value a_0. Otherwise the main feature of the variation of the speed of wave propagation is the same as that for large value of Z but small value of R_p. The damping factor per wave length due to thermal radiation has two maxima and one minimum for large Z, i.e., there are two anti-resonant system and one resonance which may be described in terms of the corresponding relaxation times. The phase velocity of the waves becomes larger than the radiation speed of sound if X and θ are large.

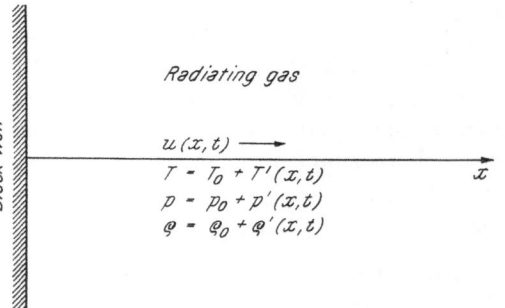

Fig. 8.2. Semi-infinite domain of a radiating gas

Since the frequency-dependent feature of the sound wave and heat wave in a radiating gas is of extremely interest, we are going to discuss this case in detail by considering a simple case according to reference 1.

We consider the case of an inviscid, non-heat-conducting but radiating perfect gas. We assume that the radiation pressure and the radiation energy density are small but the radiative heat flux is not small. In other words, we consider the case of very small radiation pressure number but finite radiative heat flux or finite effective radiative heat conductivity. In calculating the radiative heat flux, we assume that the scattering of radiation is negligible and that the condition of local thermodynamic equilibrium holds. According to our previous results, we know that there are two different waves in the present case, i.e., the sound wave and the heat wave. We are considering the variation of these waves with frequencies. We consider that the radiating gas is in a semi-infinite domain bounded one side by an infinite plane black wall. The distance perpendicular to the wall is x (Fig. 8.2). Initially the gas is at rest with a uniform state of temperature T_0, pressure p_0 and density ρ_0. The temperature of the wall is also T_0 initially. We may introduce small disturbances into the radiating gas by moving the wall which is at constant temperature or by varying the temperature of the fixed wall or both. We write for the perturbed flow

$$u = u'(x, t), \; T = T_0 + T'(x, t), \; p = p_0 + p'(x, t), \; \rho = \rho_0 + \rho'(x, t) \quad (8.35)$$

Substituting equation (8.35) into equations (5.3), (5.4) and (5.7) for an inviscid, non-heat-conducting and radiating fluid without body force and with negligibly small radiation pressure number, we have the following linearized equations after the second and higher order terms of the perturbed quantities are neglected:

$$\frac{\partial \rho'}{\partial t} + \rho_0 \frac{\partial u'}{\partial x} = 0 \qquad (8.36\,\text{a})$$

$$\rho_0 \frac{\partial u'}{\partial t} + \frac{\partial p'}{\partial x} = 0 \qquad (8.36\,\text{b})$$

$$\rho_0 C_p \frac{\partial T'}{\partial t} - \frac{\partial p'}{\partial t} = Q_R' \qquad (8.36\,\text{c})$$

where Q_R' is the perturbed quantity of the radiating heat transfer term $\nabla \cdot \vec{q}_R$. The expression of Q_R' can be derived from the one dimensional radiative transfer analysis of chapter V, § 10 except that we use x instead of y as the spatial co-ordinate now, and one of the plate $(y = L)$ is at infinity. For a black wall $(x = 0)$ the non-linear expression of the radiative heat transfer of our problem is [cf. eq. (5.46)].

$$Q_R = \frac{\partial q_R}{\partial x} = \int_0^\infty \frac{\partial q_{R\nu}}{\partial x} \, d\nu \qquad (8.37\,\text{a})$$

and

$$\frac{\partial q_{R\nu}}{\partial x} = \frac{2\pi}{L_{R\nu}} \left[B_\nu(\tau_w)\,\varepsilon_2(\tau_\nu) + \int_0^{\tau_\nu} B_\nu(t_\nu)\,\varepsilon_1(\tau_\nu - t_\nu)\,d t_\nu + \right.$$
$$\left. + \int_{\tau_\nu}^\infty B_\nu(\tau_\nu)\,\varepsilon_1(t_\nu - \tau_\nu)\,d t - 2B_\nu(T) \right] \qquad (8.37\,\text{b})$$

where T_w is the temperature of the wall which may be different from the initial temperature T_0 and $L_{R\nu} = 1/(\rho\,k_\nu')$ is the mean free path of radiation at the frequency ν. $B_\nu(T)$ is the Planck function given by equation (4.22). We are going to define the Planck average mean free path of radiation and the average optical thickness as follows:

$$L_{Rp} = \frac{\displaystyle\int_0^\infty B_\nu(T)\,d\nu}{\displaystyle\int_0^\infty B_\nu(T)(1/L_{R\nu})\,d\nu} \qquad (8.38)$$

and

$$\tau = \int_{x_w(t)}^x (1/L_{Rp})\,d x \qquad (8.39)$$

In order to find the perturbed quantity Q_R' from equation (8.37 a), BALDWIN (1) found that the following relation is useful so that the integration with respect to ν can be carried out:

$$d\tau_\nu = \frac{L_{Rp}}{L_{R\nu}}\,d\tau \qquad (8.40)$$

or

$$\tau_\nu = \tau_\nu(\tau, t) = \int_0^\tau \frac{L_{Rp}}{L_{R\nu}}\,d\bar{\tau} \qquad (8.41)$$

If we expand the Planck function $B_\nu(T)$ in power series of T' etc. and use the relation (8.40) and neglect the higher order terms, we have finally

$$Q_R' = 8\,\sigma\,T_0^3\,\frac{1}{L_{R0}}\Bigg[F(\tau)\{T_w' - T'(\tau=0)\} - \int\limits_0^\tau F(\tau-\bar\tau)\frac{\partial\,T'(\bar\tau)}{\partial\,\bar\tau}\,\mathrm{d}\bar\tau +$$

$$+ \int\limits_\tau^\infty F(\bar\tau-\tau)\frac{\partial\,T'(\bar\tau)}{\partial\,\bar\tau}\,\mathrm{d}\bar\tau\Bigg] \tag{8.42}$$

where

$$F(\tau; T_0) = \frac{\pi}{4\,\sigma\,T_0^3}\int\limits_0^\infty \frac{L_{R0}}{L_{R\nu0}}\frac{\mathrm{d}\,B_\nu(T_0)}{\mathrm{d}\,T_0}\,\varepsilon_2\left(\frac{L_{R0}}{L_{R\nu0}}\tau\right)\mathrm{d}\nu \tag{8.43}$$

where subscript 0 refers to the value in the undisturbed state and $L_{R0} = L_{Rp0}$. For a gray gas, $F(\tau) = \varepsilon_2(\tau)$. From the definition (8.41), we have to the lowest order, the relation between the optical thickness and the spatial coordinate x as follows:

$$\tau = \frac{1}{L_{R0}}[x - x_w(t)] \tag{8.44}$$

where $x_w(t)$ is the coordinate of the wall which may be a function of time t. With the help of equation (8.44), the first order radiative heat transfer Q_R' of equation (8.42) may be expressed in terms of x and t:

$$Q_R' = 8\,\sigma\,T_0^3\,\frac{1}{L_{R0}}\Bigg[F\left(\frac{x}{L_{R0}}\right)\{T_w'(t) - T'(t,0)\} -$$

$$- \int\limits_0^x F\left(\frac{x-\bar x}{L_{R0}}\right)\frac{\partial\,T'}{\partial\,\bar x}\,\mathrm{d}\bar x + \int\limits_x^\infty F\left(\frac{\bar x-x}{L_{R0}}\right)\frac{\partial\,T'}{\partial\,\bar x}\,\mathrm{d}\bar x\Bigg] \tag{8.45}$$

From equation (8.36b), we define a potential ϕ such that

$$u' = \frac{\partial\,\phi}{\partial\,x},\quad p' = -\,\rho_0\frac{\partial\,\phi}{\partial\,t} \tag{8.46}$$

Then the rest of equations (8.36) gives two equations for ϕ and T':

$$\frac{\partial^2\,\phi}{\partial\,t^2} + a_0^2\frac{\partial^2\,\phi}{\partial\,x^2} = -\frac{\gamma-1}{\rho_0}Q_R' \tag{8.47a}$$

$$\frac{\partial\,T'}{\partial\,t} = -\frac{1}{R}\left(\frac{\partial^2\,\phi}{\partial\,t^2} - \frac{a_0^2}{\gamma}\cdot\frac{\partial^2\,\phi}{\partial\,x^2}\right) \tag{8.47b}$$

where Q_R' is given by equation (8.45) and $a_0 = (\gamma\,R\,T_0)^{1/2}$ isentropic soundspeed.

We may eliminate T' from equations (8.47) and obtain a differential equation for ϕ:

$$\frac{\partial}{\partial t}\left(\frac{\partial^2\,\phi}{\partial\,t^2} - a_0^2\frac{\partial^2\,\phi}{\partial\,x^2}\right) = \frac{8\,(\gamma-1)\,\sigma\,T_0^3}{R\,L_{R0}\,\rho_0}\Bigg\{- F\left(\frac{x}{L_{R0}}\right)\left[R\frac{\mathrm{d}\,T_w'}{\mathrm{d}\,t} +\right.$$

$$+ \left.\left(\frac{\partial^2\,\phi}{\partial\,t^2} - \frac{a_0^2}{\gamma}\frac{\partial^2\,\phi}{\partial\,x^2}\right)_{x=0}\right] - \int\limits_0^x F\left(\frac{x-\bar x}{L_{R0}}\right)\frac{\partial}{\partial\,\bar x}\left(\frac{\partial^2\,\phi}{\partial\,t^2} - \frac{a_0^2}{\gamma}\frac{\partial^2\,\phi}{\partial\,\bar x^2}\right)\mathrm{d}\bar x + \tag{8.48}$$

$$+ \int\limits_x^\infty F\left(\frac{\bar x-x}{L_{R0}}\right)\frac{\partial}{\partial\,\bar x}\left(\frac{\partial^2\,\phi}{\partial\,t^2} - \frac{a_0^2}{\gamma}\frac{\partial^2\,\phi}{\partial\,\bar x^2}\right)\mathrm{d}\bar x\Bigg\}$$

The boundary conditions associated with equation (8.48) are

$$\frac{\partial \phi}{\partial x}(t, 0) = u_w(t) = \text{wall velocity} = \text{a given function of } t$$

$$T_w'(t) = T_w(t) - T_0 = \text{given function of } t \tag{8.49}$$

$$\phi(t, \infty) = \text{finite quantity for all time } t$$

For given functions $u_w(t)$ and $T_w'(t)$, we may solve for ϕ from equation (8.48). BALDWIN solved two cases: One is the case of an oscillating piston so that both $u_w(t)$ and $T_w'(t)$ are sinusoidal functions of time with radian frequency $\omega = 2\pi\nu$ and arbitrary phase and the other is the case of impulse motion of the wall, so that both $u_w(t)$ and $T_w'(t)$ are step functions of t.

In order to obtain a close form solution, some approximations on the function $F(z)$ have to be made. For a gray gas, $F(z) = \varepsilon_2(z)$. Hence we may approximate $F(z)$ by an exponential function. For the general case BALDWIN wrote

$$F(z) = m_1 \exp(m_2 z) \tag{8.50}$$

where m_1 and m_2 are constant so chosen that the approximation is exact in the Rosseland limit of optical thick case and that it is close to the expression (8.43) by a least square fit. BALDWIN found that n satisfies the relation:

$$m_2 \int_0^\infty \frac{(\mathrm{d} B_\nu/\mathrm{d} T_0) L_{R0}}{(4/\pi) T_0^3 \sigma L_{R\nu 0}} \mathrm{d}\nu - 4 \int_0^\infty \frac{\mathrm{d} B_\nu/\mathrm{d} T}{\frac{4}{\pi} \sigma T_0^3 \left(1 + \frac{L_{R\nu 0}}{L_{R0}} n\right)} \mathrm{d}\nu = 0 \tag{8.51}$$

The value of m_2 may be solved from equation (8.51) numerically. For gray gas $m_2 - [4/(1+m_2)] = 0$ or $m_2 = 1.562$. After m_2 is obtained, the value of m_1 is given by the relation

$$m_1 = \frac{m_2^2}{3} \int_0^x \frac{L_{R0}}{L_{R\nu 0}} \frac{\mathrm{d} B_\nu/\mathrm{d} T_0}{\frac{4}{\pi} \sigma T_0^3} \mathrm{d}\nu \tag{8.52}$$

for gray gas $m_1 = m_2^2/3$.

With the approximation (8.50), the solution of an oscillating wall with frequency of oscillation ω is

$$u(t, x) = \frac{a_0}{\gamma} Re \left\{ \left[c_1 C_1 \exp\left(\frac{c_1 \omega x}{a_0}\right) + c_2 C_2 \exp\left(\frac{c_2 \omega x}{a_0}\right) \right] \exp(i\omega t) \right\} \tag{8.53}$$

where the complex quantities c_1, c_2, C_1 and C_2 satisfy the following relations:

$$Re\left\{ (c_1 C_1 + c_2 C_2) \exp(i\omega t) \right\} = \frac{\gamma}{a_0} u_w(t) \tag{8.54a}$$

$$Re\left\{ -i\left[\left(1 + \frac{c_1^2}{\gamma}\right)\left(\frac{\beta}{\beta + c_1}\right) C_1 + \left(1 + \frac{c_2^2}{\gamma}\right)\left(\frac{\beta}{\beta + c_2}\right) C_2 \right] \exp(i\omega t) \right\} = \frac{T_w'(t)}{T_0} \tag{8.54b}$$

$$\begin{Bmatrix} c_1 \\ c_2 \end{Bmatrix} = -\left[\frac{-(1 - \beta^2 - iK\beta) \mp \sqrt{(1 - \beta^2 - iK\beta)^2 + 4\beta^2(1 - iK\beta/\gamma)}}{2(1 - iK\beta/\gamma)} \right]^{\frac{1}{2}} \tag{8.54c}$$

$$\beta = \frac{m_2 \, a_0}{L_{R0}}, \quad K = \frac{16 \, (\gamma - 1) \, T_0^3}{R \, \rho_0 \, a_0} \frac{m_1}{m_2} \tag{8.54 d}$$

where Re means the real part of the complex function.

For given sinusoidal functions $u_w \, (t)$ and $T_w' \, (t)$ which are both proportional to $\exp{(i \omega t)}$, we may calculate the complex quantities c_1, c_2, C_1 and C_2. Equation (8.53) then represents two damped sinusoidal travelling waves. One corresponds to sound wave and the other, heat wave. Fig. 8.3 shows the wave

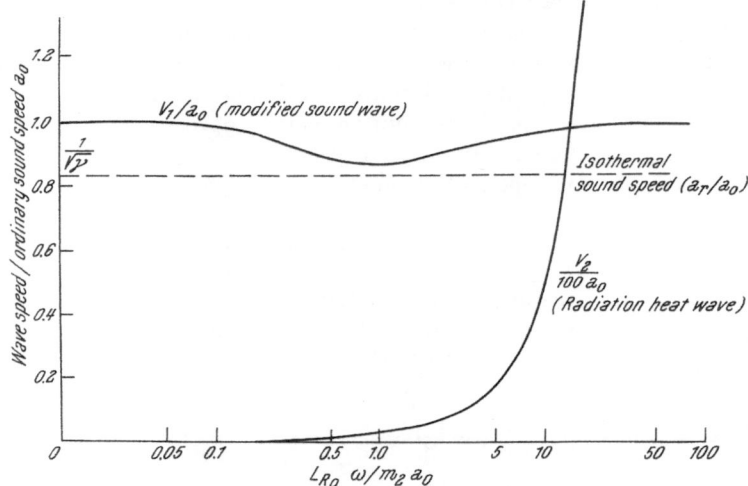

Fig. 8.3. Radiation wave speeds *vs* frequency of oscillation.
(Fig. 2 of reference 1 by B. S. BALDWIN jr., Courtesy of NASA)

speeds for these two waves as functions of the frequency at $\gamma = 7/5$ and $K = 4$. The parameter K represents the radiative flux number. For the first wave, the speed of propagation equals to the isentropic sound speed at both low frequency and high frequency range. At intermediate frequency, the speed of propagation approaches the isothermal sound speed $a_T = a_0/\sqrt{\gamma}$. This is the radiating sound wave. For the second wave, the speed of propagation increases with the frequency. This is the radiating heat wave. It is interesting to compare these results with the optically thick case, i.e., equation (8.22) and (8.23). For a first approximation, these two results agree. For the radiating sound wave, the speed of propagation is close to isentropic sound speed and for the radiating heat wave, the speed of propagation increase with the frequency.

4. Shock waves in an optically thick medium. It is well known that waves of finite amplitude may develop into shock waves across which large changes of the velocity and state variables occur. Since shock waves are important in high speed flow where the thermal radiation is also important, it is interesting to see what are the effects of thermal radiation on the shock waves. We first consider a normal shock wave in an ionized and radiating gas. We shall choose the coordinate system such that the shock wave is stationary (Fig. 8.4). In our system, the gas flow is parallel to the x-axis and has the x-component of velocity u only. We assume that the magnetic field strength $\vec{H}$ is planar and perpendicular to the velocity field and has the y-component only $H = H_y$. Both the velocity and the magnetic field strength are uniform in front of the shock and far behind the shock and there are large variations of all the variables u, H, p, ρ and T in the shock transition region. In general the shock transition region is very narrow and then for first approximation it may be considered as a surface of discontinuity in many flow problems. The fundamental equations which govern the flow field with shock wave shown in Fig. 8.4 are as follows:

$$\rho \, u = \text{constant} = m \tag{8.55a}$$

$$m u + p_t + \mu_e \tfrac{1}{2} H^2 - \frac{4}{3} \mu \frac{\mathrm{d} u}{\mathrm{d} x} = \text{ constant } = m C_1 \tag{8.55b}$$

$$m h_R + u p_t - \frac{4}{3} \mu u \frac{\mathrm{d} u}{\mathrm{d} x} - K^* \frac{\mathrm{d} T}{\mathrm{d} x} + E \mu_e H = \text{constant} = m C_2 \tag{8.55c}$$

$$u H - \nu_H \frac{\mathrm{d} H}{\mathrm{d} x} = \text{ constant } = E \tag{8.55d}$$

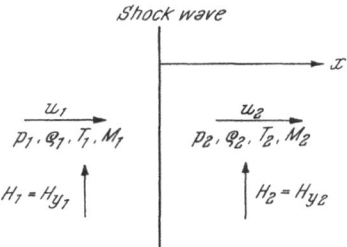

Fig. 8.4. Normal shock in a transverse magnetic field

Here we assume that the gas is optically thick so that these equations are derived from equation (5.23) and (8.1) to (8.6) with the usual magnetogasdynamic approximations that charge separation can be neglected. The total pressure p_t is the sum of gas pressure p and the radiation pressure p_R, H is the y-component of the magnetic field; $h_R = \tfrac{1}{2} u^2 + C_v T + E_R/\rho$ and C_v is the specific heat of the gas at constant volume; K^* is the effective coefficient of heat conductivity with radiation [cf. eq. (8.10)]. The rest symbols have their usual meanings.

We shall discuss the shock transition region in the next section. Here we consider only the Rankine-Hugoniot relations between the two uniform states separated by the shock transition. In these uniform states, we have

$$\frac{\mathrm{d} u}{\mathrm{d} x} = \frac{\mathrm{d} H}{\mathrm{d} x} = \frac{\mathrm{d} T}{\mathrm{d} x} = 0 \tag{8.56}$$

If we substitute the conditions (8.56) into equations (8.55), we obtain the generalized Rankine-Hugoniot relations in radiation magnetogasdynamics. Before we write these relations, we shall introduce the following non-dimensional variables:

$$\xi = \frac{u}{u_1}, \quad T^* = \frac{R T}{u_1^2}, \quad h = H/(2 m u_1/\mu_e)^{\tfrac{1}{2}}$$

$$P = T_1^* = 1/\gamma \, M_1^2, \ M_1 = \frac{u_1}{a_1}, \quad a_1 = (\gamma \, R \, T_1)^{\tfrac{1}{2}} \tag{8.57}$$

$$Q = a_R u_1^6/(R^4 \rho_1)$$

and

$$R_n = p_R/p = \frac{1}{3} Q \, \xi \, T^{*3} = \text{radiation pressure number} \tag{8.58}$$

where M is the Mach number and the subscript 1 refers to the values in the uniform state in front of the shock.

Substituting equations (8.56) and (8.57) into equations (8.55), we obtain the following non-dimensional form of the generalized Rankine-Hugoniot relations:

$$h\,\xi = h_1 \tag{8.59a}$$

$$\xi^2 + T^* + \frac{1}{3}\,Q\,\xi\,T^{*4} - \left(1 + P + h_1{}^2 - h^2 + \frac{1}{3}\,Q\,P^4\right)\xi = 0 \tag{8.59b}$$

$$\frac{T^*}{\gamma-1} - \frac{1}{2}\,\xi^2 + (1 + P + h_1{}^2 - h^2)\,\xi + Q\,\xi\,T^{*4} +$$

$$+ \frac{1}{3}\,Q\,P^4\,(\xi-4) - \left[\frac{1}{2} + \frac{\gamma\,P}{\gamma-1} + 2\,(h_1{}^2 - h_1\,h)\right] = 0 \tag{8.59c}$$

If the radiation effects are negligible, the terms containing Q can be omitted and equations (8.59) reduce to those of ordinary magnetogasdynamics.

If we eliminate h from equations (8.59) and introduce the radiation pressure number R_p as a parameter, we have

$$\xi^2 + (1 + R_p) - \left[1 + P + h_1{}^2\left(1 - \frac{1}{\xi^2}\right) + R_{p1}\,P\right]\xi = 0 \tag{8.60a}$$

$$-\tfrac{1}{2}\,\xi^2 + \left(\frac{1}{\gamma-1} + 3\,R_p\right)T^* + \left[1 + P + h_1{}^2\left(1 - \frac{1}{\xi^2}\right) + R_{p1}\,P\right]\xi -$$

$$- \left[\frac{1}{2} + \frac{\gamma\,P}{\gamma-1} + 2\left(1 - \frac{1}{\xi^2}\right)h_1{}^2 + 4\,R_{p1}\,P\right] = 0 \tag{8.60b}$$

Evidently we consider R_p and R_{p1} as parameters which account for the radiation effects on the shock wave.

If we eliminate T^* from equations (8.60), we have

$$(\xi - 1)\,\{\xi^2 - (7\,R_p + r^2)^{-1}$$

$$[(8\,R_p + r^2 + 1)\,(h_1{}^2 + f\,R_{p1}\,P + f\,P) + R_{p1} + 1]\,\xi - \tag{8.61}$$

$$- (7\,R_p + r^2)^{-1}\,(4\,R_p + r^2 - 3)\,h_1{}^2\} = (\xi - 1)\,(\xi - \xi_2)\,(\xi + \xi_3) = 0$$

where
$$r^2 = (\gamma + 1)\,/(\gamma - 1)$$

$$f = f\,(R_p) = [\xi - g\,(R_p)]/(\xi - 1)$$

with
$$g\,(R_p) = \frac{(R_p + 1)\,(8\,R_{p1} + r^2 + 1)}{(R_{p1} + 1)\,(8\,R_p + r^2 + 1)}$$

There are three roots of equation (8.61). The negative root $-\xi_3$ has no physical significance because ξ represents the magnitude of the velocity of our flow field which is a positive quantity. The root "unity" represents the velocity of the original flow, i.e., the case of no shock. The only interesting solution is the root ξ_2 which represents the velocity of the gas behind a normal shock wave.

The formal expression for ξ_2 is

$$\xi_2 = \frac{1}{2}\left[\frac{\gamma_e-1}{\gamma_e+1} + \frac{2\,\gamma_e\,(P_e+h_1{}^2)}{\gamma_e+1}\right] + \frac{1}{2}\left\{\left[\frac{\gamma_e+1}{\gamma_e+1} + \frac{2\,\gamma_e\,(P_e+h_1{}^2)}{\gamma_e+1}\right]^2 + 8\frac{2-\gamma_e}{\gamma_e+1}h_1{}^2\right\}^{\frac{1}{2}}$$

(8.62)

where $\qquad \gamma_e = \dfrac{4\,(\gamma-1)\,R_{p2}+\gamma}{3\,(\gamma-1)\,R_{p2}+1}$ = effective ratio of specific

heats in radiation gasdynamics (8.63)

$$P_e = (R_{p1}+1)\,f\,(R_{p2})\,P = \text{effective value of } P \text{ in}$$
radiation gasdynamics (8.64)

where subscript 2 refers to the value in the uniform state behind the shock wave. The value of γ_e is γ when $R_{p2}=0$, i.e., no radiation effect and $\gamma_e = 4/3$ if $R_{p2}=\infty$, i.e., very large radiation effect.

Since both γ_e and P_e are functions of R_{p2} and R_{p2} depends on ξ_2, we have to find the value of ξ_2 for a given set of initial conditions P, h_1 and R_{p1} by the method of successive approximation as we shall discuss below. The function $g\,(R_{p2})$ depends on γ, R_{p1} and R_{p2}. When the difference between R_{p1} and R_{p2} is small, $g\,(R_{p2})$ is approximately equal to unity and hence $f\,(R_{p2})$ is also approximately equal to unity. The function $g\,(R_{p2})$ decreases as R_{p2} increases. The limiting value of $g\,(R_{p2})$ for very large value of R_{p2} is a constant, i.e.,

$$g_0 = [8\,R_{p1}+r^2+1]/[8\,(R_{p1}+1)]$$

(8.65)

for given values of γ and R_{p1}. In general, g_0 does not differ greatly from unity.

It is interesting to find the value of ξ_2 for a few limiting cases:

(a) Low temperature case. If the temperatures both in front of the shock and behind it are not too high, we have $R_{p1} \simeq R_{p2} \simeq 0$. Hence $\gamma_e = \gamma$ and $P_e = P$. Equation (8.62) becomes the normal shock relation in ordinary magnetogasdynamics.

(b) Weak shock in high-temperature gas. If the temperature of the gas is initially very high, R_{p1} is then not negligible. If in addition, the shock wave strength is weak, R_{p2} will be approximately equal to R_{p1}. Hence in equation (8.62), we may write $\gamma_e \simeq \gamma_{e1}$ and $P_e \simeq P_{e1}$. The effects of radiation on the uniform state behind a weak shock in this case are:

(i) The value of γ is replaced by the effective value γ_{e1}, i.e.,

$$\gamma_{e1} = \frac{4\,(\gamma-1)\,R_{p1}+\gamma}{3\,(\gamma-1)\,R_{p1}+1}$$

(8.66)

and (ii) the value of P changes into the effective value P_{e1}, i.e., the gas pressure is replaced by the total pressure which is the sum of the gas pressure and the radiation pressure.

If there is no magnetic field strength and the shock strength is infinitesimally small, we have

$$u_1{}^2 = \frac{p_1}{\rho_1}\gamma_{e1}\,(R_{p1}+1) = \gamma_{e1}\frac{p_1+p_{R1}}{\rho_1} = C_R{}^2$$

(8.67)

This formula (8.67) is another way to define a radiation sound speed C_R which is practically identical to that given by equation (8.20).

(c) Very strong shock in a cold gas. In this case, $R_{p1} \ll 1$ and $R_{p2} \gg 1$. Hence if $\gamma = 5/3$, we have $\gamma_e = 4/3$ and $P_e = (9/16)\,P$. Equation (8.62) becomes

$$\xi_2 = \frac{1}{2}\left[\frac{1}{7} + \frac{8}{7}\left(\frac{9\,P}{16} + h_1{}^2\right)\right] + \frac{1}{2}\left\{\left[\frac{1}{7} + \frac{8}{7}\left(\frac{9\,P}{16} + h_1{}^2\right)\right]^2 + \frac{32}{7}\,h_1{}^2\right\}^{\frac{1}{2}} \quad (8.68)$$

If there is no magnetic field, $h_1 = 0$, equation (8.68) reduces

$$\xi_2 = \frac{1}{7} + \frac{9\,P}{14} = \frac{1}{7} + \frac{27}{70\,M_1{}^2} \quad (8.68\,a)$$

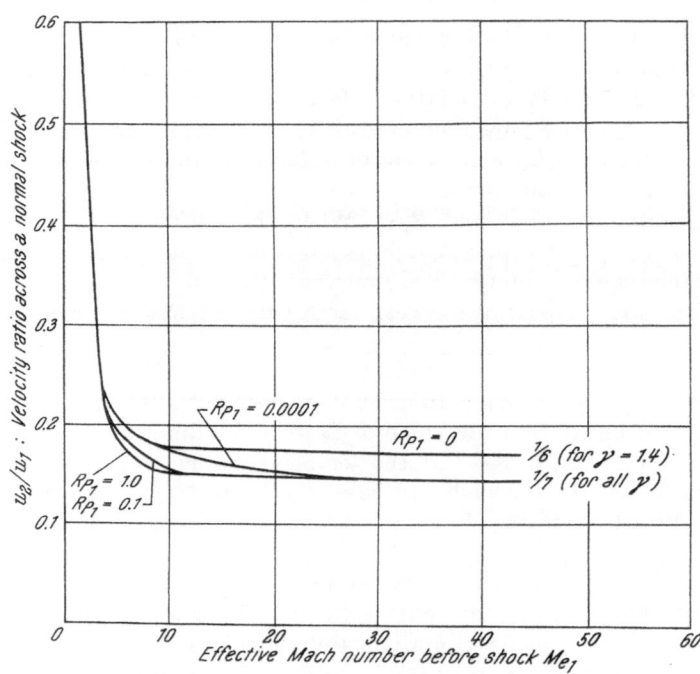

Fig. 8.5. Velocity ratio across a normal shock in an ideal gas
with and without radiation effect

For very strong shock, Mach number $M_1 \gg 1$, we have the limiting value $\xi_2 = {}^1\!/_7$ for all values of γ (see Fig. 8.5).

(d) Shock wave in a very hot plasma. In this case $R_{p1} \gg 1$. For simplicity, we consider the case without magnetic field, i.e., $h_1 = 0$. The procedure can be extended to the case with a magnetic field in a straightforward manner. The formal solution of ξ_2 is then

$$\xi_2 = \frac{\gamma_e - 1}{\gamma_e + 1} + \frac{2\gamma_e P_e}{\gamma_e + 1} \quad (8.69)$$

For the limiting case $R_{p1} \gg 1$, equation (8.69) becomes

$$\xi_2 = \frac{1}{7} + \frac{1}{6\,M_{e1}^2} \quad (8.69\,a)$$

where M_{e1} is the effective shock Mach number, defined by the equation

$$M_{e1}^2 = \frac{\gamma\,[1 + 12\,(\gamma - 1)\,M_1^2\,R_{p1}]}{\gamma + 20\,(\gamma - 1)\,R_{p1} + 16\,(\gamma - 1)\,R_{p1}^2} \tag{8.70}$$

When $R_{p1} \gg 1$, we have

$$M_{e1}^2 = \frac{3\,\gamma\,M_1^2}{4\,R_{p1}} \tag{8.70a}$$

Equation (8.69a) with the help of equation (8.58) may be written as

$$T^{*4} + A^{-1}\,T^* - A^{-1}\,B = 0 \tag{8.71}$$

where

$$A^{-1} = P^3/(R_{p1}\,\xi) > 0$$

and

$$B = [(R_{p1} + 1)\,P + 1]\,\xi - \xi^2 > 0$$

For $R_{p1} \gg 1$, $A^{-1}\,T^* \ll T^{*4}$. Hence we have

$$\frac{T_2}{T_1} = \frac{T^*}{P} \simeq \frac{(A^{-1}\,B)^{\frac{1}{4}}}{P} \simeq \left[1 + \frac{8}{7}\,(M_{e1}^2 - 1)\right]^{\frac{1}{4}} \tag{8.72}$$

For very large M_{e1}, equation (8.72) becomes

$$T_2/T_1 = 1.033\,M_{e1}^{\frac{1}{2}} \tag{8.72a}$$

Without radiation effect, it is well known that at high shock Mach number, the temperature ratio across a shock increases with the square of the shock Mach number M_1. Here we have shown that if the radiation effects are included, at very high shock Mach numbers, the temperature ratio across a shock wave increases only with the square root of the effective Mach number M_{e1}.

For finite R_{p1} (not large compared to unity, but still with $A^{-1}\,T^* \ll T^{*4}$, we have

$$\frac{T_2}{T_1} = \frac{T^*}{P} = \frac{(A^{-1}\,B)^{\frac{1}{4}}}{P} \simeq \left\{\frac{R_{p1}+1}{R_{p1}} + \frac{6\,(M_{e1}^2 - 1)\,[\gamma + 20\,(\gamma - 1)\,R_{p1} + 16\,(\gamma - 1)\,R_{p1}^2]}{7\,R_{p1}\,[1 + 12\,(\gamma - 1)\,R_{p1}]}\right\}^{\frac{1}{4}} \tag{8.73}$$

It was found that equation (8.73) gives very accurate results whenever $R_{p1} \gtrsim 1$.

The accurate values of the temperature ratio T_2/T_1 may be obtained by successive approximations. The results are shown in Fig. 8.6.

5. Shock wave structure in an optically thick medium. Equation (8.59) gives the uniform states in front of and behind a shock wave. Actually there is a transition region in which the flow variables change gradually from the value in front of the shock to that behind the shock. This transition region is usually very thin and in many flow problems we may consider it as a surface of discontinuity. It is interesting to know what are the main effects of radiation on the flow variables in the transition region and how the radiation affects the thickness of the shock transition region. In order to answer these questions we have to solve the system of differential equations (8.55) with the initial and final conditions given by equations (8.59). The general solution of equations (8.55) has not been worked out. However from our knowledge of shock wave structure in ordinary gasdynamics, some general conclusions may be drawn. We shall consider the case without magnetic field only. Equations (8.55b) and (8.55c) in non-dimensional form are

$$\frac{4}{3}\frac{\mu}{m}\xi\frac{\mathrm{d}\xi}{\mathrm{d}x} = T^* - T_\infty^*(\xi, R_p) \tag{8.74a}$$

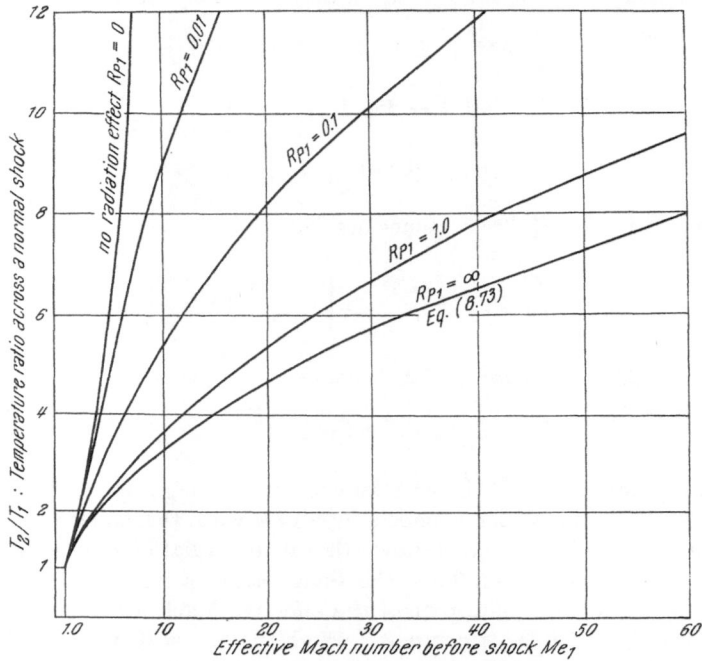

Fig. 8.6. Temperature ratio across a normal shock in an ideal gas with and without radiation effect

$$\frac{K^*}{m\,C_p}\frac{\mathrm{d}T^*}{\mathrm{d}x} = T^* - T_0^*(\xi, R_p) \tag{8.74b}$$

where
$$T_\infty^* = -\xi^2 - \frac{1}{3}Q\,\xi\,T^{*4} + \left(1 + P + \frac{1}{3}Q\,P^4\right)\xi \tag{8.75a}$$

$$T_0^* = -(\gamma-1)\left[-\frac{1}{2}\xi^2 + (1+P)\xi + Q\xi T^{*4} + \frac{1}{3}Q\,P^4(\gamma-4) - \frac{1}{2} - \frac{\gamma P}{\gamma-1}\right] \tag{8.75b}$$

Eliminating x from equation (8.74), we have

$$\frac{\mathrm{d}T^*}{\mathrm{d}(\frac{1}{2}\xi^2)} = \frac{4}{3}\frac{P_{rR}}{\gamma}\frac{T^* - T_0^*}{T^* - T_\infty^*} \tag{8.76}$$

where $P_{rR} = \mu\,C_p/K^* = $ effective Prandtl number with radiation effect.

Equation (8.76) may be regarded as a differential equation for T^* in terms of ξ. Regardless of whether or not the effective Prandtl number P_{rR} is a constant, if $0 < P_{rR} < \infty$, equation (8.76) has a singularity whenever the numerator and the denominator of the right-hand side of equation (8.76) vanish simultaneously. The conditions $T^* = T_0^*$ and $T^* = T^*_{\infty}$ lead to the Rankine-Hugoniot relation of shock in radiation gasdynamics (8.61) with $h_1 = 0$. Hence the two uniform states defined by equation (8.61) are two singular points of equation (8.76). We are interested in the integral curves of equation (8.76) in the $\xi - T^*$ plane joining the two uniform states $\xi = 1$ and $\xi = \xi_2$. Since $T^* = T_0^*$, $d\,T^*/d\,(\tfrac{1}{2}\xi^2) = 0$ and $T^* = T^*_{\infty}$, $dT^*/d\,(\tfrac{1}{2}\xi^2) = \infty$, the integral curve must lie between the two

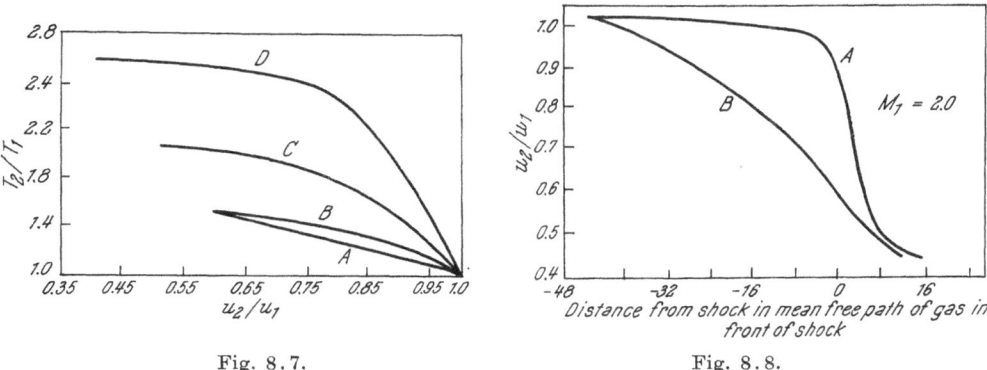

Fig. 8.7. Fig. 8.8.

Fig. 8.7. Variation of temperature with velocity in a shock transition region. A: $M_1 = 1.5$ without radiation; B: $M_1 = 1.5$ with radiation; C: $M_1 = 2.0$ with radiation; D: $M_1 = 2.5$ with radiation. (Fig. 1 of reference 16 by H. K. SEN and A. W. GUESS, Courtesy of American Institute of Physics, Physical Review)

Fig. 8.8. Velocity distribution in a shock transition region. A: without radiation; B: with radiation. (Fig. 2 of reference 16 by H. K. SEN and A. W. GUESS, Courtesy of American Institute of Physics, Physical Review)

curves $T^* = T_0^*$ and $T^* = T^*_{\infty}$ in the $\xi - T^*$ plane. In general, we have

$$T^* = T^* (\xi) \qquad (8.77)$$

from equation (8.76) by numerical or graphical integration. The general case for finite value of R_{p1} has not been calculated, GUESS and SEN calculated the case for $R_{p1} = 0$ but finite radiative heat transfer and the results are shown in Fig. 8.7 and 8.8. The main result is that the radiation effect broadens the shock transition region. For finite value of R_p, we know from Figs. 8.5 and 8.6, the change of velocity ratio across the shock is small but the ratio of temperatures across the shock reduces. This will further increases the thickness of the shock transition region.

After $T^* = T^* (\xi)$ is obtained, the x-coordinate in the shock transition region may be obtained by simple quadrature, i.e.,

$$\frac{\rho_1 u_1}{\mu_1} x = \frac{4}{3} \int \frac{(\mu/\mu_1)}{T^* - T^*_{\infty}}\, d\left(\frac{1}{2}\xi^2\right) \qquad (8.78)$$

6. Shock wave in a medium of finite mean free path of radiation. In the last two sections, we consider only the case of very small mean free path of radiation so that all the radiation terms may be expressed as simple functions of temperature and some mean absorption coefficient independent of frequency. For the case of finite mean free path of radiation, all the radiation terms must be expressed in terms of integral forms as we have discussed in chapter V, §§ 2 to 6. The general case has not been investigated yet. The only case which has been studied is the case of a gray gas with negligibly small radiation pressure number but finite radiation flux number. The fundamental equations for the shock transition region are [cf. eq. (8.55)]

$$\rho\, u = \text{constant} = m \qquad (8.79\,\text{a})$$

$$-\frac{4}{3}\,\mu\,\frac{\mathrm{d}u}{\mathrm{d}x} + p + mu = \text{constant} = m\,C_1 \qquad (8.79\,\text{b})$$

$$-\varkappa\frac{\mathrm{d}T}{\mathrm{d}x} - \frac{4}{3}\,\mu\,u\,\frac{\mathrm{d}u}{\mathrm{d}x} + m\left(\frac{1}{2}\,u^2 + C_p\,T\right) - q_R = \text{constant} = m\,C_2 \qquad (8.79\text{c})$$

where q_R is the radiative heat transfer as given by equation (5.35) or for a gray gas given by equation (8.80):

$$\frac{\mathrm{d}\,q_R}{\mathrm{d}\,x} = \frac{c a_R}{2\,L_R}\left\{\int_{-\infty}^{\infty} T^4\,(t)\,\varepsilon_1\,(|t - \tau|)\,\mathrm{d}t - 2\,T^4\,(\tau)\right\} =$$

$$= \frac{2\,c\,a_R}{L_R}\left\{\int_{\tau}^{\infty} T^3\,(t)\,\varepsilon_2\,(t - \tau)\frac{\mathrm{d}T}{\mathrm{d}t}\,\mathrm{d}t - \int_{-\infty}^{\tau} T^3\,(t)\,\varepsilon_2\,(\tau - t)\frac{\mathrm{d}T}{\mathrm{d}t}\,\mathrm{d}t\right\} \qquad (8.80)$$

where

$$\tau = \int_{0}^{x} (1/L_R)\,\mathrm{d}x = \text{optical thickness}$$

and L_R is the mean free path of radiation. In equation (8.80), we assume that the radiative heat transfer at infinity would not affect the value of radiative heat transfer at x. We have to solve simultaneously equations (8.79) and (8.80) or its equivalent expression with the boundary values given by the two uniform states $(u_1,\ T_1)$ and $(u_2,\ T_2)$ as determine by the Rankine-Hugoniot relations. The range of this transition region is from $x = -\infty$ where $u = u_1$ and $T = T_1$ to $x = +\infty$ where $u = u_2$ and $T = T_2$.

The main feature of the present problem is the introduction of a new diffusive transport property, the mean free path of radiation L_R, in addition to the viscosity and heat conductivity in the shock structure problem of ordinary gasdynamics in which the mean free path of collision of gas particles L_f plays an important role. Since the relative values of L_R and L_f may vary greatly, new phenomena would occur, particularly when L_R is much larger than L_f. The problem is very similar to the shock structure in a chemically reacting medium in which the relaxation phenomena are important and in which a relaxation length L_r may be introduced to represent the distance to attain the chemical equilibrium condition in the shock transition region. If L_r is much larger than L_f, a considerable portion of the shock transition region, which is usually referred to as an inviscid and non-heat-conducting tail, is determined by the new diffusive property due to the chemical reaction only. If we neglect completely the viscosity and heat conductivity in a chemically reacting gas, we first obtain a shock discontinuity in temperature and velocity as if it is a non-chemically reacting gas and the kinetic temperature immediately behind this shock discontinuity is higher than the final equilibrium temperature. We then have a long transition region, the tail, in which the temperature decreases from the overshot value to its final equilibrium value as if the gas is an inviscid and non-heat-conducting medium. Such a shock is referred to as partly dispersed shock. If the shock is sufficiently weak, we may obtain continuous solution of the shock structure from the Rankine-Hugoniot conditions by neglecting the viscosity and heat conductivity and with the relaxation phenomena only. This is known as a fully dispersed shock.

Similar situation has been obtained for the shock structure in a radiating gas. If we consider the effect of radiative heat transfer only, i.e., equations (8.79) and (8.80), in a shock structure problem, the main new feature is the radiation mean free path L_R. If L_R is much larger than L_f, we would have some portion of the shock transition region in which the flow field will be determined essentially by L_R. Hence we would have phenomena similar to the partly dispersed shock and fully dispersed shock in a chemically reacting medium. However, in the shock structure in a chemically reacting gas, the chemical reaction is important only behind the normal shock and hence we have only the rear tail in the shock transition region. One the other hand, the radiative transfer phenomena may be important both in front of and behind the main shock region, we would expect that the inviscid and non-heat-conducting tails occur in front of and behind the main shock transition region. For a partly dispersed shock, the front and the rear inviscid and non-heat-conducting tails may join together by a surface of discontinuity, the ideal shock. For a fully dispersed shock, the front and the rear inviscid and non-heat-conducting tails would merge together without any discontinuity between them. In the region of the discontinuity surface, the effects of viscosity and heat conductivity must be considered and then the discontinuity surface will be replaced by a sharp transition region in the actual case. Because in real air, L_R is usually much larger than L_f, the dispersed solution of the shock structure in a radiating gas is important under a wide range of conditions. We shall consider it first:

(i) Radiation resisted shock wave or radiation dispersed shock. In this case, we neglect the viscous and heat-conducting terms in equations (8.79) and solve the resultant equations with the expression of q_R given by equation (8.80). Since equations (8.79) with the expression (8.80) are integro-differential equations, it is difficult to solve them. It is advisable to transform the equations into a system involving only differential equations.

CLARKE (3) solved this problem approximately by the following approximation:

$$\varepsilon_2(t) = \frac{m_2{}^2}{3} \exp(-m_2 t) \tag{8.81}$$

where $m_2 = 1.562$. This approximation has been discussed in section 3. From equations (8.79) to (8.81) we may obtain a single differential equation of u in terms of the optical thickness τ as follows:

$$\frac{d^2}{d\tau^2}\left[\left(u - \frac{a^2}{u}\right)\frac{du}{d\tau}\right] + \frac{2 m_2{}^2}{3}\frac{8(\gamma - 1)\sigma}{mR}\frac{d}{d\tau}\left[T^3\left(u - \frac{a_T{}^2}{u}\right)\frac{du}{d\tau}\right] - m_2{}^2\left(u - \frac{a^2}{u}\right)\frac{du}{d\tau} = 0$$

(8.82)

where

$$a_T = (RT)^{\frac{1}{2}} = \frac{a}{\sqrt{\gamma}} = \text{isothermal sound speed}$$

(8.83)

$$T = (C_1 - u)\,u$$

(8.84)

and

$$\sigma = \frac{1}{4}(c\,a_R).$$

If we integrate equation (8.82) with resepct to τ and substitute the conditions of the uniform state, we obtain the following equation:

$$-\frac{d^2}{d\tau^2}F(v) + C(1-v)^3 v^3 (2v - 1)\frac{dv}{d\tau} + m_2{}^2 F(v) = 0$$

(8.85)

where

$$v = \frac{u}{C_1} = \frac{2\gamma}{\gamma + 1}\frac{u}{u_1 + u_2}$$

(8.86)

$$F(v) = (v_1 - v)(v - v_2)$$

(8.87)

and

$$C = \frac{32\,m_2{}^2\,(\gamma - 1)\,C_1{}^6}{3\,m\,(\gamma + 1)\,R^4}$$

(8.88)

Since u_1 and u_2 and then v_1 and v_2 are determined by the Rankine-Hugoniot relation, for an ideal gas with constant γ, we have that v lies between the limits 1 and $(\gamma - 1)/(\gamma + 1)$ and $v = \frac{1}{2}$ when $u = a_T$ and that $v = \gamma/(\gamma + 1) = v^*$ when $u = a = \frac{1}{2}(u_1 + u_2)$.

Now if we write

$$z = \frac{dv}{d\tau}$$

(8.89)

equation (8.85) reduces to a first order differential equation:

$$-\frac{dz}{dv} = \frac{2z^2 + C(v - v^2)^3(2v - 1)z + m_2{}^2(v_1 - v)(v - v_2)}{2(v - v^*)z}$$

(8.90)

Equation (8.90) is similar to our previous shock structure equation (8.76). Equation (8.90) has also two singular points at the two uniform states, i.e., $z = 0$, $v = v_1$ and $z = 0$, $v = v_2$. CLARKE discussed these singular points in reference 3. We also want to find the integral curve joining the two singular points. In general this integral curve should be obtained by numerical integration. In order to show some essential features we consider a limiting case as follows:

First we introduce a new variable y such that

$$y = \tau/C \tag{8.91}$$

equation (8.85) becomes

$$-\epsilon \frac{d^2 F(v)}{dy^2} + (v - v^2)^3 (2v - 1) \frac{dv}{dy} + m_2^2 F(v) = 0 \tag{8.92}$$

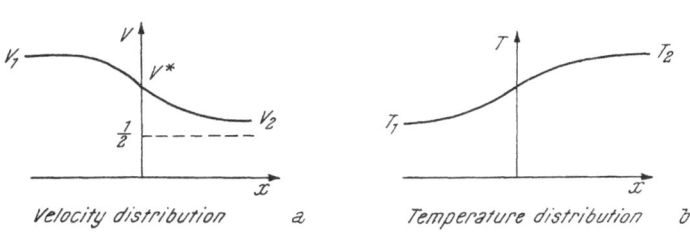

Fig. 8.9. Fully dispersed shock in a radiating gas

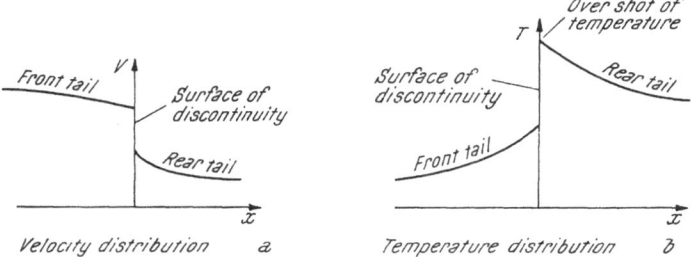

Fig. 8.10. A solution of partly dispersed shock in a radiating gas

where $\epsilon = C^{-2}$. For the limiting case $\epsilon \to 0$ for a fixed y, equation (8.92) reduces to

$$\frac{(v - v^2)^3 (2v - 1)}{(v_1 - v)(v - v_2)} \frac{dv}{dy} = -m_2^2 \tag{8.93}$$

From equation (8.93), we see that the equation (8.92) has a continuous solution with finite dv/dy everywhere from $v = v_1$ to $v = v_2$, provided that $v > \frac{1}{2}$. Fig. 8.9 shows sketches of the continuous solution. Even though we assume $\mu = \varkappa = 0$, both the velocity and the temperature vary continuously from the values of state 1 to those of state 2. However, if $v_2 \leq \frac{1}{2}$, the solution shows that the velocity first decreases smoothly from the value $v = v_1$ until a value $v = v_c > v_2$, the slope dv/dy becomes infinite. Hence we will not be able to obtain a smooth solution for this case. Physically, it means that neat this critical point where $dv/dy = \infty$, we should consider the viscosity and heat conductivity. In general, the shock transition region in radiation gasdynamics behaves as follows:

The velocity first decreases gradually until near some critical value if the shock strength is large and then there will be a sharp transition region until it reaches near the value of v_2 or u_2 and finally it will gradually reaches the final uniform state u_2 (Fig. 8.10).

The dispersed shock in a radiating gas was first investigated by PROKOF'EV (*13*) of USSR in 1952 who found the fully dispersed shock but did not notice the partly dispersed shock. In 1957, Zel'dovich (*21*) pointed out the existence of the partly dispersed shock in a radiating gas and gave the sufficient condition for discontinuous profile to appear within Prokof'ev model that the shock Mach number should satisfy the condition:

$$M_1{}^2 > M_y{}^2 = (2\,\gamma - 1)/[\gamma\,(2 - \gamma)] \tag{8.94}$$

($M_y{}^2 \leq 4.20$ for $\gamma \leq {}^5/_3$) or that the upstream temperature is sufficiently low. In USA, CLARKE (*3*) independently studied Prokof'ev problem in 1962 and obtained same results. Clarke did not find the partly dispersed shock with a surface of discontinuity within the shock transition region because of an error in joining the solution of the front and the rear tails. Clarke's problem had been reexamined by HEASLET and BALDWIN (*5*) who defined an approximate boundary between fully dispersed and partly dispersed shocks in a radiating gas and presenting both continuous and discontinuous shock profiles in various regimes. Heaslet and Baldwin found that (i) for weak shock and strong radiation, continuous solution, i.e., fully dispersed shock occurs and (ii) for strong shock and weak radiation, partly dispersed shock occurs, in which the surface of discontinuity depends on the total shock strength and the degree of radiative heat transfer.

(ii) Radiation dispersed shock in a transverse magnetic field. MITCHNER and VINOKUR (*7*) considered the above problem with and without a transverse magnetic field. In general, if we consider the effects of electromagnetic field on shock structure, we have to introduce another diffusive property, the electrical conductivity of the medium. This diffusive property would have its own influence on the smoothing effect of the shock transition region. However, Mitchner and Vinokur did not consider the effect of the electrical conductivity by studying the case of infinite electrical conductivity only. In their treatment a differential approximation for the radiative transfer term has been used than that of equation (8.81). Their approximation is to replace the integral form of radiative transfer (8.80) by the following system of differential equations:

$$\frac{d\,q_R}{d\tau} = c\,(a_R\,T^4 - E_R) \tag{8.95a}$$

$$E_R = \frac{1}{2}\,\int\limits_{-\infty}^{\infty} dt.\,a_R\,T^4\,(t)\,\varepsilon_1\,(|\tau - t|) \tag{8.95b}$$

$$\frac{d\,p_R}{d\tau} = -\,\frac{1}{c}\,q_R \tag{8.95c}$$

$$p_R = \frac{1}{2}\,\int\limits_{-\infty}^{\infty} dt \cdot a_R\,T^4\,(t)\,\varepsilon_3\,(|\,\tau - t\,|) \tag{8.95d}$$

$$p_R = \frac{1}{3}\,E_R \tag{8.95e}$$

Equation (8.95e) is known as the Milne-Eddington approximation, which is exact for the thermodynamic equilibrium condition.

For the non-magnetic case, Mitchner and Vinokur define more precisely the limits of the existence for a fully dispersed shock. The effect of increasing magnetic field strength is to make more difficult the role of radiation as a smoothing agent.

(iii) Viscous solution of the shock structure in a radiating gas. Near the surface of discontinuity, we have to consider the effect of viscosity and heat conductivity. This problem has been studied by TRAUGOTT (17). In Traugott's analysis, the integral expression of radiative heat transfer (8.80) is replaced by the following equation:

$$\frac{d^2 q_R}{d \tau^2} - 3 q_R + 16 \sigma T^3 \frac{dT}{d\tau} = 0 \tag{8.96}$$

It is interesting to notice that for optically thick case, the first term of equation (8.96) is negligible and we have the Rosseland diffusion limit while for the optically thin case, the second term of equation (8.96) may be neglected and we have the usual expression of optically thin case.

Traugott found the solutions of equation (8.79) together with equation (8.96) from the Rankine-Hugoniot conditions under various conditions. In his results, the fully dispersed and the partly dispersed shocks are included. When the viscosity and heat conductivity are included, the surface of discontinuity in the inviscid analysis is replaced by a sharp transition region. For strong radiative transfer, the thickness of the shock transition increases and the temperature distribution across the shock transition region tends to be isothermal but in certain cases, there is a temperature overshot in the shock. Traugott also showed that the variation of the mean free path of radiation with temperature has a great influence on the temperature distribution in the shock transition region. For instance, for the same shock Mach number, if we use the equation (7.49) with positive m_1, the temperature overshot disappears while if we use the expression (7.50), a large temperature overshot occurs.

Scala and Simpson studied the shock structure with radiation and chemical reaction for two extreme cases: one is the optically thick case with Rosseland diffusion expression (5.21) and the other is the optically thin case with expression similar to (6.43) or (6.45). For the optically thick case, the result is similar to those obtained by Sen and Guess at low Mach numbers and the radiation effect increases the shock thickness. There is no temperature overshot. The major influence of the radiation occurs in the front tail region. They calculated three cases of shock Mach number of 12.5; 25 and 50 and found that in this Mach number range, the rate of increase of shock thickness due to radiation decreases as Mach number increases. When $M_1 = 50$, the shock thickness changes slightly by the effect of radiation.

In the optically thin case, Scala and Simpson also found the temperature overshot in the shock transition region and they found that the major effect of radiation is in the rear tail of the shock. When the effects of dissociation equilibrium are included, the chemical energy required for dissociation acts to decrease the temperature throughout the shock transition region. Hence for same shock Mach number, the radiation has larger influence in the non-dissociating gas than in the dissociating gas.

7. Flow field behind shock waves. In the last two sections, we consider only the flow field in an isolated shock wave. In the actual condition, we should consider the shock wave associated with some bodies. Since the thermal radiation affects both the stresses and the energy in the flow field, the flow field in a radiating gas will differ greatly from the corresponding one without radiation effect. However at present time, in most practical flow problems, the radiation pressure number is usually negligibly small but the radiative heat transfer is not negligible. It is interest to examine the deviation of the flow field with the effect of radiative heat transfer only. In this section, we consider the case of very high Reynolds number and in the region outside the boundary layer or shock rapid transition region, we may assume that the gas is inviscid and non-heat conducting. We shall con-

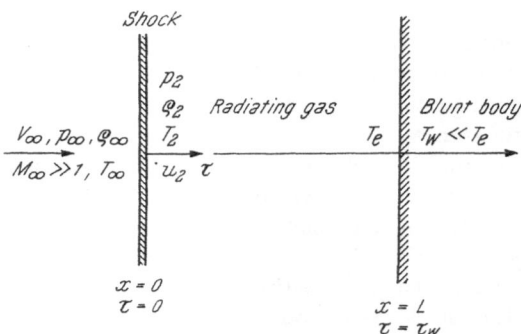

Fig. 8.11. Idealized model of the radiating region behind a hypersonic normal shock

sider two typical cases: One is the flow field behind a bow shock of a blunt nosed body and the other is the flow field behind an oblique shock.

(i) Inviscid flow field behind a detached shock in front of a blunt body. We consider the case of a hypersonic flow over a blunt body. Near the stagnation point, we may idealize the situation by a one-dimensional model shown in Fig. 8.11. The body

surface may be replaced by a flat wall. There is a detached straight shock at a distance L from the wall. In front of the shock there is a uniform hypersonic flow, i.e., $M_\infty \gg 1$. We assume that the thickness of the shock is small so that the shock may be replaced by a surface of discontinuity. The state variables T_2, p_2, ρ_2 and the velocity u_2 immediately behind the shock may be obtained from the values of the free stream and ordinary Rankine-Hugoniot relation. Because of the very high speed in the free stream, the temperature in the region behind the shock and in front of the body will be high and we should consider the radiative heat transfer in this region. Except in the boundary layer region near the body, viscosity and heat conductivity are negligible and equation (8.79) with $\mu = \varkappa = 0$ may be used to describe the flow conditions in this radiating gas region behind the shock, i.e., for $0 \leqq x \leqq L$, where $x = 0$ and $\tau = 0$ is the shock and $x = L$, $\tau = \tau_w$ is the surface of the wall. Now we have to find the expression of the radiative heat transfer q_R in the shock layer region. In the free stream, the temperature is low and then the thermal radiation in the free stream may be neglected. The shock front may be considered as a transparent surface, hence the absorption coefficient and the reflectivity coefficient at the shock are zero. The wall of the body is assumed to be a black body surface and the wall temperature T_w is much smaller than the gas temperature T_2. Hence the emission from the body is also negligible. As a result, we need to consider the thermal radiation in the shock layer $0 \leqq x \leqq L$ only. Furthermore we assume that the gas is gray. The radiative heat transfer

q_R is given by equation (5.45) with $q_R(0) = q_R(\tau_w) = 0$, $\tau_2 = \tau_w$ i.e.,

$$q_R(\tau) = \int_\tau^{\tau_w} 2\,\sigma\,T^4\,\varepsilon_2\,(t-\tau)\,\mathrm{d}t - \int_0^\tau 2\,\sigma\,T^4\,\varepsilon_2\,(\tau-t)\,\mathrm{d}t \qquad (8.97)$$

and

$$\frac{\mathrm{d}q_R(\tau)}{\mathrm{d}\tau} = -4\,\sigma\,T^4 + \int_0^{\tau_w} 2\,\sigma\,T^4\,\varepsilon_1\,(|t-\tau|)\,\mathrm{d}t \qquad (8.98)$$

For a shock wave moving at hypersonic velocity relative to the undisturbed air, $M_\infty \gg 1$, the following approximations may be used:

(a) $\quad h_2 \gg \frac{1}{2}\,u_2^2$

where h is the enthalpy of the gas.

(b) $\quad -\dfrac{1}{p}\dfrac{\mathrm{d}p}{\mathrm{d}x} \simeq \dfrac{1}{u}\dfrac{\mathrm{d}u}{\mathrm{d}x} \ll \dfrac{1}{L}$

An important consequence of approximation (b) is that the flow behind the hypersonic normal shock wave is essentially a constant pressure flow.

Under these approximations, equation (8.79 c) with $\mu = \varkappa = 0$ and (8.98) give fundamental equation for the temperature distribution in the shock layer as follows:

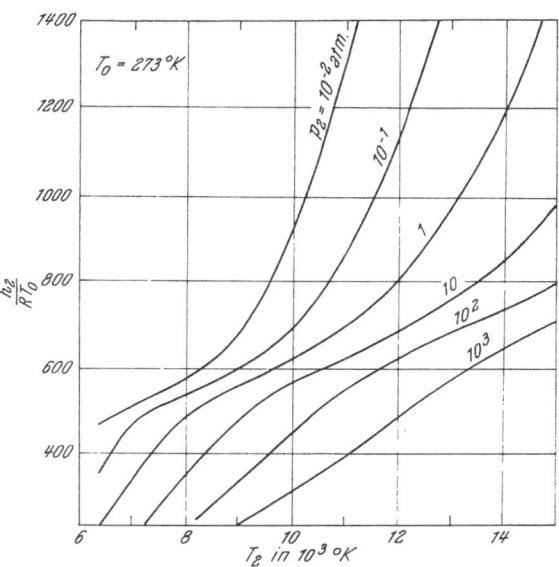

Fig. 8.12. Enthalpy of air at constant pressure. (Fig. 2 of reference 20 by K. K. YOSHIKAWA and D. R. CHAPMAN, Courtesy of NASA)

$$-\rho_\infty\,u_\infty\,\mathrm{d}h = \left[4\,\sigma\,T^4 - \int_0^{\tau_w} 2\,\sigma\,T^4\,\varepsilon_1\,(|t-\tau|)\,\mathrm{d}t\right]\mathrm{d}\tau \qquad (8.99)$$

Chapman and Yoshikawa solved equation (8.99) for the hypersonic shock wave in air. For high temperature air, we have to consider the effects of relaxation and dissociation. The specific heat at constant pressure is not a constant. Immediately behind the shock, we may assume that the air is in an equilibrium condition. Hence there is a definite relation between the enthalpy and the temperature and the pressure. Fig. 8.12 gives the enthalpy of air as a function of temperature T_2 and various pressure p_2. The pressure p_2 and the density ρ_2 immediately behind the shock wave are functions of the velocity of the flow in the free stream and the altitude which are shown in Fig. 8.13. The temperature T_2 and the density ρ_2 immediately behind the shock wave are functions of the velocity of the flow in the free stream and the altitude which are shown in Fig. 8.14. The proper mean value of the absorption coefficient of radiation for the expressions (8.98) and (8.99) is the Planck mean absorption coefficient which is a function of the temperature and pressure of the air and which is given in Fig. 8.15. With the data given in Figs. 8.12 to 8.15., Chapman and Yoshikawa calculated the temperature distribution in the shock layer for $T_2 = 15,000$ °K and $p_2 = 10$ atmospheric pressure in Fig. 8.16. For optically thin case such as $\tau_w = 0.2$, the tem-

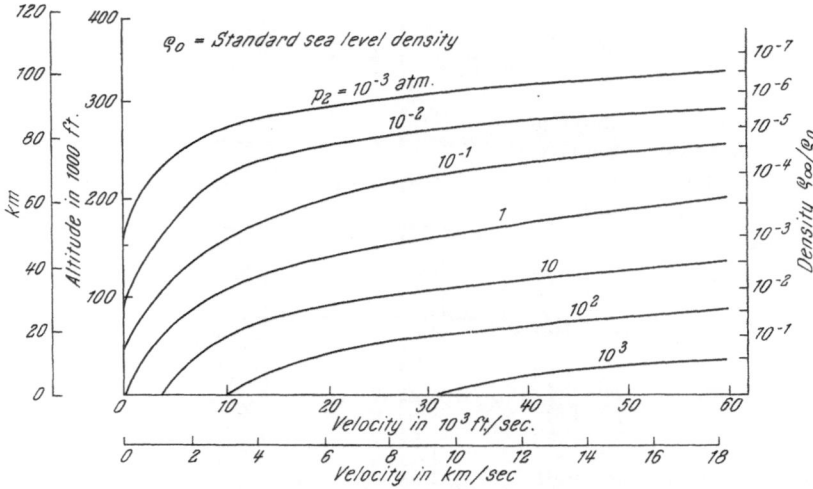

Fig. 8.13. Pressure behind a normal shock
(Fig. 5 in reference 20 by K. K. YOSHIKAWA and D. R. CHAPMAN, Courtesy of NASA)

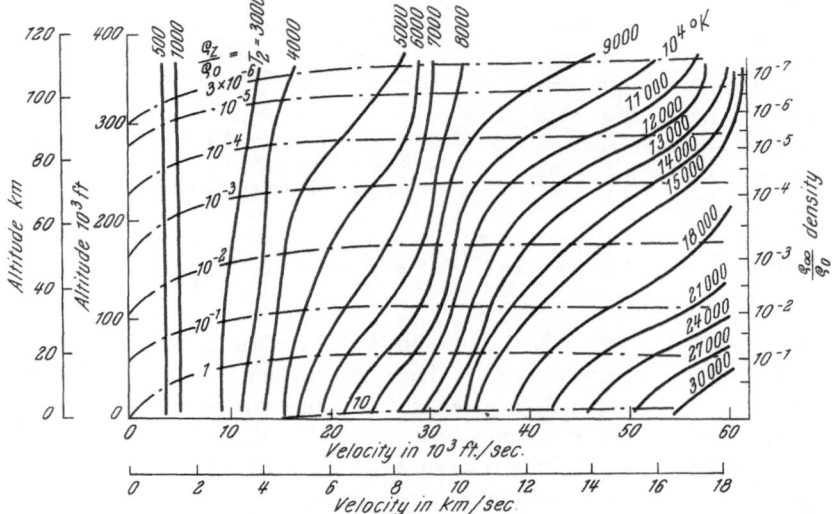

Fig. 8.14. Temperature and density behind a normal shock in standard atmosphere.
(Fig. 6 of reference 20 by K. K. YOSHIKAWA and D. R. CHAPMAN, Courtesy of NASA)

perature T drops approximately linear from T_2 to a value T_e which is the gas temperature at the wall which may be different from the wall temperature or the temperature of the body. On the other hand, for optically thick case $\tau_w = \infty$, the temperature on the major portion of the shock layer is constant as shown in Fig. 8.16.

Chapman and Yoshikawa pointed out that there are three characteristic lengths in the present problem. One is the shock standoff distance L and the

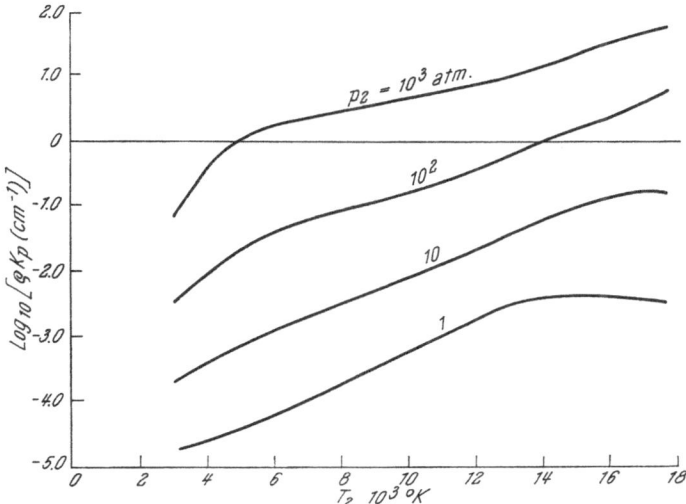

Fig. 8.15. Planck mean absorption coefficient ($\rho\, K_p$) of air.
(Fig. 7 of reference 20 by K. K. Yoshikawa and D. R. Chapman, Courtesy of NASA)

other is the radiation decay length L_d which is defined as

$$L_d = \frac{\rho_\infty\, u_\infty\, h_2\, L_R}{4\,\sigma\, T_2{}^4} = \frac{\frac{1}{2}\rho_\infty\, u_\infty{}^3\, L_R}{4\,\sigma\, T_2{}^4} \tag{8.100}$$

which is the length required to loss all energy radiation of constant intensity behind a normal shock wave. The third length is the mean free path of radiation L_R which characterizes the absorption of radiation. For high density case, $L_d \gg L_R$ so that the absorption deminates decay and for low density case $L_d \ll L_R$ so that the decay dominates absorption. The relative magnitude of shock standoff distance with respect to L_R and L_d shows the general 'characteristics of the radiating gas. Chapman and Yoshikawa suggested that if

$$L \leqq \frac{L_c}{10} \tag{8.101 a}$$

the radiation heat flux would be

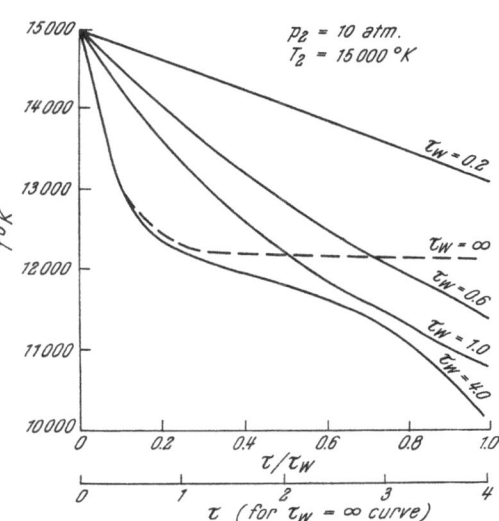

Fig. 8.16. Temperature distribution as function of optical thickness. (Fig. 11 of reference 20 by K. K. Yoshikawa and D. R. Chapman, Courtesy of NASA)

8*

that of an isothermal gas (cf. chapter IX) while

$$L \geqq \frac{L_c}{10} \qquad (8.101\,\text{b})$$

the radiation heat flux would be essentially that of a black body radiation where L_c is the smaller one of L_d and L_R.

The above simple analysis may be improved by considering the effect of curvature of the body and the flow streamlines. Since in the hypersonic flow which we are interested in, the shock layer is thin and the radius of curvature of the body is large, it is convenient to use a curvilinear coordinate system with the surface of the body as the x-axis and an orthogonal system to the x-axis as the y-coordinate as shown in Fig. 8.17. In terms of this curvilinear coordinate system, we have the following equations for the flow in the shock layer:

(a) Equation of continuity:

$$\frac{\partial r^\delta \rho u}{\partial x} + \frac{\partial K^* r^\delta \rho v}{\partial y} = 0 \qquad (8.102\,\text{a})$$

where $K^* = 1 + K y$, K is the curvature of the body, $\delta = 0$ for two dimensional flow and $\delta = 1$ for axi-symmetrical flow and r is the distance from the axis of symmetry.

(b) x-wise equation of motion

$$\rho u \frac{\partial u}{\partial x} + K^* \rho v \frac{\partial u}{\partial y} + K^* \rho u v = -\frac{\partial p}{\partial x} \qquad (8.102\,\text{b})$$

(c) y-wise equation of motion

$$\rho \frac{u}{K^*} \frac{\partial v}{\partial x} + \rho v \frac{\partial v}{\partial y} - \rho u^2 = -\frac{\partial p}{\partial y} \qquad (8.102\,\text{c})$$

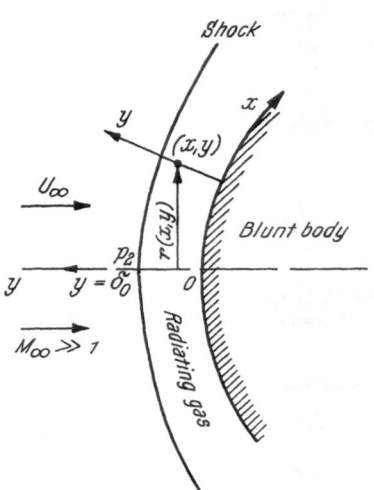

Fig. 8.17. Detached shock in front of a blunt body in a hypersonic flow

(d) energy equation

$$\rho \frac{u}{K^*} \frac{\partial h}{\partial x} + \rho v \frac{\partial h}{\partial y} + Q_R = 0 \qquad (8.102\,\text{d})$$

where h is the enthalpy of the gas.

Equations (8.102) should be solved for the boundary conditions at the shock front and at the surface of the body $y = 0$. Because of the complicated geometrical configuration, the integral expression of Q_R is very complicated for the actual case, which has not been studied yet. Only some simple approximate cases have been discussed. One of the simplified approximation is the one dimensional model discussed in Fig. 8.11 in which the curvature of the body is assumed to be zero and all the variables are assumed to be function of y only as far as the evaluation of the radiative heat transfer term Q_R is concerned (4, 20). Another simplified approximation is to use the following simple optically thin expression for the radiative heat transfer, i.e.,

$$Q_R = 4\,\pi\,\rho\,K_p\,B \qquad (8.103)$$

In reference 19, the problem with the expression (8.103) has been solved by the integral method around a sphere at flight velocities 30,000 fps to 60,000 fps at altitudes 190,000 and 200,000 ft. The results have been compared with the solution without the radiative heat transfer term. It was found that the shock shape and surface pressure distribution are in fair agreement up to the sonic line. The differences in velocity profiles are insignificant but there are significant differences in the static and total enthalpy profiles.

(ii) Perturbed flow fields of oblique shock waves. We consider a two-dimensional steady flow field with oblique shock waves, e.g. the flow over a wedge or an ogive. For simplicity, we consider a wedge in a supersonic flow of a perfect gas (Fig. 8.18). Without radiation effect, we have a straight shock at $y_s = x_s$

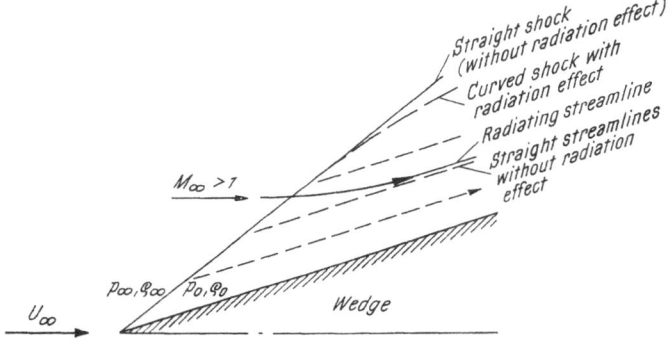

Fig. 8.18. Flow in a radiating gas over a wedge with attached shock wave

tan α where α is the shock angle and subscript s refers to the value at the shock front. We have uniform flow fields both in front of the shock (subscript 1) and behind the shock (subscript 2). If we choose the coordinate systems ahead and behind the shock such that the y-component of the gas velocity is zero in the radiationless case, the first order perturbed equations of radiation gasdynamics are:

$$\rho_0 \frac{\partial u}{\partial x} + \rho_0 \frac{\partial v}{\partial y} + U_0 \frac{\partial \rho}{\partial x} = 0 \qquad (8.104\,\text{a})$$

$$\rho_0 U_0 \frac{\partial u}{\partial x} = -\frac{\partial p}{\partial x} \qquad (8.104\,\text{b})$$

$$\rho_0 U_0 \frac{\partial v}{\partial x} = -\frac{\partial p}{\partial y} \qquad (8.104\,\text{c})$$

$$\rho_0 U_0 \frac{d U_m}{d x} + \rho_0 \left(\frac{\partial u}{\partial x} + \frac{\partial v}{\partial y}\right) + Q_R = 0 \qquad (8.104\,\text{d})$$

where subscript of refers to the value in the uniform state without radiation and the variables without subscript are the perturbed quantities. $U_m = C_v T$ is the perturbed internal energy and Q_R is the perturbed quantity of the radiative heat transfer, i.e., $\nabla \cdot \vec{q}_R$. The exact expression of Q_R depends on the radiating properties of the gas as we will show later.

To find the solution of equation (8.104), we need the boundary conditions at the shock front as well as those at the surface of the wedge. At the surface of the wedge, the normal component of the velocity vanishes, i.e.,

$$y = 0: \quad v = 0 \tag{8.105}$$

On the shock front, there are definite relations between the perturbed flow variables u_s, v_s, p_s etc. and the perturbed shock angle α which are obtained from the perturbation of the Rankine-Hugoniot relations. In general, these relations may be obtained in the following forms:

$$\alpha = A_s v_s + A; \qquad \frac{p_s}{p_0} = B_s v_s + B$$
$$\tag{8.106}$$
$$\frac{\rho_s}{\rho} = C_s v_s + C; \qquad \frac{u_s}{U_0} = D_s v_s + D$$

where A, B, C, and D depend on the radiation effects in front of the shock. In most problems, the radiation effects in front of the shock are negligible and then A, B, C and D are all zero. The factors A_s, B_s, C_s and D_s depend on the properties of the gas. For a perfect gas of constant ratio of specific heats γ, we have

$$A_s = \left(\frac{\rho_0}{\rho_\infty} - 1\right)^{-1}\left(1 + \frac{D_s}{\tan\alpha_0}\right) \tag{8.107 a}$$

$$D_s = \frac{\left(\frac{1}{U_0}\right)\left[1 - \frac{1}{2}\left(1 - \frac{\rho_\infty}{\rho_0}\right)(1 + M^2\gamma\sin^2\alpha_0)\right]\cot\alpha_0}{1 + \frac{1}{2}M^2(\gamma - 1) - \frac{1}{2}\left(1 + \frac{\rho_\infty}{\rho_0}\cot^2\alpha_0\right)(1 + M^2\gamma\sin^2\alpha_0)} \tag{8.107 b}$$

$$B_s = -\frac{1}{2}\left(\frac{\rho_0 U_0}{P_0}\right)\sin^2\alpha_0\left[D_s\left(\tan\alpha_0 + \frac{\rho_\infty}{\rho_0}\cot\alpha_0\right) + \frac{\rho_\infty}{\rho_0} - 1\right] \tag{8.107 c}$$

$$C_s = \frac{1}{U_0}\left(1 + \frac{\rho_\infty}{\rho_0}\right)\cot\alpha_0 + \left[\frac{\rho_\infty}{\rho_0}\csc^2\alpha_0 - \left(1 + \frac{\rho_\infty}{\rho_0}\right)D_s\right]\frac{1}{U_0} \tag{8.107 d}$$

The solutions of equations (8.104) should satisfy the relations (8.107) at the shock front. Equations (8.104) may be reduced into a single equation. For instance, for the variable v, we have

$$(M^2 - 1)\frac{\partial^2 v}{\partial x^2} - \frac{\partial^2 v}{\partial y^2} = \frac{1}{\rho_0 C_p T_0}\frac{\partial Q_R}{\partial y} \tag{8.108}$$

where M is the Mach number of the undisturbed flow. Since Q_R depends on the undisturbed flow and may be considered as a known function of x and y, equation (8.102) is in inhomogeneous wave equation if $M > 1$ and is a Poisson's equation if $M < 1$. In reference 8, several general cases have been solved for equation (8.108). In order to illustrate the radiation effect, we consider the case of a transparent gas so that Q_R is a constant, i.e., from equation (8.103):

$$Q_R = Q_{R0} = 4\,\rho_0\,K_p\,\sigma\,T_0{}^4 \tag{8.109}$$

Now we introduce the new variables:

$$u^* = u - Q_{R0}{}^*\,(\tan\alpha_0)\cdot y - f_1\,(y) \tag{8.110a}$$

and

$$v^* = v + Q_{R0}{}^*\,(y - x\tan\alpha_0) \tag{8.110b}$$

where $Q_{R0}{}^* = Q_{R0}/(\rho_0\,C_p\,T_0)$ and $f_1\,(y)$ is an arbitrary function of y only. The variables u^* and v^* satisfy the following relations of a potential ϕ:

$$u^* = \frac{\partial\phi}{\partial x}\quad;\quad v^* = \frac{\partial\phi}{\partial y} \tag{8.111}$$

and the potential ϕ satisfies the equation:

$$(M^2 - 1)\,\frac{\partial^2\phi}{\partial x^2} - \frac{\partial^2\phi}{\partial y^2} = 0 \tag{8.112}$$

If the Mach number of the undisturbed flow behind the shock is greater than unity, we have

$$\phi = \phi_1\,(x + \beta\,y) + \phi_2\,(x - \beta\,y) \tag{8.113}$$

where $\beta^2 = M^2 - 1$ and ϕ_1 and ϕ_2 are arbitrary functions to be determined by the boundary conditions (8.105) and (8.106). It is easy to show that the following results are obtained from equations (8.105), (8.106), (8.111) and (8.113):

$$\frac{v}{U_0} = \frac{Q^*{}_{R0}\cdot y}{1 + \beta^2\,B_s{}^*\tan\alpha_0}\,;\qquad \frac{p}{\rho_0\,U_0{}^2} = -\frac{x\,Q^*{}_{R0}\,B_s{}^*\tan\alpha_0}{1 + \beta^2\,B_s{}^*\tan\alpha_0}$$

$$\frac{u}{U_0} = \frac{x\,Q^*{}_{R0}\,B_s{}^*\tan\alpha_0}{1 + \beta^2\,B_s{}^*\tan\alpha_0} - \frac{y\,Q^*{}_{R0}\,(B_s{}^* + D_s{}^*)}{1 + \beta^2\,B_s{}^*\tan\alpha_0}$$

$$\frac{T}{T_0} = -\frac{Q^*{}_{R0}\,[M^2\,(\gamma - 1)\,B_s{}^*\tan\alpha_0 + 1]\,x}{1 + \beta^2\,B_s{}^*\tan\alpha_0} + \tag{8.114}$$

$$+\,\frac{Q^*{}_{R0}\,\{1 + [\beta^2\,B_s{}^* - M^2\,(\gamma - 1)\,(B_s{}^* + D_s{}^*)]\tan\alpha_0\}\,y}{\tan\alpha_0\,(1 + \beta^2\,B_s{}^*\tan\alpha_0)}$$

$$\frac{\rho}{\rho_0} = -\frac{Q^*{}_{R0}\,(B_s{}^*\tan\alpha_0 - 1)\,x}{1 + \beta^2\,B_s\,{}^*\tan\alpha_0} -$$

$$-\,\frac{Q^*{}_{R0}\,\{1 + \beta^2\,B_s{}^* - M^2\,(\gamma - 1)\,(B_s{}^* + D_s{}^*)\tan\alpha_0\}\,y}{\tan\alpha_0\,(1 + \beta^2\,B_s{}^*\tan\alpha_0)}$$

where $B_s{}^* = B_s\,(p_0/\rho_0\,U_0)$ and $D_s{}^* = D_s\,U_0$

ZHIGULEV (22) et al. gave the following example:

For a wedge of semi-angle of 35° traveling at Mach number $M_\infty = 60$ at an altitude of 60 km with $\rho_0 = 3.49 \times 10^{-7}$ gr/cm³, we have the following data:

$$\alpha_0 = 7°; \quad p_0 = 0.4 \text{ atm}; \quad \rho_0 = 4.8 \times 10^{-6} \text{ gr/cm}^3; \quad T_0 = 12{,}200 °\text{K};$$

$$u_0{}^2 = 1.57 \times 10^{12} \text{ cm}^2/\text{sec}^2, \quad Q_{R0} = 6.4 \times 10^9 \text{ erg/cm}^3 \text{ sec}$$

$$Q^*{}_{R0} L = 11.4\%; \quad \rho_0 K_p L = 0.15; \quad \gamma = 1.4; \quad L = 1.0 \text{ meter}$$

where L is a characteristic length of the wedge and is taken as one meter.

With the above data, equation (8.114) gives the following relations:

$$\frac{v}{U_0} = -0.075 \frac{y}{L}; \qquad \frac{u}{U_0} = -0.002 \frac{x}{L} + 0.12 \frac{y}{L}$$

$$\frac{P}{\rho_0 U_0{}^2} = -0.002 \frac{x}{L}, \qquad \frac{T}{T_0} = -0.121 \frac{x}{L} + 1.27 \frac{y}{L}$$

$$\frac{\rho}{\rho_0} = 0.07 \frac{x}{L} - 1.27 \frac{y}{L}$$

With $x = 1$, $\alpha/\alpha_0 = 6.5\%$ and at the shock wave the streamlines have an inclination $\theta = -0.01 (x/L)$. From these results, we may say that thermal radiation does affect the flow field in a hypersonic flow around a body. In general, both the temperature and the pressure at body surface drop while its density increases. Little has been done for the flow field where the temperature is so high that the radiation pressure and radiation energy density should be considered.

References

1. BALDWIN, B. S., Jr.: The propagation of plane acoustic waves in a radiating gas. NASA TR R-138, 1962.

2. BETHE, H. A., K. FUCHS, J. O. HIRSCHFELDER, J. L. MAGEE, R. E. PEIERLS, and J. VON NEUMANN: Blast Wave. Chapter 3: Thermal radiation phenomena. Los Alamos Scientific Lab. Report LA-2000, March, 1958.

3. CLARKE, J. F.: Radiation-resisted shock waves. Phys. of Fluids, vol. 5, No. 11, pp. 1347–1361, Nov. 1962.

4. GOULARD, R.: Preliminary estimates of radiative transfer effects on detached shock layers. AIAA Journal vol. 2, No. 3, pp. 494–502, March 1964.

5. HEASLET, M. A., and B. S. BALDWIN: Predictions of the structure of radiation-resisted shock waves. Phys. Fluids, vol. 6, No. 6, pp. 781–791, June 1963.

6. MARSHAK, R. L.: Effect of radiation on shock wave structure. Phys. Fluids, vol. 1, No. 1, pp. 24–29, Jan. 1958.

7. MITCHNER, M., and V. VINOKUR: Radiation Smoothing of shocks with and without a magnetic field. Phys. Fluids, vol. 6, No. 12, pp. 1682–1692, Dec. 1963.

8. OLFE, D. B.: Radiation perturbed flow fields of normal and oblique shock waves. AIAA preprint No. 64–69, Jan. 1964.

9. PAI, S. I., and A. I. SPETH: The wave motions of small amplitude in radiation-electromagneto-gasdynamics. Proc. 6th Midwest Conference on Fluid Mech. Univ. of Texas Press, pp. 446–466, 1959.

10. PAI, S. I., and A. I. SPETH: Shock waves in radiation Magneto-gasdynamics. Phys. Fluids, vol. 4, No. 10, pp. 1232–1237, Oct. 1961.

11. POMERANTZ, J.: The influence of the absorption of radiation in shock tube phenomena. U. S. Naval Ordnance Lab. Navord Report 6136, 1958.

12. PROKOF'EV, V. A.: Propagation of forced plane compression waves of small amplitude in a viscous gas when radiation is taken into account. ARS Journal, vol. 31, No. 7, p. 988, July 1961.

13. PROKOF'EV, V. A.: Shock wave in a radiating gas. Uch. Zap. Mos. Gos. Univ. Mekh. 172, p. 79, 1952.

14. SACHS, R. G.: Some properties of very intense shock waves. Phys. Rev. Vol. 69, No. 9–10, pp. 514–522, May 1946.

15. SCALA, S. M., and D. H. SAMPSON: Heat transfer in hypersonic flow with radiation and chemical reaction. Tech. Inf. Ser. R 63 SD 46, Space Sci. Lab. General Electric Co., Phil. Pa., March 1963. Also Supersonic Flow, Chemical Processes and Radiative Transfer, Pergamon Press, pp. 319–354, 1964.

16. SEN, H. K., and A. W. GUESS: Radiation effects in shock wave structure Phys. Rev. Vol. 108, No. 3, pp. 560–564, Nov. 1, 1957.

17. TRAUGOTT, S. C.: Shock structure in a radiating, heat-conducting and viscous gas. Martin Co. Research Report RR 57, May 1964.

18. VINCENTI, W. G., and B. S. BALDWIN, Jr.: Effects of thermal radiation on the propagation of plane acoustic waves. Jour. Fluid Mech. vol. 12, pt. 3, p. 449, March 1962.

19. WILSON, K. H., and H. HOSHIZAKI: Inviscid, non-adiabatic flow about blunt bodies. AIAA preprint 64–70, Jan. 1964.

20. YOSHIKAWA, K. K., and D. R. CHAPMAN: Radiative heat transfer and absorption behind a hypersonic normal shock wave. NASA TN D-1424, Sept. 1962.

21. ZEL'DOVICH, IA. B.: Soviet Phys. JETPS vol. 5, p. 919, 1957.

22. ZHIGULEV, V. N., YE. A. ROMISHEVSKII, and V. K. VERTUSHKIN: Role of radiation in modern. AIAA Journal vol. 1, No. 6, pp. 1473–1483, June 1963.

Chapter IX

Heat Transfer in Radiation Gasdynamics

1. Introduction. The importance of heat transfer by radiation in engineering problems has been noticed for a long time such as the heat transfer in an industrial furnace. It has been studied by Kirchhoff more than 100 years ago. In such earlier investigations, we consider the heat transfer by radiation alone without the interaction with heat conduction and convection. The essential feature of engineering radiative transfer problem different from those of astrophysics is the interreflection caused by the presence of walls. The problem may become very complicated due to the complex geometry of the system. It would be of interest to give a brief review of the general methods used in the analysis of engineering radiation problem which would be useful for our study of radiation gasdynamics. In section 2, we deal with the radiative heat transfer in a non-absorbing medium while in section 3, we deal with the radiative heat transfer in an absorbing medium. It should be noticed that these analyses are important in the determination of heat transfer of space vehicles because in the outer space the heat conduction and heat convection are negligible in comparison with radiation.

In radiation gasdynamics, we have to consider the coupling of thermal radiation with heat conduction and heat convection. We shall first consider the coupling of heat conduction and thermal radiation in an absorbing medium between two parallel plates in section 4. One of the most important applications of such coupling problems is the heat balance in the atmosphere of earth or other planets which will be discussed in section 5. In section 6, we shall study the most general cases of flow between two parallel plates in radiation magneto-gasdynamics. We study not only the coupling between heat convection, heat conduction and thermal radiation but also the influence of electromagnetic fields on the resultant flow. Because such problems may occur in hypersonic flight, nuclear reactors, fission and fusion reactions, the temperature of the medium may be so high that the gas is ionized and the electromagnetic fields may play an important role.

Because the fundamental equations of radiation gasdynamics are non-linear-differential equations, it is very difficult to solve the flow problems in a very general case. Many approximations used in ordinary gasdynamics may be used in radiation gasdynamics. One of these approximations is the boundary layer approximation which is useful for high Reynolds number cases. Since in hypersonic flow, the Reynolds number is very high, we may use the boundary layer approximation to solve heat transfer problem. We shall discuss boundary layer flows in radiation gasdynamics in section 7. One of the most interesting current problems is the stagnation point radiative heat transfer for manned vehicles entering the earth's atmosphere at super-orbiting speed. We shall discuss this

problem in section 8. Finally some other radiative heat transfer problems will
be discussed in section 9.

2. Radiative heat transfer in a non-absorbing medium. In outer space,
the atmosphere is practically zero and the only way of heat transfer is by radiation. Since there is no medium, there will be no absorption of radiation in the
outer space through which the radiation rays travel. We do not have to consider
the absorption of radiation by the medium but the interaction of the rays on
the surface of the body. Hence the analysis of radiative heat transfer in
outer space is just the same as that
for a non-absorbing medium in which both
the heat conduction and heat convection
are negligible.

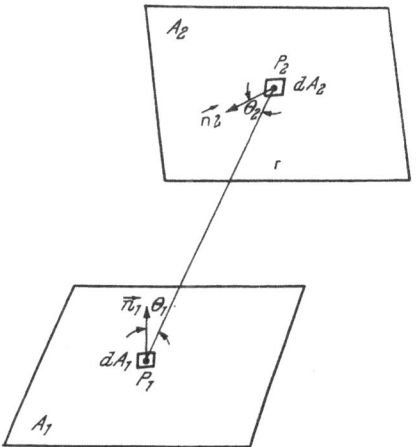

We consider two diffusively radiating
surface A_1 and A_2 and find the radiative
heat transfer between them. Let dA_1 and
dA_2 be the surface elements on these surfaces at points P_1 and P_2 respectively
(see Fig. 9.1). The distance between
P_1 and P_2 is r. We assume that both surfaces are gray radiators with constant emissivity and reflectivity. Furthermore we
assume that the temperature of each surface is a constant so that the emissive
power of the surface is a constant depending
on its temperature. For instance, the black
body emissive power is [cf. eqs. (4.29)
and (6.40)]:

Fig. 9.1. Radiative heat transfer between two diffuse walls

$$E_b = \sigma\, T^4 \tag{9.1}$$

The radiosity of the surface is

$$R_a = e_w\, E + r_w\, G \tag{9.2}$$

where e_w is emissivity coefficient (6.30), r_w is the reflectivity coefficient (6.23)
of the surface. E is the emissive power and G is the irradiation of the surface.
The radiative energy emitted from dA_1 and striking the surface dA_2 is

$$d\,E_{12} = \frac{R_1 \cos\theta_1 \cos\theta_2}{\pi\, r^2}\, d\,A_1\, d\,A_2 = R_1\, d\,F_{12} \tag{9.3}$$

Similar expression may be obtained for the radiative energy emitted from dA_2
and striking dA_1, i.e., $dE_{21} = R_2\, dF_{12}$. Integration over two surfaces gives
the total radiative energy from one surface to the other as follows:

$$E_{12} = R_1\, A_1\, F_{12} \tag{9.4}$$

where

$$A_1 F_{12} = A_2 F_{21} = \int_{A_1 A_2} \frac{\cos\theta_1 \cos\theta_2}{\pi\, r^2}\, d\,A_1\, d\,A_2 \tag{9.5}$$

and F_{ij} is known as angle factor of A_i versus A_j. Sometimes it is called the
geometrical factor or configuration factor.

There are three known methods to evaluate equation (9.4). They are known as accounting method, network method and calculus method. We are going to discuss them briefly below:

(a) Accounting method. It is the oldest method which had been first used by Kirchhoff in 1899. This method is based on the direct accounting of all the reflected radiation rays. We have already used it in chapter VI, § 6. In our present case, $\tau = 0$ and the total radiative flux per unit area from an infinite plate 1 to a parallel infinite plate 2 is (cf. Fig. 6.4).

$$q_{R12} = \sum_{n=0}^{x} e_{w1} e_{w2} r^n_{w1} r^n_{w2} \sigma T_1^4 = \frac{\sigma T_1^4}{\left(\dfrac{1}{e_{w1}}\right) + \left(\dfrac{1}{e_{w2}}\right) - 1} \tag{9.6}$$

Thus the net radiative heat transfer is simply $(q_{R12} - q_{R21})$ as given by equation (6.40).

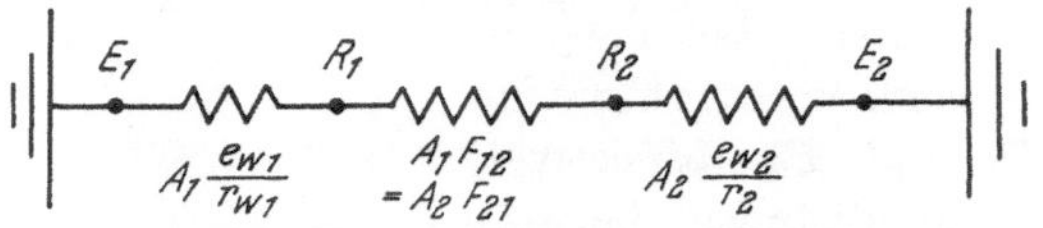

Fig. 9.2. Circuit diagram of radiative heat transfer between two surfaces A_1 and A_2

For complicated geometry, the expression $A_1 F_{12}$ may be complicated. In reference 8, JAKOB gave a number of examples for the values of $A_1 F_{12}$.

(b) Network method. The net radiative heat flux any opaque surface i is

$$q_{R_{net},\, i} = J_i - G_i = \frac{e_{wi}}{r_{wi}} (E_i - J_i) \tag{9.7}$$

According to the network method, we may divide the whole radiative system into n-elements and each element may be considered as a node in a network. The analogy of radiation network with the electrical network is evident. Particularly by Kirchhoff's law, we have

$$\sum_{i=1}^{n} q_{R_{net},\, i} = 0 \tag{9.8}$$

It is interesting to see the result of network method for the case of two infinite parallel walls as shown in equation (9.6) (cf. Fig. 6.4).

$$q_{R_{net},\,1} = -\,q_{R_{net},\,2} = R_1 - G_1 = R_1 - R_2 = E_{b1} - \left(\frac{1}{e_{w1}}\right) q_{R_{net},\,1} -$$

$$-\,B_{b2} - \left(\frac{1}{e_{w2}} - 1\right) q_{R_{net},\,2} \tag{9.9}$$

or

$$q_{R_{net},\,1} = (E_{b1} - E_{b2}) \Big/ \left(\frac{1}{e_{w1}} + \frac{1}{e_{w2}} - 1\right) \tag{9.9a}$$

Equation (9.9a) is exactly the same as that given by the accounting method [cf. eq. (9.6)]. It should be noticed that the irradiation G_1 is equal to the radiosity R_2.

Since the network method of radiative transfer is identical to the network method of electrical circuit, the well known method of solving the electrical network may be used. For instance, the circuit diagram of problem given by equation (9.9) is shown in Fig. 9.2. For complicated system, matrix method may be used to solve the problem (cf. reference 17).

(c) Calculus method or integral equation method. The irradiation at point P_2 on the surface A_2 (Fig. 9.1) due to the radiosity from the area dA_1 at P_1 according to equation (9.3) may be written as

$$dG_{21} = dE_{12} = R_1 K(P_1, P_2) dA_1 \qquad (9.10)$$

where $K(P_1, P_2)$ known as the shape factor is the integrand of equation (9.5). The irradiation on P_2 is given by the integral of equation (9.10) over the whole area A_1. From equations (9.2) and (9.10), We have the radiosity at any point P on the surface A_2 is

$$R(P) = e_{w2}(P) E_2(P) + \int_{A_1} R(P_1) K(P_1, P) dA_1 \qquad (9.11)$$

The net radiative heat transfer at the point P is then, by equation (9.7)

$$q_{R\,net}(P) = e_w(P) E(P) + \int_{A_1} R(P_1) K(P_1, P) dA_1 \qquad (9.12)$$

Hence we have to solve the integral equation (9.12). Since most of the problems of radiation gasdynamics are solved by this integral equation method, we shall discuss this method in detail later. Some general methods will be discussed in section 4.

3. Radiative heat transfer in an absorbing medium. If we consider the radiative heat transfer in an absorbing medium in which the heat conduction and the heat convection are negligible, the only difference from the results of last section is that the specific intensity of radiation decreases as the ray travels in the medium according to equation (3.1). If we neglect the scattering and the emission of radiation in the medium and assume that the mass absorption coefficient $\rho k_\nu = K =$ constant the specific intensity of radiation at P_2 is then

$$I_\nu(P_2) = I_\nu(P_1) \exp(-Kr) \qquad (9.13)$$

The radiative energy at dA_2 due to the radiation ray from dA_1 is

$$dE_{12} = \frac{R_1 \exp(-Kr) \cos\theta_1 \cos\theta_2}{\pi r^2} dA_1 dA_2 \qquad (9.14)$$

The only difference of equation (9.14) from equation (9.3) is the factor $\exp(-Kr)$. If K is a constant, we may introduce a transmissivity T_{12} such that

$$E_{12} = T_{12} A_1 F_{12} R_1 \qquad (9.15)$$

where $\qquad T_{12} = \dfrac{1}{A_1 F_{12}} \displaystyle\int\limits_{A_1}\int\limits_{A_2} \dfrac{\exp(-Kr) \cos\theta_1 \cos\theta_2}{\pi r^2} dA_1 dA_2 \qquad (9.16)$

Equation (9.16) is of similar form as equation (9.4) except the factor transmisivity T_{12}. We may use the same methods discussed in the last section to solve problems with absorbing medium from equation (9.15). If K is not a constant, more complicated method should be used because in general K is a function of temperature and the temperature distribution depends on the radiative heat transfer.

4. Heat transfer by simultaneous heat conduction and radiation in an absorbing medium. For simplicity, we consider a system consisting of two diffuse, non-black, infinite, isothermal and parallel plates separated by a finite distance. The space between the plates is filled with a thermal radiation absorbing and emitting medium. There is no flow of the medium. Hence we need to consider the energy equation (5.7) only. Since there is no energy input and all the variables are function of the distance perpendicular to the plates, i.e., y, only, the energy equation (5.7) becomes:

$$\frac{\mathrm{d}}{\mathrm{d}y}\left(\varkappa\,\frac{\mathrm{d}T}{\mathrm{d}y}\right) = -\frac{\mathrm{d}}{\mathrm{d}y}\,(q_{Ry}) \tag{9.17}$$

where q_{Ry} is the y-wise radiative heat flux given by equation (5.43) if we assume that the flow is local thermodynamic equilibrium and scattering is negligible. Equation (9.17) is an integro-differential equation. Because the coefficient of heat conductivity $\varkappa$ is in general a function of temperature T and the mass absorption coefficient $\rho\,k_\nu'$ or the mean free path of radiation $L_R = 1/(\rho\,k_\nu')$ is a function of temperature and density as well as of frequency ν, equation (9.17) is highly non-linear. It is very complicated to solve equation (9.17). In order to obtain some essential features of this problem, we follow reference 26 by assuming that both the coefficient of heat conductivity and the mean free path of radiation are constant. Then we may use equation (5.46) for $\mathrm{d}\,q_{Ry}/\mathrm{d}\,y$. Under these assumptions, equation (9.17) in non-dimensional form becomes:

$$\frac{1}{R_{F0}}\frac{\mathrm{d}^2 T^*}{\mathrm{d}\tau^2} = T^{*4} - \tfrac{1}{2}\left\{R_a^*\,(\tau_2)\,\varepsilon_2\,(\tau_2 - \tau) + R_a^*\,(0)\,\varepsilon_2\,(\tau) + \right.$$
$$\left. + \int_0^\tau T^{*4}\,(t)\,\varepsilon_1\,(|t - \tau|)\,\mathrm{d}t \right. \tag{9.18}$$

where $R_{F0} = \dfrac{4\sigma\,T_0^3\,L_R}{\varkappa} = $ radiation flux number [cf. equation (7.42)]

$T^* = T/T_0$, T_0 is a reference temperature, $R_a^* = $ radiosity$/\sigma\,T_0^4$, $\tau = y/L_R$ is the optical thickness of the coordinate y. At the lower plate $y = 0$, $\tau = 0$ and at the upper plate $y = L$, $\tau = \tau_2 = L/L_R$. Hence $1/\tau_2$ may be considered as the Knudsen number of radiation of our problem [cf. eq. (7.16a)]. The boundary conditions of our problem are:

$$y = 0 \ \text{ or } \ \tau = 0 \ : \ T^*\,(\tau) = T^*\,(0) = 1$$
$$y = L \ \text{ or } \ \tau = \tau_2 \ : \ T^*\,(\tau) = T^*\,(\tau_2) = T_w^* = \text{constant} = \frac{T_2}{T_0} \tag{9.19}$$

where we take the temperature at the lower wall as our reference temperature and both walls are maintained at constant but different temperature.

Even with our simplified assumptions, equation (9.18) is still a non-linear integro-differential equation. There is no simple mathematical method for solving this equation with the boundary equations (9.19) in a closed form solution. We have to solve it by approximate or iterative methods. Some of these methods will be briefly discussed as follows:

(i) Linearization of the dependent variable.

If the temperature difference between the two plates is small, we expect that the temperature in the whole field can deviate only slightly from the temperature of the walls or from some average temperature which is a constant. Under this condition, we have:

$$T^4 = T_a{}^4 \left[1 + \left(\frac{T - T_a}{T_a} \right) \right]^4 = 4\,T\,(\tau)\,T_a{}^3 - 3\,T_a{}^4 + 0 \left[\left(\frac{T - T_a}{T_a} \right)^2 \right] \qquad (9.20)$$

If we substitute equation (9.20) into equation (9.18) and neglect terms of second or higher order of $(T - T_a)/T_a$, where T_a is the average temperature, we will have a linear integral equation which may be easily solved.

(ii) Iterative method from a differential equation.

If the radiative flux number R_{F0} is small, equation (9.18) may be written as

$$\frac{d^2 T^*}{d\tau^2} = R_{F0}\,G\,(T^*;\tau) \qquad (9.21)$$

For first approximation, we may neglect the term on the right-hand side and have

$$\frac{d^2 T^*}{d\tau^2} = 0 \qquad (9.22)$$

We solve equation (9.22) for the boundary condition (9.19) and obtain

$$T^*\,(\tau) = T^*\,(0) + \frac{\tau}{\tau_2}\,[T^*\,(\tau_2) - T^*\,(0)] \qquad (9.23)$$

Substituting equation (9.23) in the right hand side of equation (9.21) we have

$$\frac{d^2 T^*}{d\tau^2} = R_{f0}\,G\,(\tau) \qquad (9.24)$$

Now we may solve equation (9.24) for the boundary condition (9.19). Since $G\,(T^*;\tau) = G\,(\tau)$ is in general a very complicated function of τ, we have to use numerical integration. If we call the solution of equation (9.24) as $T_1^*\,(\tau)$, we substitute $T_1^*\,(\tau)$ into the expression $G\,(T^*,\tau)$ of equation (9.21) and obtain a new section for the second approximation of T^* which is symbolically the same as equation (9.24). The procedure may be repeated until the desired accuracy is obtained.

(iii) Iterative method from an integral equation.

In this method, we first integrate equation (9.21) twice with respect to τ from 0 to τ. The result of this integration is a non-linear Fredholm integral equation of the second kind as follows:

$$T^*\,(\tau) = R_{F0}\,F_0\,(\tau) + R_{F0} \int_0^\tau F\,(t)\,dt \qquad (9.25)$$

where

$$F_0 = C_1 + C_2\,\tau - \frac{1}{2} \int_0^\tau dt_2 \int_0^{t_2} [R^*\,(\tau_2)\,\varepsilon_2\,(\tau_2 - t) + R^*\,(0)\,\varepsilon_2\,(t)]\,dt \qquad (9.26)$$

and

$$F\,(t) = \int_0^t \left[T^{*4}\,(t_1) - \frac{1}{2} \int_0^{t_2} T^{*4}\,(t_0)\,\varepsilon_1\,(|t_0 - t_1|)\,dt_0 \right] dt_1 \qquad (9.27)$$

Equation (9.25) may be solved by successive approximation. First we assume a distribution $T_1^*(t)$ and substitute this distribution in the function $F(t)$ of equation (9.25). We may calculate $T^*(\tau) = T_2^*(\tau)$ from equation (9.25) and the boundary conditions (9.19) which determine the arbitrary constants C_1 and C_2. The procedure may be repeated by substituting $T_2^*(t)$ in the expression $F(t)$ until the desired accuracy is reached.

The above three methods are used to solve the accurate equation (9.18). However under certain conditions, it is possible to simplify equation (9.18)

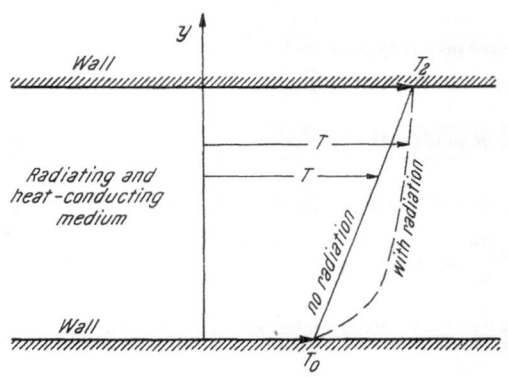

before we try to solve it. The resultant equation after simplification will be much easier to solve then the original equation. The following are some of such simplifications:

(a) Rosseland approximation. For optically thick case, we have shown in chapter V, section 8, that the radiative heat transfer can be expressed in terms of the gradient of temperature and an effective coefficient of heat conductivity by radiation (5.22). Under such conditions, equation (9.17) becomes simply

Fig. 9.3. Temperature distributions between two parallel plates with and without radiation

$$\frac{\mathrm{d}}{\mathrm{d}y}\left(K^* \frac{\mathrm{d}T}{\mathrm{d}y}\right) = 0 \qquad (9.28)$$

where

$$K^* = \varkappa + \varkappa_R = \varkappa(T) + 4\,D_R\,a_R\,T^3 \qquad (9.29)$$

The solution of equation (9.28) with the boundary conditions (9.19) is

$$\frac{y}{L} = \frac{\displaystyle\int_{T_0}^{T} K^*\,\mathrm{d}T}{\displaystyle\int_{T_0}^{T_w} K^*\,\mathrm{d}T}$$

In general K^* may be considered as a known function of temperature T. If we assume $\varkappa$ and D_R are constant, equation (9.30) gives

$$\frac{y}{L} = \frac{\varkappa(T - T_0) + D_R\,a_R\,(T^4 - T_0^4)}{\varkappa(T_2 - T_0) + D_R\,a_R\,(T_2^4 - T_0^4)} \qquad (9.30\,\text{a})$$

Equation (9.30a) shows that if the radiative heat transfer is negligible, we have the linear distribution of temperature and if the heat conduction is negligible, $y = A_1\,T^4 - A_2$ so that in the major portion of the gap, the temperature is closer to the higher temperature T_2 than the linear distribution (Fig. 9.3). In general the temperature distribution is between these two extreme cases.

(b) Optically thin case. For optically thin case, the optical thickness is small. We may use the first few terms of the power series in "t" for the exponential integral (5.56) and the integro-differential equation (9.18) may be greatly simplified. We shall discuss this method in next section. For the case of extremely optically thin case, the radiative heat flux will be a constant as shown in equation (6.40). Then $dq_R/dy = 0$. If the coefficient of heat conductivity $\varkappa$ is a constant, we have again the linear temperature profile.

(c) Milne-Eddington approximation. In one dimensional radiative heat transfer problem, accurate results can be obtained by assuming a simple form for the integrated intensity of radiation I [eq. (2.3)]. We introduce the following average values:

$$J_0 = \frac{1}{4\pi} \int I \, d\omega; \quad H_0 = \frac{1}{4\pi} \int I \cos\theta \, d\omega; \quad K_0 = \frac{1}{4\pi} \int I \cos^2\theta \, d\omega \qquad (9.31)$$

For a gray gas with local thermodynamic equilibrium, the integration of the radiative transfer equation (5.30) gives

$$\frac{d H_0}{dy} = \frac{1}{L_R} (B - J_0) \qquad (9.32)$$

$$\frac{d K_0}{dy} = -\frac{H_0}{L_R} \qquad (9.33)$$

The Milne-Eddington approximation is that $K_0 \simeq J_0/3$. Then

$$\frac{d^2 J_0}{dy^2} = -\frac{3}{L_R}\frac{d H_0}{dy} = \frac{3}{L_R^2}(J_0 - B) \qquad (9.34)$$

The energy equation (9.17) for constant coefficient of heat conductivity becomes

$$\varkappa \frac{d^2 T}{dy^2} = 4\pi \frac{d H_0}{dy} \qquad (9.35\,a)$$

with

$$\frac{d^2 H_0}{dy^2} = \frac{1}{L_R}\left(\frac{3 H_0}{L_R} + \frac{4}{\pi}\sigma T^3 \frac{d T}{dy}\right) \qquad (9.35\,b)$$

Since equations (9.35) are still non-linear, they can be solved by numerical integration only. Equation (9.35 b) is essentially the same as equation (8.96).

(d) Exponential approximation. Another approximation which has been often used is to replace the exponential integral by an exponential such as given by equation (8.81). With this approximation, equation (9.18) may be reduced to a differential equation.

Some numerical examples were given by VISKANTA and GROSH (26). Since the expression for $F(t)$ of equation (9.27) in reference 26 was not correct, their results will not be given here. But qualitatively, the main effect of radiation is to make the temperature distribution more uniform in the major portion of the flow field as shown in Fig. 9.3. Quantitatively the optical thickness is the pertinent parameter in the heat transfer of radiation and the corresponding non-dimensional parameter is the radiation flux number R_{F0} given in equation (9.18).

5. Radiative processes in the atmosphere. One of the most important application of heat transfer by simultaneous heat conduction and radiation is the determination of the temperature distribution in the atmosphere of the earth or other planets. The sun is the main source of energy that determines the atmospheric temperature which is determined by the absorption of the radiation from the sun and the emission of the radiation of the earth into the free space as well as the heat conduction and convection in the atmosphere and the conduction and reflection on the earth surface. The detailed discussion is very complicated. Special treatize should be referred to (3, 9, 22). We shall discuss only briefly the general principle of the heat balance in an atmosphere and the general temperature distribution in the earth's atmosphere here. The problem is very involved because of the large variation of the composition of the atmosphere with altitude. It would be of interest to give a brief description of the earth atmosphere before we discuss the radiative process and the heat balance in the atmosphere.

The earth atmosphere consists of various types of gases, ions and electrons. The composition varies with altitude as well as time and places. However it is accurate enough to divide the earth atmosphere into various regions in which different laws of variation of temperature or pressure with altitude hold. There are two major regions: One is known as the homosphere which covers the altitude approximately below 85 km. and in which the composition of the earth atmosphere remains almost the same, and the other is known as the heterosphere which covers the altitude above 85 km. and in which the composition of the atmosphere changes greatly with altitude. In ordinary meteorology, only the homosphere has been extensively studied (3, 9). However for space flight, the heterosphere becomes important (22). In each of these two main regions, we may further divide them into several sub-regions according to the variation of temperature with altitude as follows:

(I) *Homosphere,* in which we have three sub-regions: troposphere, stratosphere and mesosphere.

(i) *Troposphere.* At low altitude, from sea level to an altitude of about 10 km., the composition of the air is approximately constant and the temperature decreases with altitude. It is known as troposphere. The lapse rate in the troposphere is almost a constant even though its value varies from place to place and time to time. Dry atmospheric air at sea level consists of 78.03% of molecular nitrogen N_2, 20.99% of molecular oxygen O_2, 0.93% of argon A, 0.030% of carbon dioxide CO_2, and 0.01% of molecular hydrogen H_2 by volume and a slight amount of Neon N_e, helium H_e, krypton K_r and Xenon X_e. In many flow problem of gasdynamics, the physical properties of air are computed by assuming that air is simply composed of 78.12% of N_2, 20.95% of O_2 and 0.93% of A at sea level by volume. For low altitude, a standard atmosphere has been adopted in order to faciliate the comparison of test data in different atmospheric conditions. The standard atmosphere for the troposphere is based on the following assumptions:

(a) The air is a perfect gas with gas constant $R = 53.33$ ft./°F.

(b) The pressure of air at sea level is $p_0 = 29.921$ inch. Hg.

(c) The temperature of air at sea level is $T_0 = 59$ °F.

(d) In the troposphere, the lapse rate is a constant of a value of -0.003566 °F/ft.

(e) The troposphere ends when the temperature reaches $T_s = -67$ °F.

Another very important species in the atmosphere of troposphere is the water vapor which is the most important species in the atmosphere as far as the absorption of long wave length of radiation originating at the earth's surface is concerned but which has little influence on the other physical properties of the air in the calculation of the flow field of ordinary gasdynamics. Thus, in ordinary aerodynamics, the water vapor in the atmosphere is usually not considered but in the study of radiation processes and heat balance in the troposphere, it is the most important species.

At the end of the troposphere, there is a region known as tropopause which is a short region between the troposphere and stratosphere. The height of the tropopause depends greatly on the latitude. It may be below 10 km. in the polar region and above 15 km. in the equatorial belt.

(ii) *Stratosphere*. The stratosphere covers the region from tropopause up to an altitude of about 50 km. In this region, the temperature does not decrease with increase of altitude. First the temperature remains constant and then increases with altitude. In the old definition of standard atmosphere, we assume that the temperature remains constant in the stratosphere. But from recent data we know that the temperature in the stratosphere increases with altitude to a peak of the order of 270 °K. Hence we may divide the stratosphere into two sub-regions. In the lower region, the temperature of the atmosphere is constant up to a level A and in the upper region the temperature increases with altitude to the stratopause which is also known as level B.

(iii) *Mesosphere*. It is the region above stratopause up to an altitude of about 85 km. in which the temperature decreases again with increase of altitude. The minimum temperature in this region is of the order of 150 °K. The end of mesosphere is known as mesopause. In the mesosphere, the photochemical action is very important, i.e., the solar radiation will cause the chemical reactions in the atmosphere.

(II) *Heterosphere* in which the composition of the air varies greatly with altitude and in which the dissociation and ionization of air are important. The science which studies the upper atmosphere of heterosphere is known as Aeronomy. Heterosphere may be divided into various subregions: thermosphere, ionosphere, metasphere, protosphere and exosphere.

(i) *Thermosphere and ionosphere*. Above mesosphere, the temperature of the air again increases with altitude up to a maximum value of the order of 1,500 °K to 2,000 °K. This region is known as the thermosphere in which dissociation and ionization of the air are important and the composition of the air is no longer constant. At lower end of the thermosphere of the altitude of about 100 km., the dissociation of oxygen takes place. Hence we cannot assume that the composition of the air remains constant. At an altitude around 500 km.

most of the molecular oxygen dissociate and we may assume that the major constituents of the air are molecular nitrogen and atomic oxygen. At higher altitudes, the dissociation of molecular and the recombination of atomic oxygen are not the only chemical reactions. There are many other chemical reactions take place. For instance, the atomic oxygen may attach to molecular oxygen to form ozone. The molecular nitrogen may dissociate into atomic nitrogen. The atomic oxygen and the atomic nitrogen may form nitric oxide. The water vapor in the air may dissociate into atomic oxygen and hydroxyl (OH). The carbon dioxide may dissociate into atomic carbon and carbon monoxide and many other chemical reactions may take place.

From mesopause and up, ionization of air also takes place. Even below mesopause at about 80 km., there are some free electrons in the atmosphere which is known as D-layer with free electrons of a density of 10^2 to 10^4 electrons per cubic centimeter. As the altitude increases, the electron density increases too until a maximum of the order of 10^5/cc which is known as the E-layer and which is at about 120 km. After a slight drop of electron density with the increase of altitude, the electron density increases again with altitude to a greater maximum of 10^6/cc which is known as the F-layer and which is at about 300 km. As the altitude further increases, the electron density decreases. The region in which there are considerable amount of free electrons is known as Ionosphere. In the ionosphere, the diffusion as well as photochemical and photoelectric actions are very important. The thermosphere extends to a few hundred kilometers above the F-layer of ionosphere until the temperature of the atmosphere reaches a maximum of 1500 °K to 2,000 °K.

(ii) *Exosphere — metasphere and protosphere.*

The temperature of the atmosphere above the thermosphere remains almost constant for a considerable altitude. This isothermal region was called the exosphere because it was thought that in this region the laws of gas kinetics no longer applied. In this region, the particles will suffer little collisions and then as they move upward, they may escape from the earth gravitational field. Hence the name "exosphere" is used. In this region we should not consider the air as a continuum and should consider the discrete character of the air particles. Hence we have to use the rarefied gasdynamics to study the flow of these particles, particularly using the free molecular flow analysis as we shall discuss in chapter. The radiation is a very important mode of heat transfer in this region.

In the lower portion of the exosphere, the air is still mainly unionized. We should use the free molecule flow of neutral particles with gravitational force as the main body force to study the dynamic process of the atmosphere. This region is known as metasphere. In the upper portion of the exosphere, the gas particles of the atmosphere are almost ionized and the protons are more abundant than the neutral hydrogen. This region is known as the protosphere. In the protosphere, we should consider both the gravitational and the electromagnetic forces to study the dynamic processes of the atmosphere. In the exosphere, we should find out the molecular distribution function of the various species and then determine the statistical average of various properties of the atmosphere.

Because of the different laws of variation of temperature with altitude at various regions in the atmosphere, the relative importance of radiation, heat conduction and heat convection in the atmosphere is different in these regions. First let us consider the earth surface temperature. Average over all seasons and altitudes, the mean temperature of the earth's surface is almost constant. Hence we may assume that under the average condition, the earth is in a state of thermal equilibrium, i.e., the mean absorption of the effective solar radiation is equal to the mean effective terrestrial emission of radiation to the free space. As observed at the earth surface, the solar radiation at shorter wave lengths than 0.29 micron is totally absorbed by the atmosphere. Hence in meteorology, we usually refer to the solar radiation as short-wave radiation as the term used in the heat balance of the atmosphere E_{Rs} (z) where z is the altitude. This short-wave radiation flux consists of the direct sun light plus the scattered and multi-reflected light from the air cloud and earth. The radiation emitted by the earth to free space is usually referred to as the long-wave radiation in meteorology. The water vapor is the decisive factor in the absorption of the long-wave radiation E_{RL} (z) in the atmosphere. In the determination of the atmospheric temperature distribution, we should consider the energy balance between the difference of E_{Rs} and E_{RL} and other terms such as the heat conduction and heat convection and chemical reactions. For instance, near the earth surface, we need to consider the heat conduction into the soil, the heat conduction into the atmosphere and the latent heat of phase transformation of water in addition to the radiation terms. Such calculations have been discussed in details in standard textbooks of meteorology (cf. references 3 and 9) and we shall not discuss them here. At upper atmosphere, the problem is more complicated because of the large number of chemical reactions as well as ionizations. Many of these problems are still not solved (cf. reference 22).

6. Flow between two parallel plates in radiation magnetogasdynamics. In the most general case, we should consider the combined effects of heat convection, heat conduction and thermal radiation. We are going to consider a simple case which will demonstrate fully the interaction of thermal radiation and other gasdynamic effects. We consider an ionized gas flowing between two parallel plates: One of the plates is at rest and the other is of uniform motion with a velocity U (Fig. 9.4). There is a constant pressure gradient in the flow direction, i.e., x-direction. There is an externally applied transverse magnetic field $H_y = H_0 =$ constant in the direction perpendicular to the plates. The temperatures of the two plates are kept at constant values T_0 and T_1 respectively. The x-component of the magnetic field is assumed to be zero at the lower plate and equal to a finite value or zero at the upper plate depending on the externally applied electric field $E_z = E_0 =$ constant in the direction perpendicular to both the plates and the direction of the flow. The ionized gas is assumed to be viscous, heat-conducting, thermal radiating and electrically conducting. The flow is assumed to be steady and laminar. Furthermore, we assume that all variables are functions of the distance perpendicular to the plates only, i.e., they are functions of y only.

The fundamental equations of our problem are as follows:

(i) Equation of state [cf. eq. (5.2)].

$$p = \rho R T \tag{9.36}$$

(ii) Equation of motion [cf. eq. (5.4)].

The x-wise equation of motion is

$$\frac{d}{dy}\left(\mu\frac{du}{dy}\right) + B_y\frac{dH_x}{dy} = \frac{d}{dx}(p + p_R) \tag{9.37}$$

where we assume that only the x-wise velocity component u is different from zero and it is a function of y only. $B_y = \mu_e H_0$ is the constant external magnetic induction. Our problem is approximately the situation of the flow field far from the entrance of a two dimensional channel. For incompressible fluid, the situation holds even if there is a constant x-wise pressure gradient, i.e., $dp/dx = p_x =$ $=$ constant. However it is not true for the case of compressible fluid, because

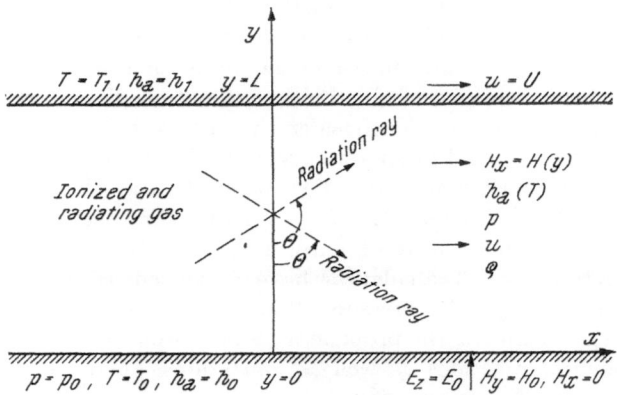

Fig. 9.4. Couette flow in radiation magnetogasdynamics

if the pressure is a function of x, the temperature, density as well as the velocity will be functions of x too. Then we can not assume that u is a function of y only. For compressible fluid with $u = u(y)$, dp/dx must be equal to zero. Furthermore, if $dp/dx \neq 0$, dT/dx will not be zero. As a results, $dp_R/dx = (dT/dx)$ $(dp_R/dT) \neq 0$. We shall treat two special cases of equation (9.37): One is the case that $u = u(y)$ and $d(p + p_R)/dx = 0$ which is called plane Couette flow and the other is the case that $u = u(y)$ and $d(p + p_R)/dx \neq 0$ but may be approximated by a simple expression as we shall show later. This case is called the plane Poiseuille flow.

The y-wise equation of motion is

$$p + p_R + \tfrac{1}{2}\mu_e H_x^2 = \text{constant} \tag{9.38}$$

For the thermal radiation term, we assume that the one dimensional model of chapter V § 10 holds. Hence the only radiation stress component different from zero is the radiation pressure p_R which is related to the specific intensity $I\nu$ by the integral:

$$p_R = \frac{2\pi}{3c}\int_0^\infty d\nu \int_0^\pi I_\nu(\theta, y)\sin\theta\, d\theta \tag{9.39}$$

where θ is the angle between the direction of the ray of radiation and the minus y-axis as shown in Fig. 9.4 and c is the velocity of light.

(iii) The x-wise magnetic field equation [cf. equations (8.1) & (8.3)]

$$H_y \frac{d\,u}{d\,y} + \frac{d}{dy}\left(\nu_H \frac{d\,H_x}{d\,y}\right) = 0 \tag{9.40}$$

where the ordinary magnetogasdynamic approximations have been used so that the terms with excess electric charge ρ_e are neglected.

(iv) Energy equations [cf. equation (5.7)]

$$\rho\,u\left[\frac{dC_p T}{dx} + \frac{d\,(E_R/\rho)}{dx}\right] + \frac{d\,q_{Rx}}{dx} = \frac{d}{dy}\left(u\,\mu\,\frac{d\,u}{dy}\right) + \frac{d}{dy}\left(\varkappa\,\frac{dT}{dy}\right) + \vec{E}.\vec{J} + \frac{d\,q_{Ry}}{dy} \tag{9.41}$$

For plane Couette flow, the terms on the left-hand side of equation (9.41) are zero because all the terms are independent of x. For the plane Poiseuille case, strictly speaking they are not zero. But far from the entrance of the channel, these terms are very small. They may be neglected in some cases or may be approximated by simpler expressions in other cases as we shall discuss later. The electric field $\vec{E}$ in the present case is a constant and has a z-component only, i.e., $E_z = E_{z0}$. The electric current density $\vec{J}$ also has a z-component only, i.e.,

$$J_z = -\frac{d\,H_x}{d\,y} \tag{9.42}$$

The radiative flux q_{Ry} is given by equation (5.43).

(A) Plane Couette flow in radiation magnetogasdynamics (19).

In this case, we assume that $dp/dx = dp_R/dx = dT/dx = 0$, we may integrate equation (9.37) with respect to y and obtain the following non-dimensional equation:

$$\mu\,\frac{d\,u}{d\,y} + R_e\,R_H\,H_x = \left(\frac{d\,u}{dy}\right)_0 = \text{constant} \tag{9.43\,a}$$

where the velocity u is expressed in terms of the uniform velocity of the upper wall U, the distance y is expressed in terms of the gap between the plates L, the magnetic field is expressed in terms of the uniform y-wise externally applied magnetic field H_0, all other quantities in equations (9.43) will be expressed in terms of their corresponding values at the lower wall, e.g., the coefficient of viscosity is expressed of μ_0 where subscript 0 refers to the value at the lower wall $y = 0$.

$$R_e = LU\rho_0/\mu_0 = \text{Reynolds number of the Couette flow}$$

$$R_H = (\mu_e\,H_0{}^2)/(\rho_0\,U^2) = \text{Magnetic pressure number} = \frac{\text{magnetic pressure}}{\text{dynamic pressure}}$$

Equation (9.38) in the non-dimensional form is

$$p + R_{p0}\,p_R + \tfrac{1}{2}\,\gamma\,M^2\,R_H\,H_x{}^2 = 1 + R_{p0} \tag{9.43\ b}$$

where the radiation pressure is expressed in terms of $p_{R0} = \frac{1}{3} a_R T_0^4$ and γ is the ratio of the specific heats and

$R_{p0} = (\frac{1}{3} a_R T_0^4)/p_0$ = radiation pressure number at the lower wall

$$M = U/a_0 = U/(\gamma R T_0)^{\frac{1}{2}} = \text{Mach number}$$

where a_0 is the sound speed at the lower wall.

Equation (9.40) in the non-dimensional form is, after the integration with respect to y,

$$\frac{d H_x}{d y} = R_\sigma (-u + R_E) \tag{9.43 c}$$

where
$$R_\sigma = \frac{U L}{\nu H} = \text{magnetic Reynolds number}$$

and
$$R_E = \frac{E_0}{B_0 U} = \text{electric field number}$$

In the present problem the electric field number may be considered as a parameter which has a significant effect on both the velocity and temperature distributions of the flow field.

For the energy equation (9.41), the terms on the left-hand side are zero because they are all independent of x. Since E is a constant, we may integrate equation (9.41) with the help of equation (9.42) with respect to y. The resultant non-dimensional equation of energy is then

$$\frac{1}{(\gamma - 1) P_r} \frac{d T}{d y} + M^2 \mu \frac{d}{dy} (\tfrac{1}{2} u^2) + R_e R_E R_H M^2 H_x +$$

$$+ R_{p0} R_F R_e q_R (T) = \text{constant} = b \tag{9.43 d}$$

where
$$P_r = \frac{\mu_0 C_{p0}}{\varkappa_0} = \text{Prandtl number of the gas.}$$

The radiation flux term $q_R (T)$ is in general a complicated integral depending on the properties of the gas as well as those of the plates. We shall consider two special cases. One is the optically thick case, the radiation flux number is

$$R_F = R_{F1} = \frac{c L_R}{U L} = \text{radiation flux number for small } (L_R/L)$$

and
$$q_R (T) = q_{R1} (T) = \frac{4}{\gamma} T^3 \frac{d T}{d y} \tag{9.44 a}$$

The second case is the optically thin case with black plates and the exponential integrals may be replaced by its first power expansion of small optical thickness [cf. eq. (5.56)]. Hence we have

$$R_F = R_{F2} = \frac{c L}{U L_R} = \text{radiation flux number for large } (L_R/L)$$

where L_R is the Planck average mean free path of radiation. It is interesting to see that for small (L_R/L), the radiation flux is directly proportional to (L_R/L) while for large (L_R/L), the radiation flux is inversely proportional to (L_R/L). The radiation flux is then

$$q_R(T) = \frac{3}{2\gamma} \left\{ (T_1{}^4 + 1)\tau_a + \int_{\tau_a}^{\tau_{a2}} T^4(t)\,dt - \int_0^{\tau_a} T^4(t)\,dt \right\} \qquad (9.44\,b)$$

where $\tau_a = \tau(L_R/L)$ and $\tau = \int_0^{y^*} \rho\,k_\nu'\,dy =$ optical thickness and y^* is the dimensional y.

Finally the non-dimensional form of equation of state (9.36) is

$$p = \rho\,T \qquad (9.43\,e)$$

Our fundamental equations (9.43) should be solved for the boundary conditions:

$$y = 0 : u = 0,\ T = 1,\ H_x = 0 \qquad (9.45)$$
$$y = 1 : u = 1,\ T = T_1,\ H_x \text{ is finite}$$

Actually we have three first order differential equations (9.43a), (9.43c) and (9.43d) but we also have two arbitrary constants in equations (9.43a) and (9.43d). Hence we need the five conditions of equation (9.45). The system of equations (9.43) is highly non-linear because the properties of the gas such as μ, $\varkappa$ etc. are functions of the temperature T. In order to bring out some essential features of the present problem, we consider the simple case of a fluid of constant properties so that $\mu = 1$, $\varkappa = 1$ and $\tau_a = y$. Then we may separate the variables into two groups: One consists of u and H_x which are governed by equations (9.43a) and (9.43c) and the other consists of p, ρ, and T which are governed by equations (9.43b), (9.43d) and (9.43e).

Equations (9.43a) and (9.43c) give the distribution of u and H_x as follows:

and

$$u = \frac{1 + R_E(\cosh R_h - 1)}{\sinh R_h} \sinh R_h\,y - R_E(\cosh R_h\,y - 1) \qquad (9.46)$$

$$H_x = \frac{R_E}{R_h} \sinh R_h\,y - \frac{1 + R_E(\cosh R_h - 1)}{R_h \sinh R_h} (\cosh R_h\,y - 1) \qquad (9.47)$$

where

$$R_h = (R_\sigma R_H R_e)^{\frac{1}{2}} = \text{Hartmann number}$$

Fig. 9.5 shows some typical distributions of velocity u and Fig. 9.6 shows some typical distributions of the magnetic field H_x. It is interesting to notice that the velocity and the magnetic field distributions depend greatly on the electric field number R_E as well as the Hartmann number R_h. Only when $R_E = \frac{1}{2}$, H_x is zero at $y = 1$. This is the case for insulated plates.

After u and H_x are obtained, equation (9.43d) gives the temperature distribution. In general, equation (9.43d) is an integro-differential equation with an integral expression for the radiation flux term. We shall calculate only the two limiting cases given by the expressions (9.44).

For the optically thick case with the expression (9.44a), equation (9.43d) becomes a differential equation. This differential equation with the boundary

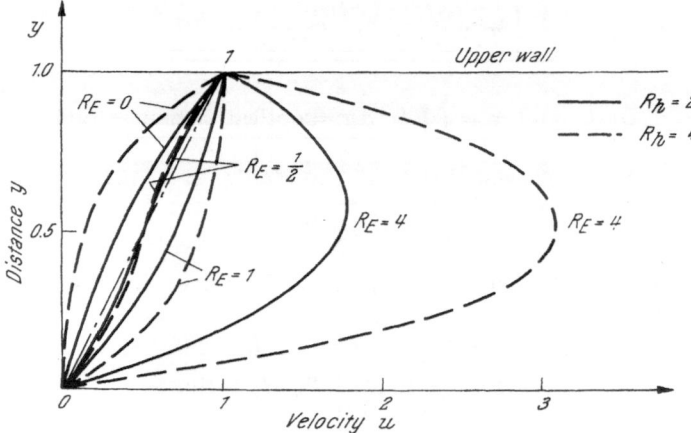

Fig. 9.5. Velocity distributions of plane Couette flow in magnetogasdynamics

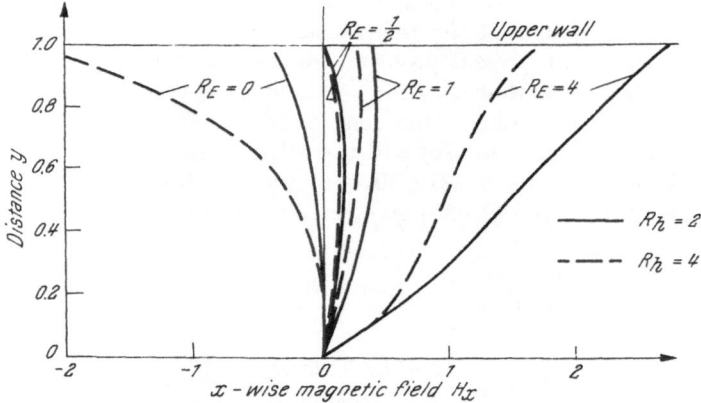

Fig. 9.6. X-wise magnetic field distributions of plane Couette flow in magnetogasdynamics

conditions (9.45) for temperature gives

$$A\left[T^4 - 1 - y(T_1^4 - 1)\right] + \frac{\gamma}{(\gamma-1)\,P_r}\left[T - 1 - y(T_1 - 1)\right] =$$

$$= \gamma\,M^2\,(R_E/R_\sigma)\,R_h^2\left[y\int_0^1 H_x\,dy - \int_0^y H_x\,dy\right] + \tfrac{1}{2}\gamma\,M^2\,(y - u^2) \tag{9.48}$$

where $A = Rp_0\,R_F\,R_e$. Fig. 9.7 shows typical temperature distributions with and without radiation effect based on equation (9.48).

For the optically thin case with expression (9.44 b), equation (9.43 d) gives the following relation:

$$\tfrac{3}{2} A \left\{ \tfrac{1}{2} (T_1{}^4 + 1)(y^2 - y) + \int_0^y I_R dy - y \int_0^1 I_R dy \right\} +$$

$$+ \frac{\gamma}{(\gamma - 1) P_r} [T - 1 - y (T_1 - 1)] = \qquad (9.49)$$

$$= \gamma M^2 (R_E/R_\sigma) R_h{}^2 \left[y \int_0^1 H_x dy - \int_0^y H_x dy \right] + \tfrac{1}{2} \gamma M^2 (y - u^2)$$

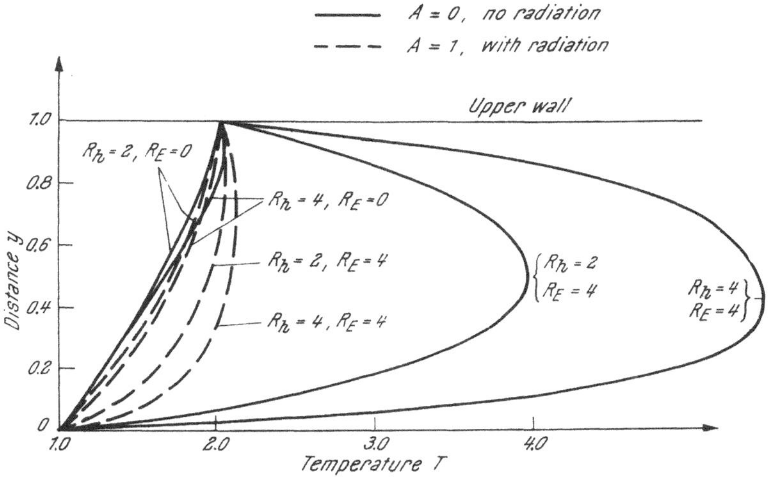

———— $A = 0$, no radiation
– – – – $A = 1$, with radiation

Fig. 9.7. Temperature distributions of plane Couette flow in radiation magnetogasdynamics for optically thick case. $(M = 1, P_r = 1, \gamma = 5/3, R_\sigma = 1)$

where

$$I_R = \int_y^1 T^4 (y) \, dy - \int_0^y T^4 (y) \, dy \qquad (9.50)$$

Fig. 9.8 shows some typical temperature distributions for the optically thin case. From Figs. 9.7 and 9.8, it is seen that the radiation has a significant effect on the temperature distributions. Without radiation effect, i.e., $A = 0$, the temperature may be enormously high in the flow field for the case of large R_E and large R_h. With radiation effects, the maximum temperature is always drops a great deal. For the optically thick case, the temperature is always lower than the corresponding case of optically thin gas. There is a tendency that the temperature is uniform near the value of the higher temperature side of one of the two plates over a major portion of the flow field and drops to the value of the other plate of lower temperature by the heat conductivity. The heat conductivity shows essentially a boundary layer effect, because we consider the Prandtl number of unity and a very large Reynolds number. There is a large drop of temperature near the walls. For one of the optically thin cases, because of the large effect due to the electromagnetic fields, we have two boundary layers: one near each wall.

(B) Plane Poiseuille flow in radiation magnetogasdynamics.

For plane Poiseuille flow, we assume that both plates are at rest and there is a constant x-wise pressure gradient. Strictly speaking, in this case all the variables p, T, ρ, u and H_x are functions of both x and y; and some of the other variables such as v and H_y may also come into the problem. No general solution for such a complete problem has been worked out yet. In order to bring out the essential features of this problem, we shall follow reference 29 by assuming that the fluid has constant properties and that both p_R, E_R and $\mathrm{d}\,q_{Rx}/\mathrm{d}x$ are negligibly small and that for fully developed flow at a distance far away from

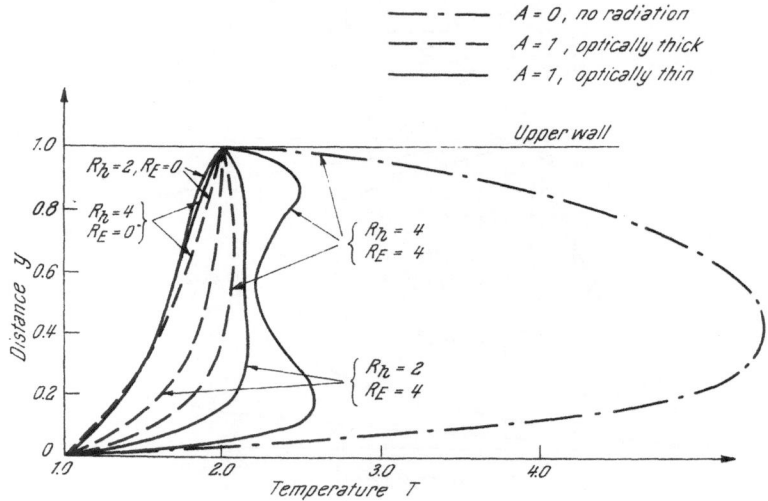

Fig. 9.8. Temperature distributions of plane Couette flow in radiation magnetogasdynamics for optically thin case. ($M = 1$, $P_r = 1$, $\gamma = 5/3$, $Ri = 1$)

the entrance of the channel, the axial temperature gradient $\partial T/\partial x$ may be replaced by the following expression:

$$\frac{\partial T}{\partial x} = \left(\frac{T_w - T}{T_w - T_m}\right)\frac{\mathrm{d}\,T_m}{\mathrm{d}x} \tag{9.51}$$

where the mean temperature over the cross section T_m is defined as follows:

$$T_m = \left(\int_0^L T u\,\mathrm{d}y\right)\bigg/\left(\int_0^L u\,\mathrm{d}y\right) \tag{9.52}$$

If q_t is the total heat flux at a section far away from the entrance and the exit of the channel, the heat balance over a section of length $\mathrm{d}x$ gives

$$\rho\,C_p\,L\,u_m\,\mathrm{d}T_m = 2\,q_t\,\mathrm{d}x \tag{9.53}$$

where
$$q_t = -\left(\varkappa\frac{\partial T}{\partial y}\right)_0 - q_{Ry\,(y=0)} = \left(\varkappa\frac{\partial T}{\partial y}\right)_L + q_{Ry\,(y=L)}$$

Finally equation (9.41) becomes

$$\frac{d}{dy}\left(\varkappa\frac{dT}{dy}\right) + \frac{dq_{Ry}}{dy} = \frac{2q_t}{L}\left(\frac{T_w - T}{T_w - T_m}\right)\left(\frac{u}{u_m}\right) - \frac{d}{dy}\left(u\,\mu\,\frac{du}{dy}\right) - \vec{E}.\vec{J} \quad (9.54)$$

Now equation (9.54) is a total differential equation of y only. In reference 29, some numerical examples had been given for the case of constant properties of the fluid and neglecting the viscous dissipation as well as the joule heating. Furthermore, the basic flow is for the case of zero external electric field $R_E = 0$. Some typical curves of the temperature distributions are shown in Fig. 9.9.

7. Boundary layer flow in radiation gasdynamics.
In many practically important problems, the configuration is more complicated than those discussed in section 6 and we have to consider the temperature and other flow variables at least as functions of two spatial variables. But in most of these problems, both the Reynolds number and the Peclet number are large, and we may use the boundary layer approximations in the fundamental equations. In this section, we consider the two dimensional boundary layer flows in radiation gasdynamics which consist of the boundary layer over a flat plate, the laminar jet mixing, the wake and the boundary layer over thin bodies etc. We take the x-axis as the

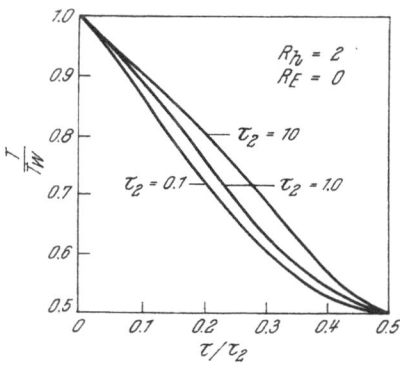

Fig. 9.9. Temperature distributions of plane Poiseuille flow in radiation magnetogasdynamics as function of optical thickness. $e_w = 1.0$, $R_F = 4\,\sigma\,L_R$ $T_w{}^3/\varkappa = 10$. (Fig. 2 of reference 29 by R. VISKANTA, Courtesy of ZAMP)

direction of the main flow and the y-axis as the direction of the boundary layer thickness and the corresponding velocity components in x- and y-directions are u and v respectively. Hence u is much larger than v and the gradient in the y-direction is much larger than that in the x-direction. The fundamental equations of the boundary layer flow in radiation gasdynamics are:

(i) Equation of state

$$p = \rho\,RT \qquad (9.55\,\mathrm{a})$$

(ii) Equation of continuity

$$\frac{\partial\rho u}{\partial x} + \frac{\partial\rho v}{\partial y} = 0 \qquad (9.55\,\mathrm{b})$$

(iii) Equations of motion

$$\rho u\frac{\partial u}{\partial x} + \rho v\frac{\partial u}{\partial y} = -\frac{\partial(p + p_R)}{\partial x} + \frac{\partial}{\partial y}\left(\mu\,\frac{du}{dy}\right) \qquad (9.55\,\mathrm{c})$$

$$p + p_R = F(x) \qquad (9.55\,\mathrm{d})$$

where the radiation pressure is given by the integral:

$$p_R = \frac{2\pi}{3c}\int_0^\infty dv\int_0^\pi I_v(\theta, x, y)\sin\theta\,d\theta \qquad (9.55\,\mathrm{e})$$

where θ is the angle between the direction of the ray of radiation and the minus y-axis. It should be noticed that in general the radiation stress is a tensor quantity [cf. eq. (2.18)]. However, within the concept of continuum theory of boundary layer flow, the radiation stress may be considered as the sum of a radiation pressure p_R [defined by equation (2.19)] and a radiation "viscous" stress tensor [defined by the difference of equations (2.18) and (2.19)]. Applying the boundary layer approximations to the radiation stress, we obtain the radiation terms in equation (9.55c) and (9.55d) because the shearing radiation viscous stresses vanish in the present problem.

By boundary layer approximations, the function $F(x)$ is a known function of x only. $F(x)$ is the sum of the gas pressure p and the radiation pressure p_R outside the boundary layer. For instance, if the free stream is a uniform flow with constant temperature, $F(x)$ is a constant.

(iv) Energy equation

$$\rho u \frac{\partial h_R}{\partial x} + \rho v \frac{\partial h_R}{\partial y} = \frac{\partial}{\partial y}\left(\varkappa \frac{\partial T}{\partial y}\right) + \frac{\partial q_{Ry}}{\partial y} + \mu\left(\frac{\partial u}{\partial y}\right)^2 + u\frac{\partial p}{\partial x} + \frac{p_R}{p}\left(u\frac{\partial \rho}{\partial x} + v\frac{\partial \rho}{\partial y}\right)$$

(9.55 f)

where

$$h_R = C_p T + \frac{E_R}{\rho} = \text{effective enthalpy with radiation effect} \quad (9.55\,\text{g})$$

The net radiation flux in the y-direction for the present case is

$$q_{Ry} = 2\pi \int\limits_0^x \mathrm{d}\nu \int\limits_0^\pi I_\nu(\theta, x, y)\ \sin\theta\ \cos\theta\ \mathrm{d}\theta \qquad (9.55\,\text{h})$$

(v) Equation of radiative transfer

$$\cos\theta\, \frac{\partial I_\nu}{\partial y} = \rho\, k'_\nu\, (I_\nu - B_\nu) \qquad (9.55\,\text{i})$$

Here we make the one-dimensional approximation that the specific intensity I_ν is essentially a function of y and its dependence on x is negligible. This is a good approximation for many practical problems. On the other hand, if we consider the two dimensional radiative transfer equation, we will have the upstream effect due to radiative transfer, because the radiative transfer should be calculated by integration over the whole space. The one-dimensional problem is very similar to the boundary layer approximation to the complete Navier-Stokes equations. In the Navier-Stokes equations, we have the upstream influence too. Little has been done for the two dimensional radiative transfer problem. In this section, we shall restrict to the one dimensional model only except for the case of very small mean free path of radiation.

We are going to solve equations (9.55) for various boundary conditions. Before we solve equations (9.55), it is interesting to consider some transformations of these equations. From equation (9.55b), we may define a streamfunction ψ such that

$$\frac{\partial \psi}{\partial y} = \rho u, \quad \frac{\partial \psi}{\partial x} = -\rho v \qquad (9.56)$$

If we use x and ψ as independent variables instaed of x and y, equation (9.55 b) is automatically satisfied.

If we use the variable x and ψ, the fundamental equations can be greatly simplified, particularly for the case of $F(x) = $ constant. Then equation (9.55 c) becomes:.

$$\frac{\partial u}{\partial x} = \frac{\partial}{\partial \psi}\left(\rho\, u\, \mu\, \frac{\partial u}{\partial \psi}\right)$$ (9.57)

Equation (9.57) is of exactly the same form as the corresponding equation of ordinary gasdynamics. The only difference of the present case from that of ordinary gasdynamics is due to fact that the density is no longer inversely proportional to the temperature T as in the case of ordinary gasdynamics where the gas pressure is a constant but the density is a function of temperature T given by equations (9.55 a) and (9.55 d).

The equation of energy (9.55 f) becomes

$$(C_p + 4 R R_p)\frac{\partial T}{\partial x} + \frac{4}{\rho}(1 + R_p)\frac{\partial p_R}{\partial x} = \frac{\partial}{\partial \psi}\left(\varkappa \rho\, u\, \frac{\partial T}{\partial \psi}\right) +$$

$$+ \frac{\partial q_{Ry}}{\partial \psi} + \mu \rho\, u\left(\frac{\partial u}{\partial \psi}\right)^2 + v\,\frac{\partial p_R}{\partial \psi}$$ (9.58)

where R_p is the local radiation pressure number.

Equation (9.58) differs considerably from the corresponding equation of ordinary gasdynamics, even though equation (9.58) is still a modified generalized heat conduction equation. In general, p_R is a function of both p and T and we cannot combine the two terms in the left-hand side of equation (9.58). However, for special cases of very small mean free path of radiation, we shall show that the two terms on the left hand side of equation (9.58) may be combined into one.

Since the density can be expressed in terms of the temperature T, we need to solve only the two integro-differential equations (9.57) and (9.58) for u and T with the initial conditions:

$$x = x_0 : u = u_0\,(\psi) \text{ and } T = T_0\,(\psi)$$ (9.59)

We are going to discuss two special cases for these equations as follows:

(A) Laminar jet mixing problems:

We consider a two dimensional free jet mixing problem which may represent a jet flow issued from a nozzle into a surrounding fluid or the mixing of two uniform streams, or a wake behind a two-dimensional body. It is convenient to treat the cases of very small mean free path of radiation separately from that of finite mean free path of radiation.

(i) Case for very small free path of radiation.

With the help of the results of chapter V, section 8 for the radiation terms, equation (9.58) reduces to

$$C_{pR}\frac{\partial T}{\partial x} = \frac{\partial}{\partial \psi}\left(K^* \rho\, u\, \frac{\partial T}{\partial \psi}\right) + \mu \rho\, u\left(\frac{\partial u}{\partial \psi}\right)^2 + v\,\frac{\partial p_R}{\partial \psi}$$ (9.58 a)

where $\qquad C_{pR} = C_p + 20\,R\,R_p + 16\,R\,R_p{}^2 = $ effective specific heat $\qquad$ (9.60a)
with radiation effect

$$K^* = \varkappa + \varkappa_R = \varkappa + 4\,a_R\,D_R\,T^3 = \text{effective coefficient of} \qquad (9.60\,b)$$
heat conductivity with radiation effect

The last term on the right-hand side of equation (9.58a), $v\,(\partial\,p_R/\partial\,\psi)$, is usually negligibly small, particularly for the extreme cases of very large and very small values of the radiation pressure number R_p. The main effects for thermal radiation are to increase the specific heat according to equation (9.60a) and the coefficient of heat conductivity according to equation (9.60b). Since C_{pR} depends on R_p only and K^* depends on both R_p and other factors such as (c/U), there are many practical cases where the increase of specific heat is negligibly small but the increase of the coefficient of heat conductivity is not negligible. The main effect of thermal radiation is then to decrease the effective Prandtl number:

$$P_{re} = \frac{C_{pR}\,\mu}{K^*} \qquad (9.61)$$

Since Prandtl number represents the square of the ratio of the spread of the velocity profile to that of the temperature profile, the radiation effect will widen the thermal jet mixing region. For instance, we consider the case of a very hot jet in a uniform stream U. Since the temperature of the whole flow field is very high, the viscous dissipation and the work done by the pressure force are negligible, equation (9.58a) becomes:

$$C_{pR}\frac{\partial\,T}{\partial\,x} = \frac{\partial}{\partial\,\psi}\left(K^*\,\rho\,u\,\frac{\partial\,T}{\partial\,\psi}\right) \qquad (9.58\,b)$$

Equation (9.58b) is of the exactly same form as equation (9.57). If we further assume that the deviations of the velocity u from the uniform speed U and of the temperature T from the uniform temperature of the surrounding stream T_0 are small in the whole flow field, the linearized solutions of equations (9.57) and (9.58b) for the boundary conditions: $x = 0$:

$$\left.\begin{array}{l} u = U + u' \\ T = T_0 + T' \end{array}\right\} \text{ at } 1 \geqq y \geqq -1$$

$$\left.\begin{array}{l} u = U \\ T = T_0 \end{array}\right\} \text{ at } y > 1 \text{ and } y < -1$$

$\qquad\qquad\qquad$ (9.62)

$$u' \ll U \text{ and } T' \ll T_0$$

are respectively

$$u = U + \tfrac{1}{2}u'\left[\theta\left(\frac{1+y}{2\,\alpha\,x^{\frac{1}{2}}}\right) + \theta\left(\frac{1-y}{2\,\alpha\,x^{\frac{1}{2}}}\right)\right] \qquad (9.63\,a)$$

$$T = T_0 + \tfrac{1}{2}T'\left[\theta\left(\frac{1+y}{2\,\alpha_1\,x^{\frac{1}{2}}}\right) + \theta\left(\frac{1-y}{2\,\alpha_1\,x^{\frac{1}{2}}}\right)\right] \qquad (9.63\,b)$$

where
$$\theta\,(y) = \frac{2}{\sqrt{\pi}} \int\limits_0^y \exp\,(-z^2)\,\mathrm{d}\,z \tag{9.64}$$

$$\alpha^2 = \mu_0/\rho_0 \ U = P_{reo}\alpha_1{}^2 \tag{9.65}$$

and ρ_0, μ_0 and P_{reo} are values of ρ, μ and P_{re} at $T = T_0$ respectively. Equation (9.63) shows the increases of the spread of the thermal jet mixing T by the thermal radiation effect because of the increase of the effective coefficient of heat conductivity.

(ii) Case of finite mean free path of radiation.

In this case, we have to use the integral expressions for the thermal radiation terms. In order to show some essential effects of thermal radiation, we shall use the one-dimensional approximation for these radiation terms as given in chapter V, section 10. Since there is no solid boundary, we should take the boundaries at plus and minus infinity. The radiation energy density and the radiation pressure in the jet mixing region are given by the following integrals:

$$E_R = 3\,p_R = \frac{2\,\pi}{c} \left\{ - \int\limits_0^\infty \mathrm{d}\nu \int\limits_{-\infty}^\tau B_\nu\,(t,\nu)\,\varepsilon_1\,(\tau - t)\,\mathrm{d}\,t\ + \right.$$
$$\left. + \int\limits_0^\infty \mathrm{d}\nu \int\limits_\tau^\infty B_\nu\,(t,\nu)\,\varepsilon_1\,(t - \tau)\,\mathrm{d}\,t \right\} \tag{9.66}$$

The y-wise radiative heat transfer is

$$q_{Ry} = 2\,\pi \left\{ - \int\limits_0^\infty \mathrm{d}\nu \int\limits_{-\infty}^\tau B_\nu\,(t,\nu)\,\varepsilon_2\,(\tau - t)\,\mathrm{d}t\ + \right.$$
$$\left. + \int\limits_0^\infty \mathrm{d}\nu \int\limits_\tau^\infty B_\nu\,(t,\nu)\,\varepsilon_2\,(t - \tau)\,\mathrm{d}t \right\} \tag{9.67}$$

In general the optical thickness τ is a function of the frequency ν, i.e.,

$$\tau = \int\limits_0^y \rho\,k_\nu{}'\,\mathrm{d}y \tag{9.68}$$

Equations (9.66) and (9.67) cannot be simplified unless we know the explicit relation between τ and ν. However, if the gas is approximately gray so that the absorption coefficient is practically independent of ν, we may assume that τ is independent of ν. Then equations (9.66) and (9.67) reduce respectively to the following forms:

$$E_R = 3\,p_R = \tfrac{1}{2}\,a_R \left\{ - \int\limits_{-\infty}^\tau T^4\,(t)\,\varepsilon_1\,(\tau - t)\,\mathrm{d}\,t + \int\limits_\tau^\infty T^4\,(t)\,\varepsilon_1\,(t - \tau)\,\mathrm{d}t \right\} \tag{9.66a}$$

$$q_{Rx} = \tfrac{1}{2}\,c\,a_R \left\{ - \int\limits_{-\infty}^\tau T^4\,(t)\,\varepsilon_2\,(\tau - t)\,\mathrm{d}\,t + \int\limits_\tau^\infty T^4\,(t)\,\varepsilon_2\,(t - \tau)\,\mathrm{d}t \right\} \tag{9.67a}$$

Then the radiative heat transfer term in the energy equation for a gray gas is

$$Q_R = \frac{\partial q_{Ry}}{\partial y} = \left(\frac{c}{L_R}\right)\left[\frac{1}{2}\int_{-\infty}^{\infty} E_{R0}(t)\,\varepsilon_1\,(|t-\tau|)\,\mathrm{d}t - E_{R0}(\tau)\right] \qquad (9.69)$$

where

$E_{R0} = a_R\,T^4$ and $L_R = 1/K_p$ is the mean free path of radiation based on Planck mean absorption coefficient K_p of equation (5.36). It is seen that for large L_R, the radiative heat transfer is inversely proportional to L_R and that we should use the Planck mean absorption coefficient instead of the Rosseland mean coefficient. In order to illustrate the nature of the above integral formula, we consider the case of very small radiation pressure number but finite radiation flux number so that the radiation pressure and radiation energy density may be neglected. Then the energy equation (9.58) for a gray gas becomes

$$\frac{\partial h_a}{\partial x} = \frac{\partial}{\partial \psi}\left(\frac{\varkappa\,\rho\,u}{C_p}\frac{\partial h_a}{\partial \psi}\right) + \rho\,u\,\mu\left(\frac{\partial u}{\partial \psi}\right)^2 + \left(\frac{c}{\rho\,u\,L_R}\right)\left\{\frac{1}{2}\int_{-\infty}^{\infty} E_{R0}(t)\,\varepsilon_1(|\tau-t|)\,\mathrm{d}t - E_{R0}(\tau)\right\}$$

$$(9.70)$$

where $h_a = C_p\,T$ is the enthalpy of the gas.

Without carrying out the numerical integration, the general effect of the radiation can be seen from equation (9.70), because the integral

$$\int_{-\infty}^{\infty} \varepsilon_1\,(|\,t-\tau\,|)\,\mathrm{d}t = 2 \qquad (9.71)$$

the last terms of equation (9.70) may be written as

$$\left(\frac{c}{\rho\,U\,L_R}\right)\left\{\frac{1}{2}\int_{-\infty}^{\infty} E_{R0}(t)\,\varepsilon_1\,(|\tau-t|)\,\mathrm{d}t - E_{R0}(\tau)\right\} = \left(\frac{c}{\rho\,u\,L_R}\right)(\overline{E}_{R0} - E_{R0}) \qquad (9.72)$$

where E_{R0} is the mean value of E_{R0} at each x-station over the whole y-direction, i.e.,

$$2\,\overline{E}_{R0} = \int_{-\infty}^{\infty} E_{R0}(t)\,\varepsilon_1\,(|\,\tau-t\,|)\,\mathrm{d}t \qquad (9.73)$$

Hence near the axis of the jet where the temperature is high, E_{R0} is larger than $\overline{E}_{R0}$, the radiation term (9.72) is negative. As a result, the decrease of the temperature or enthalpy with respect to x is faster than that without radiation effect. On the other hand, far away from the axis of the jet where the temperature is low, E_{R0} is smaller than $\overline{E}_{R0}$ and the radiation term (9.72) is positive. As a result, the decrease of temperature or enthalpy with respect to x is slower than that without radiation. Thus we conclude that the radiation effect is to decrease the maximum temperature in the mixing region and to increase the spread of the thermal mixing region.

(B) Laminar boundary layer over a flat plate.

Now we consider a uniform stream of radiating gas over a flat plate. The fundamental equations (9.55) are applicable in this problem too. Since there is no variation of total pressure in the x-direction, i.e., $F(x) = $ constant, we may use equation (9.57) and (9.58) for the present problem too. The only difference of the present case from those of jet mixing problem is in the boundary conditions. In addition to the initial conditions (9.59), we have to consider the boundary conditions on the surface of the plate, i.e.,

$$y = 0: \ u = v = 0, \ T = T_w$$
$$y = \infty: \ u = u_\infty, \quad T = T_\infty \tag{9.74}$$

Hence we solve equations (9.57) and (9.58) with the boundary conditions (9.74). The cases which have been solved are those with negligibly small radiation pressure number only. Under such a condition, equation (9.58) becomes:

$$C_p \frac{\partial T}{\partial x} = \frac{\partial}{\partial \psi} \left(\varkappa \rho u \frac{\partial T}{\partial \psi} \right) + \mu \rho u \left(\frac{\partial u}{\partial \psi} \right)^2 + \frac{\partial q_{Ry}}{\partial \psi} \tag{9.58 c}$$

If we make the assumption that $\mu \rho = \text{con-stant}$, the velocity distribution in terms of x and ψ will be exactly the same as that without radiation effect and also the same as that of an incompressible fluid. The solution is one of the similarity solutions of boundary layer equation, the well-known Blasius solution. The Blasius solution may be obtained in the following manner:

We first introduce the similarity transformation:

$$s_1 = \rho_\infty u_\infty \mu_\infty x \tag{9.75a}$$

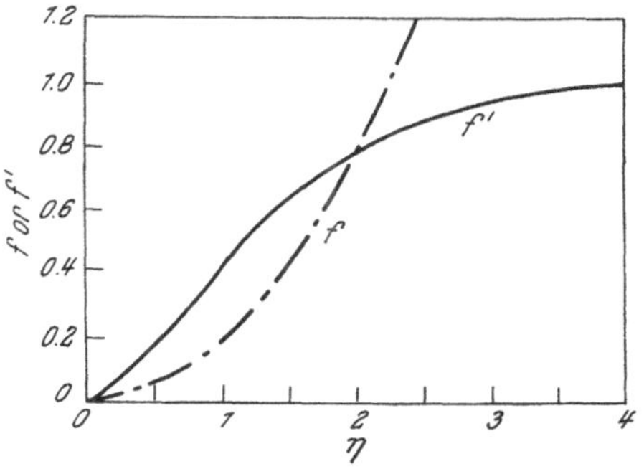

Fig. 9.10. Blasius solution of streamfunction f and velocity distribution f' in a boundary layer over a flat plate

$$\eta = \frac{\rho_\infty u_\infty}{\sqrt{s_1}} \int_0^y \frac{\rho}{\rho_\infty} \, \mathrm{d}y \tag{9.75 b}$$

$$\psi = \sqrt{s_1} \, f(\eta) \tag{9.75 c}$$

$$T^*(x, \eta) = T/T_\infty \tag{9.75 d}$$

With the variables (9.75), the equation of motion (9.75) becomes

$$f''' + \tfrac{1}{2} f f'' = 0 \tag{9.76}$$

where prime refers to the differentiation with respect to η. The boundary conditions are:

$$\eta = 0: \quad f' = f = 0$$
$$\eta \to \infty: \quad f' \to 1 \tag{9.77}$$

Equation (9.76) is the Blasius equation and the functions $f(\ )$ and $f'(\ )$ are shown in Fig. 9.10.

Without radiation effect, we found that the energy equation (9.58c) has a similarity solution too, i.e., $T^* = T^*(\eta)$. But with radiation effect, we do not have similarity solution for T^* in general. Hence we should assume first that T^* is a function of both x and η and then determine whether the similarity solution $T^*(\eta)$ exists or not. The energy equation (9.58c) in terms of x and η is

$$\frac{\partial^2 T^*}{\partial \eta^2} + \frac{Pr}{2} f \cdot \frac{\partial T^*}{\partial \eta} - s_1 Pr f' \frac{\partial T^*}{\partial s_1} + (P_r - 1) \frac{u_\infty{}^2}{C_p T_\infty} (f''^2 + f' f''') =$$

$$= - \frac{P_r L s_1}{\rho C_p T_\infty u_\infty} \left(\frac{\partial q_{Ry}}{\partial y}\right) \qquad (9.78)$$

Since there are terms with the variable s_1 in equation (9.78), there will be no similarity solution except for the optically thick case. Without radiation effect, we may assume that similarity solution exists and then both $\partial T^*/\partial s_1$ and $\partial q_{Ry}/\partial y$ vanish. Equation (9.78) becomes a total differential equation with respect to η only. With radiation term, except for the optically thick case in which the term of radiative heat transfer is a function of η only, the x-dependent terms in equation (9.78) will not vanish. Thus we do not have the similarity solution. Equation (9.78) may be integrated numerically, in order to obtain the temperature distribution. After the temperature distributions are obtained, we may calculate the heat transfers to the wall, i.e., the conductive heat transfer is

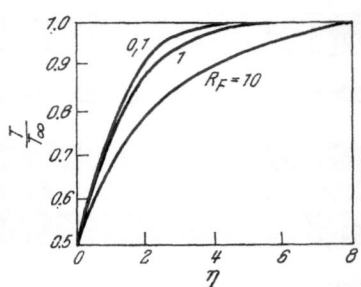

Fig. 9.11. Temperature distribution in the boundary layer over a flat plate at various radiative flux number $R_F = 4 \sigma L_R T^3/\varkappa$ for optically thick case. (Fig. 3b of reference 27 by R. VISKANTA and R. J. GROSH, Courtesy of Pergamon Press)

$$q_c = - \varkappa \left(\frac{\partial T}{\partial y}\right)_{y=0} \qquad (9.79\,a)$$

and the radiative heat transfer is

$$q_R = 2 \pi \int_0^\infty B(t)\, \varepsilon_2(t)\, dt \qquad (9.79\,b)$$

Numerical calculations have been carried out for the following two extreme cases:

(i) Optically thick case. For this case, the radiative heat transfer term becomes

$$\frac{\partial q_R}{\partial y} = \frac{\partial}{\partial y}\left(4\, a_R\, D_R\, T^3 \cdot \frac{\partial T}{\partial y}\right) \qquad (9.80)$$

or in similarity variable η, we have

$$\frac{\partial q_R}{\partial y} = \frac{\rho\, u_\infty}{\sqrt{s_1}} \frac{\partial}{\partial \eta}\left(4\, a_R\, D_R\, T^3\, \frac{\rho\, u_\infty}{\sqrt{s_1}} \frac{\partial T}{\partial \eta}\right) \qquad (9.80\,a)$$

Substituting equation (9.80a) into equation (9.78) it is evident that the radiative heat transfer will be independent of x or s_1 and similarity solution exists. Equation (9.78) with the expression (9.80a) has been integrated by VISKANTA and GROSH in reference 27 who neglected the viscous dissipation. Fig. 9.11 shows some typical temperature distributions for optically thick case with various radiative flux number.

(ii) Optically thin case. When the mean free path is very large, the radiative transfer term becomes:

$$\frac{\partial q_R}{\partial y} = -4 \rho K_p \sigma T^4 \tag{9.81}$$

In this case, there will be no similarity solution for temperature because the factor s_1 will not disappear in the radiative heat transfer term. KOH and DE-SILVA (13) integrated equation (9.78) numerically with the expression (9.81). They made calculations for free stream Mach number $M_\infty = 40$ and altitude up to 100,000 ft. They found that under these conditions, the radiative effect is negligibly small. Since the maximum temperature in sea level in their calculations was about 10,000 °F and in other of their cases, the average temperature was much less than 10,000 °K. It is expected that the radiation effect would be small. For radiation effect to be important, the free stream temperature should be of the order of 20,000 °K or higher at standard sea level conditions (cf. § 8).

8. Stagnation point heat transfer in radiation gasdynamics. One of the most interesting radiative heat transfer problems in the aerospace engineering is the heat transfer during the superorbital re-entry into the earth atmosphere. A very high temperature region will be developed behind the normal shock in front of the vehicle. In order to get the essential features of this problem, we consider a blunt body (two-dimensional or axi-symmetrical) in a hypersonic flow as shown in Fig. 9.12. Ordinary, the flow field may be divided into three distinguish regions. Region I is the shock wave region in which a very large variation of velocity and state variables occur in a very short distance. If the density of the fluid is not too small, the distance of region I is only of a few mean free path. We have discussed the flow field in this region in chapter VIII, sections 5 and 6. The main effect of radiative heat transfer is to increase this region. If the radiation stresses and the radiation energy density are not negligible, there is a large change of temperature distribution of radiation gasdynamics from those of ordinary gasdynamics. In all the cases, we expect that there is a temperature jump across this region. In many practical problems, we may assume that this region is so thin that it may be replaced by a surface of discontinuity. The relations of the variables of the flow field behind the shock with those in front of the shock are given by the Rankine-Hugoniot relations (cf. chapter VIII, § 4). In reference 23, SCALA and SAMPSON gave some numerical results in the shock wave structure for the cases of negligible radiation pressure number but finite radiation flux number for two extreme cases: One is the optically thick case in which the Rosseland expression of radiative heat transfer [eq. (5.17)] was used and in which the radiative effect was extended upstream and the other case is optically thin case in which the Planck mean absorption coefficient and equation (6.45) were used and in which the radiative effect has extended mainly downstream.

Region II may be called the inviscid shock layer region which has been studied by many authors for the case without radiation effects as well as the case of negligibly small radiation pressure number by finite radiative flux number. The main feature in this region is that both the viscous stress and the heat conduction can be neglected. We have already discussed the flow field in this region for special cases in chapter VIII, sections 6 and 7. We shall discuss it more in this section too.

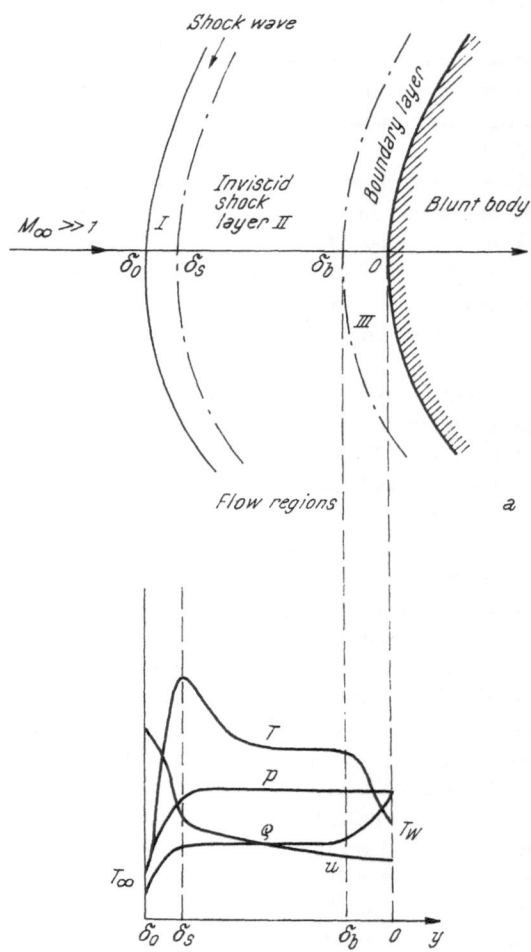

Flow regions

Variations of velocity and state variables

Fig. 9.12. Flow field near a stagnation point of a blunt body (not in scale)

Region III is the boundary layer region near a stagnation point in which the viscous shearing stress and the heat conduction cannot be neglected completely. However this region is usually thin in the y-direction and boundary layer approximations may be used. We shall discuss this region in details later in this section.

The thickness of the boundary layer region depends on the Reynolds number of the flow field. Since Reynolds number is proportional to the density of the fluid, at very high altitude where the density of the air is very low, the Reynolds number may be very small. As a result, the boundary layer thickness is so large that the boundary layer and the shock layer will merge into one single layer which may be called the viscous shock layer. We shall discuss this case too in this section.

(i) Inviscid shock layer.

The general solution of an inviscid flow field behind a detached shock is very complicated because we have to deal with anisentropic flow for which the variation of entropy is not known before the solution is obtained. Ordinarily, the solution is obtained by numerical integration of the fundamental equation. However, in hypersonic flow in which the Mach number in the free stream M_∞ is much larger than unity, some simplification may be made and the fundamental equations can be greatly simplified. We shall discuss this hypersonic flow near a stagnation point. Due to the hypersonic speed of the flow, the density of the fluid behind the shock wave is approximately constant. Hence the inviscid flow in the shock layer may be considered as a constant density flow. For constant density flow if the radiation pressure number is negligibly small, the velocity components u_i and the pressure p are independent of the temperature. We may solve the velocity and the pressure distribution without considering the temperature. After the velocity and the pressure are known, we may calculate the temperature distribution from the energy equation. Since our interest is in the heat transfer, we shall not discuss the calculation of the velocity and pressure in the inviscid shock layer region which are given in references 14 and 21. We shall consider the temperature distribution only. The coordinate system to be used (see Fig. 9.13) is a set of orthogonal curvilinear coordinates x, y and z in which x is measured parallel to the body surface with

$$K\left(x\right) = -\frac{\mathrm{d}\theta}{\mathrm{d}x} = -\frac{1}{R_b} \text{ local curvature of the body surface} \quad (9.82\,\mathrm{a})$$

y is measured normal to the body while z is perpendicular to both x and y. Hence $r\left(x, y\right)$ is the distance from an arbitrary point $P\left(x, y\right)$ normal to the axis of symmetry which is parallel to the

free stream. By symmetry, we assume that all the variables are independent of the coordinate z. We shall restrict ourselves to blunt body with large radius of curvature R_b and for the flow field near the stagnation point 0. The flow field which we are interested in lies between $y = 0$ and $y = \delta$ where δ is the detached shock distance measured along the y-axis. At the stagnation point 0, $\delta = \delta_0$. We shall assume that $\delta \ll R_b$ and $yK \ll 1$. For any physical quantity Q, we may assume that

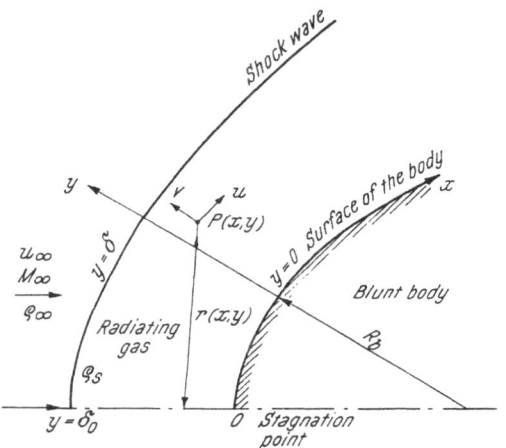

Fig. 9.13. Inviscid shock layer near a stagnation point of a blunt body in a hypersonic flow

$$\left|\frac{\partial Q}{\partial y}\right| \gg \left|KQ\right| \cong \left|\frac{Q}{R_b}\right| \quad (9.82\,\mathrm{b})$$

Furthermore we assume

$$\frac{\partial K}{\partial x} \cong \frac{1}{R_b} \quad (9.82\,\mathrm{c})$$

In the vicinity of the stagnation point 0, we may assume

$$r \cong x, \quad \frac{\partial r}{\partial x} \cong 1, \quad \frac{\partial r}{\partial y} \cong 0 \quad (9.82\,\mathrm{d})$$

With the approximations (9.82), the energy equation for an inviscid, non-heat-conducting and radiating gas with negligibly small radiation pressure number in the x-y-coordinates of Fig. 9.13 is

$$\rho u \frac{\partial h_0}{\partial x} + \rho v \frac{\partial h_0}{\partial y} = \frac{1}{x^a}\frac{\partial x^a q_{Rx}}{\partial x} + \frac{\partial q_{Ry}}{\partial y} \quad (9.83)$$

where $h_0 = h_a + \frac{1}{2}\left(u^2 + v^2\right)$ is the stagnation enthalpy; $a = 0$ for two-dimensional flow and $a = 1$ for axi-symmetrical flow. In general, we should consider both the x-wise and the y-wise radiative heat transfer given in equation (9.83). But such a general case has not been studied yet. Near the stagnation point region with $x \to 0$, we would expect that the x-wise radiative heat transfer is much smaller than the y-wise term, i.e. $(1/x_a)\, (\partial\, x_a\, q_{Rx}/\partial\, x) \ll (\partial\, q_{Ry}/\partial\, y)$. Hence in most of the analysis in literature, the term $(1/x_a)\, (\partial\, x_a\, q_{Rx}/\partial\, x)$ is neglected. We shall also neglect this x-wise radiative heat transfer term in the following analysis.

Near the stagnation point, the velocity components can be expressed in terms of a streamfunction ψ such that

$$\frac{\partial\,\psi}{\partial\,y} = -\,x^a\,u; \quad \frac{\partial\,\psi}{\partial\,x} = -\,x^a\,v \tag{9.84}$$

Here the approximation of constant density is used. The streamfunction near the stagnation point of a hypersonic flow was found in the following form (*14, 21*):

$$\psi = x^{1+a}\,f\,(y) \tag{9.85}$$

where the function $f\,(y)$ has the following forms:

For the two-dimensional case

$$f\,(y) = A_0\exp\,(A_2\,y) + A_1\exp\,(-\,A_2\,y) \tag{9.86}$$

where

$$A_0 = -\,A_1 = k_\rho\,u_\infty\,\sqrt{k_\rho\,(2-k_\rho)}/[2\,(1-k_\rho)] \text{ and } A_2 = -\,(1-k_\rho)/(k_\rho\,R_s)$$

$k_\rho = \rho_\infty/\rho_s$ is the ratio of the density across the shock, R_s is the radius of curvature of the body at the stagnation point. The shock detached distance is

$$\delta_0 = [(1-k_\rho\,R_s)/(1-k_\rho)]\sinh^{-1}[(1-k_\rho)/\sqrt{k_\rho\,(2-k_\rho)}] \tag{9.87}$$

For the axi-symmetrical case

$$f\,(y) = C_1\,y + C_2\,y^2 \tag{9.88}$$

where

$$C_1 = -\,\sqrt{1-(1-k_\rho)^2}\,(u_\infty/R_s); \quad C = [\tfrac{1}{2}\,(1-k_\rho^2]\,[k_\rho u_\infty/(k_\rho\,R_s)^2];$$
$$\delta_0 = \{[-1+\sqrt{1-(1-k_\rho)^2}]/(1-k_\rho)^2\}\,k_\rho\,R_s \tag{9.89}$$

If we substitute the expression of velocity components according to equation (9.85) into equation (9.83) and neglect the higher order terms by putting $x = 0$, we have the following equation of energy which is a function of y only:

$$\frac{\mathrm{d}q_{Ry}}{\mathrm{d}y} + (1+a)\,f\,\frac{\mathrm{d}h_a}{\mathrm{d}x} = -\,\rho\,(1+a)^3\,f^2\,\frac{\mathrm{d}f}{\mathrm{d}y} \tag{9.90}$$

where h_a is the enthalpy of the gas. In general, equation (9.90) is an integro-differential equation because q_{Ry} is an integral. The general solution of this integro-differential equation is very difficult to obtain. Usually further approximations have to be made. The following are a few simplified cases:

(a) Optically thick case.

In this case, the Rosseland heat conductivity coefficient $\varkappa_R$ may be used to express the radiative heat flux. Since both $\varkappa_R$ and the specific heat at constant pressure $C_p = \mathrm{d}h_a/\mathrm{d}T$ are functions of temperature, equation (9.90) may be written in the following form:

$$\frac{\mathrm{d}^2 T}{\mathrm{d}y^2} + A\,(T, y)\,\frac{\mathrm{d}T}{\mathrm{d}y} = -B\,(T, y) \tag{9.91}$$

where

$$A\,(T, y) = \frac{1}{\varkappa_R}\,\frac{\mathrm{d}\varkappa_R}{\mathrm{d}y} + (1 + a)\,f\,\frac{\mathrm{d}h_a}{\mathrm{d}T}$$

$$B\,(T, y) = \frac{\rho\,(1 + a)^3\,f^2}{\varkappa_R}\,\frac{\mathrm{d}f}{\mathrm{d}y}$$

Equation (9.91) is a quasi-linear differential equation which can be solved by the method of successive approximation. We may assume a function $T = T_0\,(y)$ as a first approximation. Substituting $T_0\,(y)$ in the expressions $A\,(T, y) = A\,(y)$ and $B\,(T, y) = B\,(y)$, equation (9.91) becomes a linear differential equation with variable coefficient, i.e.,

$$\frac{\mathrm{d}^2 T}{\mathrm{d}y^2} + A\,(y)\,\frac{\mathrm{d}T}{\mathrm{d}y} = -B\,(y) \tag{9.91 a}$$

Equation (9.91 a) is the well-known Pohlhausen's equation. The boundary conditions are:

$$y = 0\;:\; T = T_w \quad \left(\text{or}\;\frac{\mathrm{d}T}{\mathrm{d}y} = 0\;\text{for insulated case}\right) \tag{9.92}$$

$$y = \delta_0\;:\; T = T_s = \text{temperature immediately behind the shock wave}$$

The solution of equation (9.91 a) may be written as

$$T\,(y) - T_s = (T_w - T_e)\,T_1\,(y) + T_2\,(y) \tag{9.93}$$

where T_1 represents the complimentary solution of the homogeneous equation and T_2, the particular solution of the inhomogeneous equation. T_e is a constant to be determined later and is the temperature of the plate if it is insulated. We have then

$$T_1'' + A\,(y)\,T_1' = 0 \tag{9.94}$$

where prime refers to the differentiation with respect to y. The boundary conditions for T_1 are: $y = 0$, $T_1 = 1$ and $y = \delta_0$, $T_1 = 0$.

$$T_2'' + A\,(y)\,T_2' = -B\,(y) \tag{9.95}$$

with the boundary conditions: $y = 0$, $T_2' = 0$ and $y = \delta_0$, $T_2 = 0$.

Since both equations (9.94) and (9.95) are first order ordinary differential equation for T', they can always be integrated. But since $A\,(y)$ and $B\,(y)$ are complicated functions of y, it is advisable to ingrate these equations numerically. After the first approximation of $T\,(y)$ has been obtained, we may recalcute $A\,(y)$ and $B\,(y)$ and then the second approximation of $T\,(y)$. The process may be repeated until the desired accuracy is obtained.

The numerical solution has not been carried out for any particular case. In reference *30*, YOSHIKAWA and CHAPMAN made a further approximation by neglecting the kinetic energy terms, i.e., $B(y) = 0$. Their solution is given in Fig. 8.15 for the case $\tau_w = \infty$. It shows that the temperature drops to a constant value from T_s. In the major portion of the inviscid shock layer, the temperature would be constant. This limiting temperature T_L may be different from the wall temperature T_w. The temperature T_L will drop to the value of T_w through the boundary layer region near the stagnation point. Hence T_L may be considered as the freestream temperature in the boundary layer analysis.

(b) Optically thin case.

For this case, the optical thickness is very small and since the shock wave may be consider as a transparent surface, the radiative heat transfer term becomes

$$\frac{d\,q_{Ry}}{d\,y} = \frac{2}{L_R}\left(-2\,\sigma\,T^4 + e_w\,T_w^4\right) \tag{9.96}$$

where $L_R = 1/\rho\,K_p$ is the Planck mean free path of radiation. If the wall temperature is small, we have simply

$$\frac{d\,q_{Ry}}{d\,y} = -\frac{4\,\sigma\,T^4}{L_R} = -E(\rho, T) \quad \text{(say)} \tag{9.96 a}$$

Because the Planck mean free path of radiation L_R is a function of both density and temperature, the function E is also a function of density and temperature. For a first approximation, a power law may be used, i.e.,

$$\frac{E}{E_0} = \left(\frac{T}{T_0}\right)^{m_1}\left(\frac{\rho}{\rho_0}\right)^{m_2} \tag{9.97}$$

For air at the temperature of the order of 10,000 °K, $m_1 = 8.4$ and $m_2 = 1.0$. Another formula for E is

$$E = \alpha \exp(\beta\,h_a) \tag{9.98}$$

where α and β are functions of density only. For constant density flow, they are constant. The energy equation (9.90) becomes

$$\frac{d\,h_a}{d\,y} = \frac{E(h_a)}{(1+a)\rho f} - (1+a)^2\,ff' \tag{9.99}$$

The first order total differential equation (9.99) can be easily integrated. There is a singularity at $y = 0$ and $f = 0$. Such a singularity is simply due to the use of the simple expression (9.96a) by neglecting the wall temperature term. Since there is a boundary layer near the wall, the wall temperature T_w is not equal to the true wall temperature but the temperature at the outer edge of the boundary layer T_L which should be determined from the integration of the energy equation with the radiation terms given by equation (9.96).

In equation (9.99), if we neglect the kinetic energy term, i.e., ff' term, there will be a logarithmic singularity at $y = 0$, which has been shown in reference *11*.

(c) Case of finite mean free path of radiation.

In this case, we should use the integral expression for the radiative heat transfer term and the energy equation (9.90) becomes

$$(1 + a)\, \rho\, f\, \frac{\mathrm{d}h_a}{\mathrm{d}y} = \frac{2\,\sigma}{L_R} \left\{ 2\, T^4 - \int_0^{\tau w} T^4\,(t)\, \varepsilon_1\, (|\tau - t|)\, \mathrm{d}\,t - e_w\, T_w^4\, \varepsilon_2\,(\tau) - \right.$$
$$\left. - 2\, r_w\, \varepsilon_2\,(\tau) \int_0^{\tau w} T^4\,(t)\, \varepsilon_2\,(t)\, \mathrm{d}\,t \right\} - \rho\,(1 + a)^3\, f^2\, f' \tag{9.100}$$

where e_w and r_w are the emissivity and the reflectivity of the surface of the body respectively.

In solving equation (9.100), it should be noticed that the singularity at the stagnation point should be avoided because we do not expect that $\mathrm{d}\, h_a/\mathrm{d}\, y = \infty$ from physical point of view. Hence if we write

$$R\,(\tau) = 2\, T^4 - \int_0^{\tau w} T^4\,(t)\, \varepsilon_1\, (|\, \tau - t|)\, \mathrm{d}\,t - e_w\, T_w^4\, \varepsilon_2\,(\tau) - $$
$$- 2\, r_w\, \varepsilon_2\,(\tau) \int_0^{\tau w} T^4\,(t)\, \varepsilon_2\,(t)\, \mathrm{d}\,t \tag{9.101}$$

where τ_w is the optical thickness based on the detached shock distance. We have

$$R\,(0) = 0 \tag{9.102}$$

Equation (9.102) gives the wall temperature $T_w = T\,(0) = T_L$ for the inviscid shock layer which is the temperature at the outer edge of the boundary layer. At $y = 0$, we have

$$\frac{R\,(0)}{f} = \frac{R'\,(0)}{f'\,(0)} \tag{9.103}$$

When the kinetic energy is negligible, the result is given in Fig. 8.15 for the case of a black wall for various τ_w.

(ii) Boundary layer near a stagnation point.

We shall consider here for the case of negligibly small radiation pressure number. In the boundary layer flow, we should consider the diffusion, chemical reaction as well as thermal radiation. The fundamental equations for an axi-symmetrical flow over a body of radius $r_0\,(x)$ large in comparison with the boundary layer thickness are:

(a) Equation of continuity

$$\frac{\partial}{\partial x}\,(\rho\, u r_0) + \frac{\partial}{\partial y}\,(\rho\, v r_0) = 0 \tag{9.104 a}$$

(b) Equation of motion in the x-direction

$$\rho\, u \frac{\partial u}{\partial x} + \rho\, v \frac{\partial u}{\partial y} = -\frac{\partial p}{\partial x} + \frac{\partial}{\partial y}\left(\mu \frac{\partial u}{\partial y} \right) \tag{9.104 b}$$

(c) The diffusion equation or the conservation of mass of ith species in the mixture of gases, i.e., air:

$$\rho u \frac{\partial c_i}{\partial x} + \rho v \frac{\partial c_i}{\partial y} = \frac{\partial}{\partial y}\left[\rho \sum_j D_{ij} \frac{\partial c_j}{\partial y} + \frac{D_{iT}}{T}\frac{\partial T}{\partial y}\right] + \dot{W}_i \qquad (9.104\,\text{c})$$

where c_i is the mass fraction of the ith species, D_{ij} is the diffusion coefficient between ith and jth species, D_{iT} is the thermal diffusion coefficient of the ith species and W_i is the source term of the ith species.

(d) The energy equation of the mixture

$$\rho C_p \left(u\frac{\partial T}{\partial x} + v\frac{\partial T}{\partial y}\right) = u\frac{\partial p}{\partial x} + \mu\left(\frac{\partial u}{\partial y}\right)^2 + \frac{\partial}{\partial y}\left(\varkappa\frac{\partial T}{\partial y}\right) + \frac{\partial q_{Ry}}{\partial y} -$$

$$(9.104\,\text{d})$$

$$- \sum_i \dot{W}_i h_i - \sum_i C_{pi} m_i \frac{\partial T}{\partial y}\left\{\sum_{i \neq j} \frac{m_j}{m^2}\rho D_{ij}\frac{\partial X_j}{\partial y} - \frac{D_{iT}}{T}\frac{\partial T}{\partial y}\right\}$$

where C_{pi} is the specific heat at a constant pressure for ith species and $C_p = \sum_i c_i C_{pi}$ is the frozen specific heat of the gas mixture at constant pressure, m_i is the molecular weight of ith species and m is the mean molecular weight of the gas mixture, X_i is the mole fraction of ith species, h_i is the static enthalpy of ith species, including chemical enthalpy and q_{Ry} is the y-wise radiation heat flux.

The boundary conditions for equations (9.104) are

$$y = 0 : v = v_w, \quad u = 0,\ T = T_w,\quad c_i = c_{iw}$$
$$y = \infty : u = u_e, \quad T = T_e,\quad c_i = c_{ie}$$
$$(9.105)$$

where v_w is the normal velocity at the wall. If we inject some foreign species into the gas mixture, $v_w \neq 0$. Otherwise, v_w is usually taken as zero. The mass fraction c_{iw} at the wall depends on the condition of the surface. At the outer edge of the boundary layer $y = \infty$, we assume that the flow is in equilibrium condition, the mass fraction c_i would be its equilibrium value c_{ie}. The temperature should be determined by the inviscid shock layer solution, i.e., $T_e = T_L$. Since we consider only the solution near the stagnation point, the x-wise velocity component at the outer edge of the boundary layer is $u_e = b x$ where b is a constant.

For stagnation point flow, if we use the transformation:

$$\eta = \frac{\rho_e u_e}{\sqrt{2\xi}}\int_0^y r_0 \frac{\rho}{\rho_e}\,dy, \quad \xi = \int_0^x \rho_w \mu_w u_e r_0^2\,dx \qquad (9.106)$$

In terms of ξ and η, equation (9.104) becomes a system of three non-linear ordinary differential equations except the expression of q_{Ry}, i.e.,

$$(C f'')' + f f'' + \tfrac{1}{2}\left(\frac{\rho_e}{\rho} - f'^2\right) = 0 \qquad (9.107\,\text{a})$$

$$\frac{C}{P_r}\left(L_e c_i' + L_{eT}\frac{T^{*\prime}}{T^*}\right)' + f c_i' + \frac{\dot{W}_i}{2\rho b} = 0 \qquad (9.107\,\text{b})$$

$$\left[\frac{C_p C}{P_r} T^{*\prime}\right]' + \frac{1}{2\rho\, T_e b}\left(\frac{\mathrm{d}q_{Ry}}{\mathrm{d}y}\right) + C_p f\, T^{*\prime} - \frac{\sum_i \dot{W}_i h_i}{2\rho\, T_e b} +$$

$$+ \frac{C}{P_r}\sum_i\left[C_{Pi}\left(L_e\, c_i{}' + L_{eT}\,\frac{T^{*\prime}}{T^*}\right)\right]T^{*\prime} = 0 \qquad (9.107c)$$

where prime refers to the differentiation with respect to η and

$$C = \frac{\rho\mu}{\rho_e\,\mu_e};\quad P_r = \frac{C_p\mu}{\varkappa};\quad T^* = \frac{T}{T_e},\quad L_e = \frac{C_p D_{ij}}{\varkappa},\quad L_{eT} = \frac{C_p D_{iT}}{\varkappa}$$

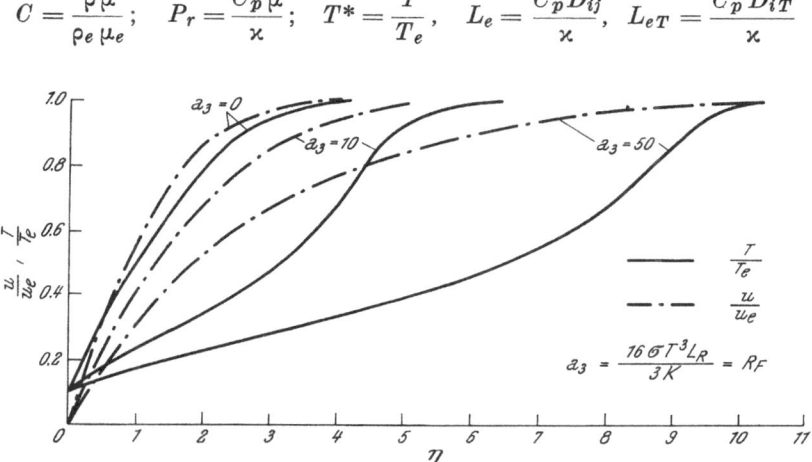

Fig. 9.14. Velocity and temperature distributions in the boundary layer near the stagnation point of a blunt body with and without thermal radiation effect. Optically thick cases. (Fig. 13 of reference 22 by S. M. SCALA and D. H. SAMPSON, Courtesy of General Electric Company)

We have to integrate equations (9.107) numerically. Fig. 9.14 shows typical velocity and temperature profiles for optically thick cases in which the Rosseland expression for radiative heat transfer is used at various modified radiation flux number [cf. eq. (7.42)]

$$a_3 = \left(\frac{16\,T^3}{3\varkappa\,K_{R}\rho}\right)_{\max} = R_{F1} \qquad (9.108\,a)$$

Fig. 9.15 shows some typical velocity and temperature profiles for optically thin case with expression (9.90a) for radiative flux at various radiation flux number [cf. eq. (7.43)]

$$a_4 = \frac{\sqrt{2}\,(K_p\,\rho)\,T_e\,\sigma\,T_e{}^{2.5}\,R_b}{\rho_e\,\sqrt{\overline{R_e}}} = R_{F2} \qquad (9.108\,b)$$

where R_b is the nose radius of the body and $\overline{R_e}$ is the gas constant of the air at $y = \infty$.

In reference 6, Howe studied the effects of the injection of a radiation absorption gas on the incident radiation field, on enthalpy profiles, and on the heat transfer to the vehicle surface. It was assumed that the energy emitted from the cold foreign gas is negligible compared with that absorbed. The fundamental equations (9.107) have been integrated for two different injection rate, i.e., v_w. It was found that the reduction of heat transfer rate is accompanied by an increase in the convective heat transfer rate. For a black surface, the total heat transfer is reduced by injection of an absorbing gas while for a totally reflecting surface, the total heat transfer is increased.

(iii) Viscous shock layer.

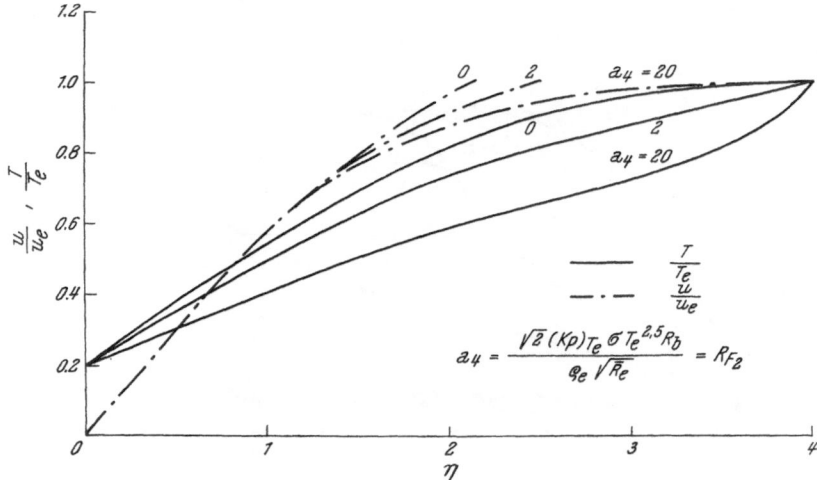

Fig. 9.15. Velocity and temperature distributions in the boundary layer near the stagnation point of a blunt body with and without thermal radiation effect. Optically thin case. (Fig. 17 of reference 22 by S. M. SCALA and D. H. SAMPSON, Courtesy of General Electric Company)

The present case is a combination of the inviscid shock layer and the boundary layer. We shall use the coordinate systems given in Fig. 9.13. We should simplify the Navier-Stokes equations by the approximation of small shock detached distance, i.e., $\delta_0/R_b \ll 1$. If we assume that x is also of the order of δ_0, i.e., near the stagnation point region, the fundamental equations of the viscous shock layer are as follows:

$$\frac{\partial \rho u r^a}{\partial x} + \frac{\partial H \rho v r^a}{\partial y} = 0 \qquad (9.109\,\text{a})$$

$$\rho u \frac{\partial u}{\partial x} + H \rho v \frac{\partial u}{\partial y} = -\frac{\partial p}{\partial x} + H \frac{\partial}{\partial y}\left(\mu \frac{\partial u}{\partial y}\right) \qquad (9.109\,\text{b})$$

$$\rho u \frac{\partial v}{\partial x} + H \rho v \frac{\partial v}{\partial y} - \frac{\rho u^2}{R} = -H \frac{\partial p}{\partial y}$$

$$\rho u \frac{\partial h_0}{\partial x} + H \rho v \frac{\partial h_0}{\partial y} = H \frac{\partial}{\partial y}\left(\varkappa \frac{\partial T}{\partial y} + \sum_i D_i h_i \frac{\partial c_i}{\partial y}\right) + \qquad (9.109\,\text{c})$$

$$+ H \rho K_p \sigma\left[\int_0^{\tau w} 2 T^4(t)\,\varepsilon_1(|t-\tau|)\,\mathrm{d}t - 4 T^4 + 2 T_w^4 \varepsilon_2(\tau)\right] \qquad (9.109\text{d})$$

where $r(x, y) = Hx$, $H = 1 + (y/R)$, $h_0 = h_a + \frac{1}{2}(u^2 + v^2)$,

$$h_a = \sum_i c_i h_i, \quad h_i(T) = \int_0^T C_{pi} \, dT + h_i^0, \quad \tau = \int_0^y \rho K_p \, dy$$

h_i^0 is the energy release due to chemical reaction.

The boundary conditions for equations (9.109) are:

$$y = 0: \quad u = 0, \quad v = v_w, \quad h_a = h_w$$

$$y = \delta: \quad u = u_s = U_\infty x/R, \quad v = v_s = -k_\rho U_\infty \left(1 - \frac{1}{2}\frac{x^2}{R^2}\right) \qquad (9.110)$$

$$p = p_s = \rho_\infty U_\infty^2 (1 - k_\rho)\left(1 - \frac{x^2}{R^2}\right), \quad h_0 = h_{0s} = \frac{1}{2} U_\infty^2$$

$$k_\rho = \rho_\infty/\rho_s.$$

In general, the y-wise velocity component v should be of the same order of magnitude as the x-wise velocity component u. But near the stagnation point and in the hypersonic flow, the y-wise velocity component v is still smaller than the x-wise velocity component u and for a first approximation, we may use the boundary layer approximation $\partial p/\partial y = 0$ to replace equation (9.109c). This approximation has been used in reference 7 from which the following numerical example is taken. For the radiation terms, we assume that the body has a black surface and the shock is a transparent surface.

In reference 7, the effect of injection of a foreign gas from the body surface has also been studied. The mass fraction of the foreign gas c_f is governed by the diffusion equation:

$$\rho u \frac{\partial c_f}{\partial x} + H \rho v \frac{\partial c_f}{\partial y} = H \frac{\partial}{\partial y}\left(\rho D_f \frac{\partial c_f}{\partial y}\right) \qquad (9.111)$$

where D_f is the diffusion coefficient of the foreign gas in air. The boundary conditions for the mass fraction of the foreign gas c_f are:

$$y = 0: \quad \left(\frac{\partial c_f}{\partial y}\right)_w = -\frac{v_w}{D_{fw}}(1 - c_{fw})$$

$$y = \delta: \quad c_f = 0 \qquad (9.112)$$

In general, the foreign gas affects the thermodynamic, transport and the radiative properties of the mixture. Since the amount of the foreign gas is small and its effects on the thermodynamic properties and transport properties are uncertain, we follow reference 7 by neglecting its effects on the thermodynamic properties and transport properties of the mixture and consider its effect on the radiative property only. The influence of the foreign gas on the Planck mean absorption coefficient may be written as follows:

$$K_p = (1 - c_f) K_{p\,(air)} + c_f K_{pf} = K_{p\,(air)}[1 + c_f(\alpha - 1)] \qquad (9.113)$$

The Planck mean absorption coefficient of air is a known function of temperature and pressure [cf. eqs. (7.49) and (7.51) and also Fig. 8.14]. The value of α

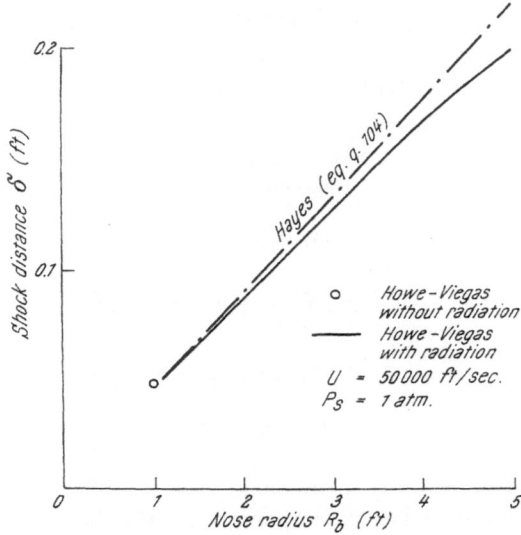

Fig. 9.16. Shock distance *vs* nose radius.
(Fig. 17 of reference 7 by J. T. Howe and J. R. Viegas, Courtesy of NASA)

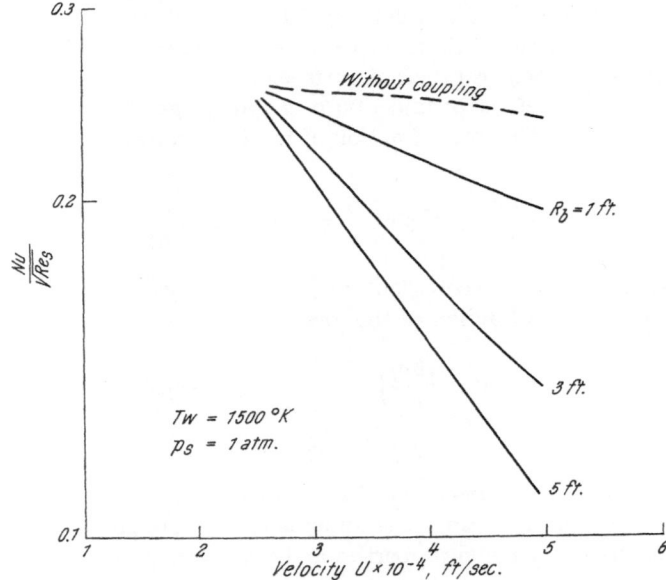

Fig. 9.17. Effect of nose radius on the heat transfer for optically thin case.
(Fig. 11 of reference 7 by J. T. HOWE and J. R. VIEGAS Courtesy of NASA)

determine the influence of the foreign gas. In reference 7, HOWE an VIEGAS solved equations (9.109) and (9.111) for various cases by numerical integration in high speed computing mechine IBM 7090. Some of their main results are given below:

(i) Shock detached distance or standoff distance. Hayes' approximate formula for the shock detached distance is

$$\delta = \frac{R_b \, k_\rho}{1 + \sqrt{2} \, k_\rho} \tag{9.114}$$

Fig. 9.16 shows the comparison of the shock detached distance with and without radiation effect.

(ii) Heat transfer. The total heat transfer rate q_T consists of the convective heat transfer q_c and the radiative heat transfer q_R, i.e.,

$$q_T = q_c + q_R = -\left(\varkappa \frac{\partial T}{\partial y}\right)_w + q_{Rw} \tag{9.115}$$

where q_{Rw} is given by the integral (6.41) with $\tau = 0$. The convective heat transfer rate may be expressed in terms of Nusselt number N_u:

$$N_u = \frac{q_c \, x \, C_{pw}}{\varkappa_w \, (h_{0s} - h_w)} \tag{9.116}$$

and the Reynolds number $R_{es} = \rho_w \, u_s \, x / \mu_w$. The convective heat transfer rate is reduced greatly by the radiation effect. Since the radiative heat transfer for an optically thin case is a strong function of the nose radius, the larger the nose radius, the larger the effect of radiation. The effect of the nose radius on the radiative heat transfer is shown in Fig. 9.17. The simple analysis of radiative heat transfer by assuming the shock to be an isoenergetic transparent gas layer (cf. Figs. 1.1 and 1.2 and also reference 12) gives a much higher value of radiative heat transfer rate than the more accurate analysis in which the coupling between the radiation and other gasdynamic variable is considered.

The effect of injection of a gas from the wall is to decrease the convective heat transfer and to increase the radiative heat transfer. But the rate of decrease of convective heat transfer is usually much larger than the increase of the radiative heat transfer. The effect of injection of foreign gas with strong absorption of radiation will have larger effect than that of air. One example given in reference 7 shows that the injection of a foreign gas that is 50 times as strong an emitter of radiant energy as air increases the radiant heat transfer by only 14%, while decreases the convective heat transfer by about 45% compared with those without injection. The total heat transfer rate was decreased by about 14%.

In conclusion, the coupling of radiation and convective heat transfer would decrease the heat transfer rate as compared with those without the coupling effect.

9. Miscellaneous problems of heat transfer in radiation gasdynamics. Since radiation gasdynamics includes all problems of gasdynamics, we may re-examine all the heat transfer problems of ordinary gasdynamics with radiation effect. In this book we shall not make such complete investigations. However in this section we consider two cases which are different from what we have studied so far. One is the case where the scattering is not negligible and the other is the unsteady flow problem.

(i) Scattering medium. So far we neglect the scattering of radiation in our analysis of flow problems and assume that both the emission and the absorption are respectivety the true emission and absorption of the medium. As a result, we use equation (5.10) for the radiative transfer. If we consider the scattering effects as well as the true emission and absorption, we should use equation (3.38) in which the emission coefficient and the absorption coefficient are given respectively by equations (3.29) and (3.5). Now the equation of radiative transfer is an integro-differential equation itself. Little has been studied for flow problems with the scattering effect. In reference 28, some discussions of the thermal radiation in an absorption and scattering medium are given. Only the gray gas has been studied because little has been known about the variation of the scattering coefficient with frequency. In this study, Viskanta and Grosh found that the only effect of isotropic scattering was to increase the optical thickness of the medium and therefore decrease the radiant heat transfer between the plates. Since our knowledge of the radiative heat transfer without scattering is still meager and we know little about the scattering coefficient, it seems to the author we should postpone the study of the flow problems with scattering effects until more knowledge of the thermal radiation effects without scattering and the information of the scattering coefficient are available.

(ii) Unsteady flow problem.

Except the wave motion of small amplitude, very little has been done on the unsteady flow with thermal radiation effects. In the wave motion of small amplitude, we usually assume that the radiative transfer corresponds to the local thermodynamic equilibrium condition. In the transient problems, the local thermodynamic equilibrium may not be reached. We shall discuss briefly the non-local thermodynamic equilibrium condition in chapter XI. Even with the assumption of local thermodynamic equilibrium, the transient effects on the radiative heat transfer may be of importance, particularly in the re-entry problem. We should study these effects in the near future.

References

1. EINSTEIN, T. H.: Radiant heat transfer to absorbing gases enclosed between parallel flat plates with flow and conduction. NASA TR R-154, 1963.

2. GEORGIEV, S., J. D. TEARS, and R. A. ALLEN: Hypervelocity radiative heat transfer. AVCO Everett Research Lab. Research note 264, August 1961.

3. GODSE, C. L., T. BERGERON, J. BJERKNES, R. C. BUNDGAARD: Dynamic Meteorology and Weather Forecasting. American Meteorology Society and Carnegie Institution of Washington, 1957.

4. GOULARD, R.: Preliminary Estimate of radiative transfer effects on detached shock layers. AIAA Jour. vol. 2, No. 3, pp. 494–502, March 1964.

5. HAYES, W. D., and R. F. PROBSTEIN: Hypersonic Flow Theory. Academic Press, New York, 1959.

6. HOWE, J. T.: Shielding of partially reflecting stagnation surfaces against radiation by transpiration of an absorbing gas. NASA TR R-95, 1961.

7. HOWE, J. T., and J. R. VIEGAS: Solutions of the ionized radiating shock layer including reabsorption and foreign species effects and stagnation region heat transfer. NASA TR R-159, 1963.

8. JAKOB, M.: Heat Transfer. Vols. I and II. John Wiley & Sons Inc., 1949 and 1957.

9. JOHNSON, J. C.: Physical Meteorology. John Wiley & Sons Inc., New York, 1954.

10. KENNET, H.: Radiation-convection interaction around a sphere in hypersonic flow. ARS Jour. vol. 32, p. 1616, 1962.

11. KENNET, H., and S. L. STRACK: Stagnation point radiative transfer. ARS Jour. vol. 31, pp. 370–373, 1961.

12. KIVEL, B.: Radiation from hot air and its effect on stagnation point heating. Jour. Aero. Sci. vol. 28, No. 2, pp. 96–102, Feb. 1961.

13. KOH, J. C. Y., and C. N. DeSILVA: Interaction between radiation and convection in the hypersonic boundary layer on a flat plate. ARS reprint 2205-61, Oct. 1961.

14. LI, T. Y., and R. E. GEIGER: Stagnation point in a blunt body in hypersonic flow. Jour. Aero. Sci. vol. 24, No. 1, pp. 25–32, Jan. 1957.

15. MÜLLER, H. G.: Energiegleichgewicht und Winde in der hohen Atmosphäre. Paper No. 64-549, ICAS 4th Congress, Aug. 1964, also report No. 6, Inst. f. Phys. der Atm. DVL, München, Germany, 1964.

16. NEMCHINOV, I. V.: Some non-stationary problems of radiative heat transfer. Report A & ES TT-4 School of Aero. & Eng. Sci. Purdue Univ., 1964.

17. OPPENHEIM, A. K.: The engineering radiation problem — an example of the interaction between engineering and mathematics. ZAMM Bd. 36, Heft 3/4, pp. 81–93, 1956.

18. PAI, S. I.: Laminar Jet mixing in radiation gasdynamics. Phys. Fluids, vol. 6, No. 10, pp. 1440–1445, Oct. 1963.

19. PAI, S. I.: Plane Couette Flow in Radiation Gasdynamics. Proc. 6th Intern. Sympo. in Ionization Phen. of Gases, Paris, France, pp. 431–436, July 1963.

20. PAI, S. I.: Viscous Flow Theory. I-Laminar Flow. D. Van Nostarnd, Inc., N. J., 1956.

21. PAI, S. I., and E. T. KORNOWSKI: Stagnation point flow of magnetized blunt body in hypersonic flow. Engineering Magnetohydrodynamics, Columbia Univ. Press, pp. 97–106, 1962.

22. RATCLIFFE, J. A. (ed.): Physics of the Upper Atmosphere. Academic Press, New York, 1962.

23. SCALA, S. M., and D. H. SAMPSON: Heat Transfer in Hypersonic flow with radiation and chemical reaction. Tech. Inf. Ser. R 63 SD 46, Space Sci. Lab. General Electric Co., Phil. Pa., March 1963. Also Supersonic Flow, Chemical Processes and Radiative Transfer, Pergamon Press, pp. 319–354, 1964

24. SFORZA, P. M.: Radiating laminar boundary layer of a grey gas over a flat plate. PIBAL report No. 812, Poly. Tech. Inst. of Brooklyn, 1963.

25. TELLEP, B. M., and D. K. Edwards: Radiant energy transfer in gaseous flows. Tech. report LMSD-288139, vol. I, Part I, No. 2, Lockheed Missiles & Space Div., 1960.

26. VISKANTA, R., and R. J. GROSH: Heat transfer by simultaneous conduction and radiation in an absorbing medium. Jour. Heat Transf. vol. 84, Ser. C. No. 1, pp. 63–72, Feb. 1962.

27. VISKANTA, R., and R. J. GROSH: Boundary layer in thermal radiation absorbing and emitting media. Intern. Jour. Heat & Mass Transf. vol. 5, pp. 795–806, Sept. 1962.

28. VISKANTA, R., and R. J. GROSH: Heat transfer in a thermal radiation absorbing and scattering medium. Purdue Univ. Report, 1960.

29. VISKANTA, R.: Effect of transverse magnetic field on heat transfer to an electrically conducting and thermal radiating fluid flowing in a parallel plate channel. ZAMP, vol. 14, No. 4, pp. 353–368, 1963.

30. YOSHIKAWA, K. K., and D. R. CHAPMAN: Radiative heat transfer and absorption behind a hypersonic normal shock wave. NASA TN D-1424, 1962.

31. YOSHIKAWA, K. K., and B. H. WICK: Radiative heat transfer during atmosphere entry at parabolic velocity. NASA TN D-1074.

Chapter X

Kinetic Theory of Radiating Gases

1. Introduction. In our discussion of radiation gasdynamics in the previous chapters, we assume implicitly that the mean free path of the gas particles is small so that the gas may be considered as a continuum in the analysis of flow problems. In many flow problems, especially those connected with space sciences, the mean free paths of the gas particles are not small and we have to consider the discrete properties of the gas. We have to use the kinetic theory of gases to study the flow problems of rarefied gases. In the kinetic theory of radiating gas, we should use the corpuscular picture to represent radiation instead of the wave picture as we discussed in chapter II. In other words, we should consider radiation as a stream of photons. The gas should be considered as a mixture of various types of particles, material particles as well as photons. However, it is usually not possible to investigate the detailed motion of all the particles in a gas flow because of the large number of particles considered. We have to use the statistical average of the motions of the particles of a gas. In such a kinetic theory, we may use a molecular distribution function for each species of the mixture of the gas. In the most successful kinetic theory of gases, one particle distribution function (*2, 3, 5, 8*) is used to describe the microscopic behavior of the system. We shall review briefly some of the essential features of the one particle molecular distribution function in section 2.

For ordinary material particles of a gas, the relativistic effects are negligible because the speed of the gas particles is negligibly small in comparison with the speed of light. However, the relativistic effect is no longer negligible for photons which move at the speed of light. Hence in the kinetic theory of photons, the relativistic fluid mechanics should be used. We shall review some essential points of relativistic fluid mechanics in section 3. The basic equation for molecular distribution function is known as Boltzmann equation. We shall discuss the ordinary Boltzmann equation in classical mechanics in section 4 which may be used for the material particles and the relativistic Boltzmann equation in section 5 which should be used for photons. We shall show that this relativistic Boltzmann equation will give the radiative transfer equation (3.31) as it should so.

Because of many physical and mathematical difficulties it is not possible at present time to study any practical problems by the Boltzmann equation even if the radiation effect is negligible. However, the Boltzmann equation serves two important aspects in the study of gasdynamics: One is that the basic equations of gasdynamics such as those derived in chapter V can be derived from the Boltzmann equation. Thus we may have some guides about the validity of the fundamental equations of gasdynamics in a macroscopic description from the analysis of Boltzmann equations. We shall discuss the relations between the fundamental equations of gasdynamics including the radiation effect with the Boltzmann equation in section 6. Another important fact is that the Boltzmann equation may give us valuable information on the transport coefficents such as the coefficients of viscosity, heat conductivity etc. The questions what would be the effect of radiation on these transport coefficients will be discussed in sections 7 and 8, particularly for the case of local thermodynamic equilibrium condition. A general discussion of rarefied radiation gasdynamics is given in section 9. The simplest type of rarefied gas flow is the free molecule flow. In section 10, we shall discuss the free molecule flow including the thermal radiation effects.

2. Molecular velocity and molecular distribution functions. A gas is composed of a large number of particles which are generally called molecules. The term of molecule in a general sense includes ordinary molecules, atoms, ions, electrons as well as photons. Thus the molecules of a gas may be all of a kind or of several kinds. The number of molecules of a gas is usually much larger than the number of kinds of molecules in a gas. For instance, at a temperature of 0 °C and one atmospheric pressure, there are about 2.69×10^{19} material molecules per cubic centimeter of a gas while ordinarily there are at most a few different kinds of molecules. For each kind of molecules, we may use a molecular distribution function F to describe its motion. In this section, we consider only the molecular distribution function for material particles so that the relativistic effect is negligible. Then the mass of each kind of molecules has a given value. The molecular distribution function $F_s (\vec{r}, \vec{p_s}, t)$ of the sth species of the material particles in a gas represents the expectation of the number of molecules of the sth species per unit volume at the position $\vec{r}$ and time t within the molecular momentum range $\vec{p_s}$ and $\vec{p_s} + \mathrm{d}\,\vec{p_s}$, i.e.,

$$\mathrm{d}\,n_s = F_s \left(\vec{r}, \quad \vec{p_s}, \quad t\right) \mathrm{d}\,\vec{p_s} \tag{10.1}$$

The average number of molecules of the sth species per unit volume at $\vec{r}$ and t is

$$n_s = \int F_s (\vec{r}, \quad \vec{p_s}, \quad t) \,\mathrm{d}\,\vec{p_s} \tag{10.2}$$

If the relativistic effect is negligible, the mass of a particle of the sth species is a constant, i.e., m_s and the momentum vector $\vec{p_s} = m_s \vec{q_s}$ may be replaced by the molecular velocity vector $\vec{q_s}$ in equations (10.1) and (10.2). However, we shall consider the cases with and without relativistic effect, hence we will use the momentum vector $\vec{p_s}$ in the molecular distribution function instead of the molecular velocity vector $\vec{q_s}$ as often used in the textbook of kinetic theory of gases (3). The $\mathrm{d}\,\vec{p_s}$ in the integral (10.2) means $\mathrm{d}\,p_{s1}\,\mathrm{d}\,p_{s2}\,\mathrm{d}\,p_{s3}$ where p_{si} is the ith component of the vector $\vec{p_s}$.

The average of any physical quantity Q of the molecules of the sth species is defined by the integral:

$$\overline{Q} = \frac{1}{n_s} \int Q F_s \mathrm{d}\,\vec{p_s} \tag{10.3}$$

For instance, the x-wise average flow velocity component of the sth species is

$$u_s = \frac{1}{n_s m_s} \int \int \int p_{s1} \, F_s \, \mathrm{d} p_{s1} \, \mathrm{d} p_{s2} \, \mathrm{d} p_{s3} \qquad (10.4)$$

where we take the axis 1 as the x-axis, i.e., $p_{s1} = p_{sx}$.

We may define other macroscopic quantities such as pressure, temperature etc. of the gas by similar expressions as we shall discuss them in section 6. From these definitions we may derive the fundamental equations of the macroscopic treatment.

3. Relativistic mechanics. When the velocity of fluid particles of which photons are considered as a special kind is not small in comparison with the velocity of light, the relativistic effect should be considered in the analysis of the fluid motion. We are going to discuss briefly the relativistic analysis of the equations of fluid motion. Of course these equations will reduce to ordinary gasdynamical equations when the velocity of the fluid q is much smaller than the velocity of light c, i.e.,

$$R_c = \frac{q^2}{c^2} \ll 1 \qquad (10.5)$$

Here we define the relativistic parameter R_c as the square of R_r defined in equation (7.17).

We shall derive the equations of relativistic fluid mechanics based on the special theory of relativity. The basic postulates of relativity are:

(i) It is impossible to measure or detect the unaccelerated translatory motion of a system through free space, and

(ii) The velocity of light in free space is the same for all observers independent of the relative velocity of the source of light and the observer.

Because of these two postulates, we have a relativistic conception of space and time. In classical mechanics, we consider only the three dimensional space. The location at a point in the space may be represented by a vector $\vec{r}$ which has three spatial components x_i, $i = 1$, 2 or 3. The time is considered as a scalar. However, in the theory of relativity, we have to use four dimensional space to describe physical quantities. The coordinates of a certain reference system x_α has four components, i.e., $\alpha = 1$, 2, 3 and 4. The first three components refer to the ordinary spatial coordinates and the fourth one $x_4 = ct$ where t is the time. Hence in the theory of relativity, we consider time as the fourth dimension. Let us consider two Cartesian coordinates (x, y, z, t) and (x', y', z', t') with a relative velocity V in the direction of the x-axis between the two systems. If at time $t = 0$, the origins of the two systems coincide, Lorentz found that the relations between these two systems are

$$x' = \frac{x - Vt}{(1 - V^2/c^2)^{\frac{1}{2}}} \; ; \quad y' = y \; ; \quad z' = z$$

$$t' = \frac{t - xV/c^2}{(1 - V^2/c^2)^{\frac{1}{2}}} \qquad (10.6)$$

Equation (10.6) is known as Lorentz transformation. From equation (10.6), we have the Lorentz invariant of the elementary length in the four dimensional space:

$$\mathrm{d}\,x^2 + \mathrm{d}\,y^2 + \mathrm{d}\,z^2 - c^2\,\mathrm{d}\,t^2 = \mathrm{d}\,x'^2 + \mathrm{d}\,y'^2 + \mathrm{d}\,z'^2 - c^2\,\mathrm{d}\,t'^2 \qquad (10.7)$$

In the classical mechanics, the term V^2/c^2 is negligible, then Lorentz transformation (10.6) will reduce to Galilean transformation and the time will be unchanged.

In relativistic mechanics, we consider four dimensional space and all vectors have four components. We have to generalize all definitions of the vectors in the three dimensional space of classical mechanics to the four dimensional space in relativistic mechanics. We would like to know, especially, what is the meaning of the fourth component of a vector in terms of the quantity in classical mechanics. Before we discuss some important 4-vectors in relativistic mechanics, we would like to discuss a little about the properties of the 4-vector in general. From now on, we use the Greek letter α, β etc. to denote the components in the four dimensional space, i.e., $\alpha = 1, 2, 3$ or 4 and the English letter i, j etc, to denote the ordinary components in the three dimensional space, i.e., $i = 1, 2$ or 3.

The Lorentz transformation (10.6) is a special case of general linear transformation:

$$x_\alpha' = A_{\alpha\beta}\, x_\beta \qquad (10.8)$$

where the summation convention is used, i.e.,

$$x_\alpha' = A_{\alpha\beta}\, x_\beta = A_{\alpha 1}\, x_1 + A_{\alpha 2}\, x_2 + A_{\alpha 3}\, x_3 + A_{\alpha 4}\, x_4 \qquad (10.8\,\mathrm{a})$$

and x_β is the coordinate in the β-axis of a rectangular system. If the unit vector along the β-axis is i_β, the 4-vector of position $\vec{x}$ is then

$$\vec{x} = i_\beta\, x_\beta \qquad (10.9)$$

Since the coordinate system is assumed to be orthogonal, the scalar product of two unit vector gives

$$i_\alpha \cdot i_\beta = \delta_{\alpha\beta} \qquad (10.10)$$

where $\delta_{\alpha\beta} = 0$ where $\alpha \neq \beta$ and $\delta_{\alpha\beta} = 1$ when $\alpha = \beta$.

Equation (10.8) gives the relations of the coordinates of a given vector x in the four dimensional space in a system X' with component x_α' to those in a system X with component x_α. Now we can easily find the matrix $A_{\alpha\beta}$ from the definition of these components, i.e.,

$$x_\alpha' = i_\alpha' \cdot \vec{x} = i_\alpha' \cdot (i_\beta\, x_\beta) = A_{\alpha\beta}\, x_\beta \qquad (10.11)$$

where

$$A_{\alpha\beta} = i_\alpha' \cdot i_\beta \qquad (10.12)$$

Comparing equation (10.11) with the Lorentz transformation (10.6), we have

$$x_1 = x, \quad x_2 = y, \quad x_3 = z, \quad x_4 = i\,c\,t$$

and

$$A = \begin{matrix} \dfrac{1}{(1-V^2/c^2)^{\frac{1}{2}}} & 0 & 0 & \dfrac{iV/c}{(1-V^2/c^2)^{\frac{1}{2}}} \\[2mm] 0 & 1 & 0 & 0 \\[2mm] 0 & 0 & 1 & 0 \\[2mm] \dfrac{-iV/c}{(1-V^2/c^2)^{\frac{1}{2}}} & 0 & 0 & \dfrac{1}{(1-V^2/c^2)^{\frac{1}{2}}} \end{matrix} \qquad (10.12\,a)$$

where $i = \sqrt{-1}$. Since the magnitude of the vector x is invariant under the rotation of the coordinate system, we have immediately equation (10.7) in the following form:

$$- d\,s^2 = x_\alpha\, x_\alpha = x_\alpha'\, x_\alpha' \qquad (10.7\,a)$$

where $d\,s$ is the elementary length in the four dimensional space.

The relation (10.8) is applicable to any 4-vector under the rotation of the coordinate system, i.e.,

$$q_\alpha' = A_{\alpha\beta}\, q_\beta; \quad q_\beta = A_{\beta\alpha}\, q_\alpha' \qquad (10.8\,b)$$

Now we would like to know what are the components of a velocity vector in the four dimensional space. In the three dimensional space, the velocity component q_i is defined as $q_i = d\,x_i/d\,t$. Since $d\,t$ is not an invariant in the theory of relativity, we should define the velocity vector slightly different from that in the classical mechanics. The velocity vector is the four dimensional space is defined in terms of the invariant element $d\,s$ defined in equation (10.7a),

$$q_\alpha = d\,x_\alpha/d\,s \qquad (10.13)$$

From equations (10.7) and (10.7a) we have

$$ds = cdt\left(1 - \frac{d\,x_i\,d\,x_i}{c^2\,d\,t^2}\right)^{\frac{1}{2}} = cdt\,(1-q^2/c^2)^{\frac{1}{2}} \qquad (10.14)$$

where $q^2 = q_i\, q_i$. From equations (10.13) and (10.14), we see that the velocity vector in the four dimensional space is a non-dimensional quantity with components:

$$q_1 = \frac{u}{c\,b_v}, \quad q_2 = \frac{v}{c\,b_v}, \quad q_3 = \frac{w}{c\,b_v}, \quad q_4 = \frac{i}{b_v} \qquad (10.15)$$

where $b_v{}^2 = 1 - (u^2 + v^2 + w^2)/c^2$ and u, v and w are respectively the velocity components in the sense of classical mechanics. Similarly we define the acceleration of a particle in the four dimensional space as $d\,q_\alpha/d\,s$.

The conservation of number of fluid particles, the equation of continuity, may be expressed in terms of the particle flux n_α which has four components and which may be expressed in terms of the velocity vector q_α as follows:

$$n_\alpha = N q_\alpha \qquad (10.16)$$

where N is a scalar which is the number density of the particles. The equation of continuity is then

$$\frac{\partial}{\partial x_\alpha}(N\,q_\alpha) = 0 \qquad (10.17)$$

Equation (10.17) reduces to the ordinary equation of continuity of fluid mechanics (5.3) if the relativistic effect if neglected, i.e., $b_v = 1$ and $m = $ constant.

One of the main features of relativistic mechanics is that the mass of a particle is not a constant but increases with the velocity of the particle q according to the Einstein formula:

$$m = \frac{m_0}{(1 - q^2/c^2)^{\frac{1}{2}}} \qquad (10.18)$$

where m_0 is the rest mass of a particle when it is not moving and m is the mass of the particle when it is moving with a velocity $\vec{q}$. The velocity q is referred to the velocity in the three dimensional space, i.e., $q^2 = q_i\,q_i = u^2 + v^2 + w^2$. It should be noticed that the velocity vector in the four dimensional space $\vec{q}_\alpha$ is essentially a unit vector, i.e.,

$$q_\alpha{}^2 = q_\alpha\,q_\alpha = \frac{\mathrm{d}x_\alpha\,\mathrm{d}x_\alpha}{\mathrm{d}s^2} = -1 \qquad (10.19)$$

Newton's second law of motion in three dimensional space is

$$F_i = \mathrm{d}\,(m\,q_i)/\mathrm{d}t = \mathrm{d}\,p_i/\mathrm{d}t \qquad (10.20)$$

where F_i is the ith component of the force on the particle of a mass m and velocity q_i and momentum $p_i = m\,q_i$. In the classical mechanics, $R_c \ll 1$ and the mass m may be assumed to be constant. Hence we may write

$$F_i = m\,\frac{\mathrm{d}q_i}{\mathrm{d}t} \qquad (10.20\,\mathrm{a})$$

However in relativistic mechanics, the mass m is no longer a constant. Thus it is convenient to use the momentum vector p_i instead of the velocity vector q_i. In the four dimensional space, the momentum vector p_α is defined by the following expression:

$$p_\alpha = m_0\,c^2\,q_\alpha \qquad (10.21)$$

Now if we use $p_i = m q_i$ as the momentum in the three dimensional space of ordinary classical mechanics, the first three component of the 4-momentum vector p are simply $c p_i$ while the 4th component is $p_4 = i\,m\,c^2 = i\,E$ where E is the total energy of the particle at velocity q. Hence the 4-momentum vector is sometimes referred to as the momentum-energy vector. We shall use the 4-momentum vector in describing the motion of a particle relativistically such as in the definition of the molecular distribution function in section 5.

4. Boltzmann equation for material particles. The main purpose of the kinetic theory of gases is to find the distribution function which is subjected to an integro-differential equation known as Boltzmann equation. In this section, we consider only the molecular distribution function for material particles so that the classical Newton mechanics is applicable. In the next section, we shall consider the molecular distribution function for photons so that relativistic mechanics should be used. The molecular distribution function of the sth species of molecules in a gas mixture $F_s(\vec{r}, \vec{p}_s, t)$ satisfies the following Boltzmann equation:

$$\frac{\partial F_s}{\partial t} + \frac{p_s^i}{m_s}\frac{\partial F_s}{\partial x^i} + \phi_s^i \frac{\partial F_s}{\partial p_s^i} = \left(\frac{\delta F_s}{\delta t}\right)_c \qquad (10.22)$$

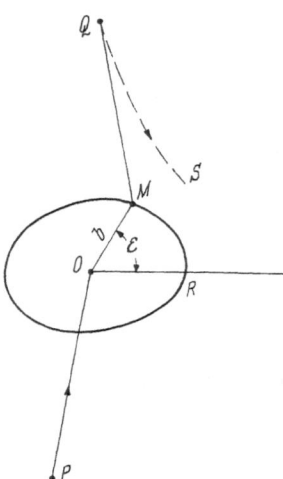

Fig. 10.1. Binary collision between gas particles

where m_s is the mass of a particle of the sth species; p_s^i is the ith component of the momentum of the sth species particles; x^i is the ith component of the space coordinate r, the location of a particle of sth species; ϕ_s^i is the ith component of the force on a particle of the sth species outside the collision. The summation convention is used. Hence the terms in the left-hand side of equation (10.22) denote the change of F_s from time t to time $t + dt$. The righthand side term represents the gain or loss of the number of particles per unit time due to collision. The exact form of the collision term $(\delta F_s/\delta t)_c$ depends on the kinetic model used in the kinetic theory. For ordinary gas, we may assume that only binary collision is important. In other words, we need only to consider the collision between two typical particles. In Fig. 10.1, we consider a particle P associated with the distribution function F and another particle Q associated with the distribution function F'. Without interaction, the particle P is moving in the direction of PO and the particle Q in the direction of QM. Because of the interaction the actual path of the particle Q is QS. If we draw a plane ROM through the point O and perpendicular to QM. The distance OM is known as the impact parameter b which is the distance of the closest approach of the two particles P and Q if there were no interaction. The angle between OM and OR is denoted by ϵ. If there is only one kind of particles in the gas, the binary collision term is given by the integral:

$$\left(\frac{\delta F}{\delta t}\right)_c = \iiint (\bar{F}\,\bar{F}' - F\,F')\,g_0\,b\,db\,d\epsilon\,d\vec{q}' \qquad (10.23)$$

where the bar refers to the distribution function after collision and g_0 is the absolute initial relative velocity of the two particles, i.e.,

$$g_0{}^2 = (q_x - q_x')^2 + (q_y - q_y')^2 + (q_z - q_z')^2 \qquad (10.24)$$

where the subscripts x, y and z refer to the corresponding component of $\vec{q}$ and $\vec{q}'$. The integral should be taken for all values of b from zero to infinity, all angle ϵ from zero to 2π and all values of q_x', q_y' and q_z' from minus infinity to plus infinity. The derivation of equation (10.23) may be found in any standard textbook of kinetic theory of gases (3).

For a gas of a mixture of N_0 species, the collision term is equal to the summation of contributions by all the N_0 species, i.e.,

$$\left(\frac{\delta F_s}{\delta t}\right)_c = \sum_{r=1}^{N_o} \iiint (\bar{F}_s \bar{F}_r - F_s F_r) g_0 b \, db \, d\epsilon \, d\vec{q}_r \qquad (10.25)$$

For charged particles, distance encounter may be important. This involves taking account of very small angle scattering and thus of a great number of acts of very small momentum transfer. Such a collision term is known as Fokker-Planck collision term.

In general it is very difficult to solve Boltzmann equation (10.22) because the equation is non-linear. However one of the exact solutions of the Boltzmann equation has been found and is known as the Maxwellian distribution function which represents a uniform steady state of the gas. It has been extensively used in various flow problems, particularly the free molecule flow.

For simplicity, we derive the Maxwellian distribution function under the following assumptions:

(i) We assume that the gas consists of one kind of particles only.

(ii) There is no body force $\phi_s{}^i$.

(iii) The state of the gas is uniform.

Under these conditions, we need to consider one molecular distribution function F which is independent of the spatial coordinates, i.e., $F = F(\vec{q}_m, t)$. From now on, we use $\vec{q}_m$ as the molecular velocity vector while $\vec{q}$ will be used as the average flow velocity vector.

The Boltzmann equation (10.22) becomes:

$$\frac{\partial F}{\partial t} = 2\pi \iint (\bar{F}\bar{F}' - FF') g_0 b \, db \, d\vec{q}_m \qquad (10.26)$$

where we assume that the particle is spherically symmetrical and the integration with respect to ϵ can be carried out immediately.

Now we define a H-function such that

$$H = \int F \cdot \log F \cdot d\vec{q}_m \qquad (10.27)$$

From equations (10.26) and (10.27), we have

$$\frac{\partial H}{\partial t} = -\frac{1}{2}\pi \iiint \log\left(\frac{F F'}{\bar{F}\bar{F}'}\right)(\bar{F}\bar{F}' - FF') g_0 b \, db \, d\vec{q}_m \, d\vec{q}_m' \qquad (10.28)$$

From equation (10.28), we have

$$\frac{\partial H}{\partial t} \leqq 0 \qquad (10.29)$$

because the sign of $\log(FF'/\bar{F}\bar{F}')$ and that of $(\bar{F}\bar{F}' - FF')$ are always the same. Equation (10.29) is known as the Boltzmann H-theorem. The H-function is associated with the entropy of the gas.

For a steady state, $\partial H / \partial t = 0$, the integrand of equation (10.28) must be zero and we have

$$\overline{F} \overline{F}' = FF' \tag{10.30}$$

and

$$\log \overline{F} + \log \overline{F}' = \log F + \log F' \tag{10.31}$$

Since the distribution function F now is a function of $\vec{q}_m$ only, it can be shown that in order to satisfy the relation (10.31) which is known as the summation invariant, the distribution function F must be of the form:

$$\log F = a_1 \, m + m \, \vec{a_2} \cdot \vec{q}_m + \tfrac{1}{2} \, m \, a_3 \, q_m{}^2 \tag{10.32}$$

where a_1, $\vec{a_2}$ and a_3 are parameters independent of $\vec{r}, \vec{q}_m$ and t. Equation (10.32) may be written in the following form:

$$F = a_0 \exp \left[- b_0 \, (\vec{q}_m - \vec{q})^2 \right] \tag{10.33}$$

where a_0 and b_0 are constant and $\vec{q}$ is the average value of $\vec{q}_m$. The constants a_0 and b_0 may be expressed in terms of the temperature of the gas T and its mass m (cf. section 6), as follows:

$$F = N \left(\frac{m}{2 \, \pi \, k \, T} \right)^{\frac{3}{2}} \exp \left[- \frac{m \left(\vec{q}_m - \vec{q} \right)^2}{2 \, k \, T} \right] = F_0 \left(N, \vec{q}, T \right) \tag{10.34}$$

where N is the number density of the gas and k is the Boltzmann constant which has the value of 1.379×10^{-16} erg/degree centigrade. Equation (10.34) is the well known Maxwellian distribution function which represent a uniform and steady state of a gas of mean flow velocity $\vec{q}$ and an absolute temperature T and a number density N.

One of the methods which has been extensively used in solving the Boltzmann equation (10.22) is the perturbation technique[3] in which we put

$$F = F_0 \, (1 + \psi_f) \tag{10.35}$$

where F_0 is the Maxwellian distribution function of the local flow velocity q and the local temperature T and the absolute value of ψ_f is much smaller than unity.

A few special terms in the Maxwellian distribution function are of special interest. We may define a random or peculiar velocity $\vec{c}_a$ with componente c_i such that

$$c_i = q_{mi} - q_i \tag{10.36}$$

The Maxwellian distribution function F_0 may be considered as a function of $\vec{c}_a$. The number of molecules per unit volume between the absolute magnitude of c_a and $c_a + d c_a$ is then

$$d \, n_c = N \left(\frac{1}{\pi \, c_m{}^2} \right) \exp \left(\frac{c_a{}^2}{c_m{}^2} \right) d \, c_a \cdot c_a{}^2 \cdot 4 \, \pi \tag{10.37}$$

where

$$c_m = (2 \, k \, T / m)^{\frac{1}{2}} = \text{the most probable speed} \tag{10.38}$$

because at $c_a = c_m$, $(d n_c / d c_a)$ is a maximum.

The Maxwellian distribution function may be written as follows:

$$F_0 = \frac{N}{(\sqrt{\pi}\,c_m)^3} \exp\left(-\frac{c_a{}^2}{c_m{}^2}\right) \tag{10.34 a}$$

The mean value of the peculair speed is

$$\overline{c_a} = \frac{4}{\sqrt{\pi}} \int_0^\infty \left(\frac{c_a}{c_m}\right)^3 \exp\left[-(c_a{}^2/c_m{}^2)\right] \mathrm{d}(c_a/c_m) \cdot c_m = \frac{2}{\sqrt{\pi}}\,c_m \tag{10.39}$$

If the gas is at rest, $\overline{c_a}$ is the mean molecular velocity. Since the sound speed of the gas at a temperature T is $a = (\gamma\,k\,T/m)^{\frac{1}{2}}$, the mean molecular speed of the gas is of the same order of magnitude as its sound speed.

Another important quantity in the kinetic theory of gases is the mean free path which is the average distance travelled by the molecules between collision. Now we consider the gas at rest under the equilibrium condition of Maxwellian distribution. The one of the two terms in equation (10.23) represents that the molecules of velocity $\overrightarrow{c_a}'$ become molecules of velocity c_a after collision and the other term represents that the molecules of velocity c_a become molecules of velocity different from c_a. In the equilibrium condition, these two terms are equal. Hence we may consider only one of these two terms to determine the average collisions. Then the number of collisions between the molecules having the velocities in the range c_a and $c_a + \mathrm{d}c_a$ and those with velocities in the range c_1 and $c_1 + \mathrm{d}c_1$ per unit volume and per unit time is

$$2\,\pi\,g_0\,F_0\,(c_a)\,F_0\,(c_1)\,S_0\,\mathrm{d}c_a\,\mathrm{d}c_1 \tag{10.40}$$

where

$$S_0 = 2\,\pi \int b\,\mathrm{d}b = \text{collision cross section} \tag{10.41}$$

The integration of expression (10.40) over all the velocities gives twice the total number of collisions per unit volume per unit time because each collision is counted twice, one as a c-molecule with a c_1-molecule and the other with the roles of these particles reversed. But since each collision terminates two free paths, the integration gives the total number of free path per unit volume and per unit time, i.e.,

$$2\,\pi \int_0^\infty\!\!\int g_0\,F_0\,(c_a)\,F_0\,(c_1)\,\mathrm{d}c_a\,\mathrm{d}c_1\,S_0 = \sqrt{2}\,N^2\,S_0\,\overline{c_a} = N\,c_a\,L_f \tag{10.42}$$

Since the total distance travelled by all molecules in unit volume per unit time is $N c_a$, we have the mean free path L_f by definition, i.e.,

$$\text{mean free path} = L_f = \frac{N\overline{c}_a}{\sqrt{2}\,N^2\,S_0\,\overline{c}_a} = \frac{1}{\sqrt{2}\,N\,S_0} \tag{10.43}$$

The mean free path may be expressed in terms of kinematic viscosity as we shall discuss later [cf. eq. (10.125a)].

For a mixture of gases, we may easily extend the above definition of the molecular speed, most probable speed and the mean free path to each species of the mixture.

Before we discuss the relation of the molecular distribution function and the macroscopic variables such as temperature, pressure, etc., we first consider briefly the Boltzmann equation for photons.

5. Boltzmann equation for photons. We shall describe the thermal radiation as being a species of gases which consists of point-like photons, each of which being characterized, at any time t, by a position vector $\vec{r}$ and a momentum $\vec{p}$. The relation of the momentum of a photon and the corresponding frequency ν of the specific intensity I_ν and its direction L is

$$p_i = \frac{h\nu}{c} L_i \qquad (10.44)$$

where the magnitude of the momentum of a photon is simply $|\vec{p}| = h\,\nu/c$.

If we denote the distribution function of photons by F_R, the number of photons per unit volume at the position $\vec{r}$ in the range of momentum $\vec{p}$ and $\vec{p} + \mathrm{d}\vec{p}$ at time t is simply

$$\mathrm{d}n_p = F_R\,(\vec{r}, \vec{p}, t)\,\mathrm{d}\vec{p} \qquad (10.45)$$

where $\mathrm{d}\vec{p}$ refers to three dimensional space, i.e., p_i.

The relation between the distribution function of photons F_R and the specific intensity of radiation I_ν is simply

$$I_\nu = \frac{h^4\,\nu^3}{c^2} F_R \qquad (10.46)$$

The symbol "c" in this section refers to the speed of light.

We have to use the relativistic mechanics to describe the motion of the photons. The 4-momentum of a photon will be p_α with

$$p_\alpha p_\alpha = 0; \quad p_4 = i\,|\vec{p}| = i\,h\,\nu/c \qquad (10.47)$$

The equation which governs F_R can be derived from the equation of radiative transfer (3.32) and may be written as

$$p_\alpha \frac{\partial F_R}{\partial x_\alpha} = \frac{|\vec{p}|}{c} \left(\frac{\delta F_R}{\delta t}\right)_c \qquad (10.48)$$

where

$$\frac{|\vec{p}|}{c} \left(\frac{\delta F_R}{\delta t}\right)_c = \frac{c}{h^3\,\nu^2} (j_\nu - k_\nu\,I_\nu) \qquad (10.49)$$

is a 4-invariant. It is evident that $(\delta F_R/\delta t)_c$ is the time variation of F_R due to the collisions of photons with material particles, i.e., due to emission and absorption processes.

According to SYNGE (12), equation (10.48) with the help of equation (10.49) may be written as

$$p_\alpha \frac{\partial F_R}{\partial x_\alpha} = E - A F_R \qquad (10.50)$$

where
$$E = \frac{c}{h^3 \nu^2}\, j_\nu; \quad A = \frac{h\nu}{c}\, k_\nu \tag{10.51}$$

and both E and A may be considered as functions of t, x_i and p_i and they are 4-invariant. It should be noticed that equation (10.48) is simply the relativistic Boltzmann equation for a photon gas which has zero rest mass and zero body force $\phi_s{}^i = 0$.

6. Conservation equations. As we have mentioned, one of the most interesting results from the Boltzmann equations is the derivation of the fundamental equations of gasdynamics from the macroscopic point of view. The first thing which we should investigate is the relations between the distribution function F_s, the molecular velocity vector $\vec{q}_{m\,s}$ with components q_{mst} and the macroscopic variables such as the flow velocity u_i, the density ρ or the number density N, the pressure p, the temperature T of a gas as well as other physical quantities such as viscous stress, heat flux etc. First we consider in this section the case · of material particles so that we shall use the three dimensional space.

By the definition of equation (10.3), we have the following relations of the variables used in the macroscopic treatment and the molecular distribution function F_s:

(i) The number density of the sth species in a mixture is

$$n_s = \int F_s\,(\vec{r},\,\vec{p}_s,\,t)\,\mathrm{d}\vec{p}_s \tag{10.52a}$$

The density of the sth species is then

$$\rho_s = m_s\,n_s \tag{10.52b}$$

The total number density of the mixture of N_0 species is

$$N = \sum_{s=1}^{N_0} n_s \tag{10.25c}$$

The density of the mixture as a whole is then

$$\rho = \sum_{s=1}^{N_0} \rho_s = m\,N \tag{10.52d}$$

where m is the mean mass of the particles in the mixture.

(ii) The flow velocity of the sth species in a mixture is from equation (10.4):

$$u_{si} = \frac{1}{n_s\,m_s} \int p_{si}\,F_s\,\mathrm{d}\vec{p}_s \tag{10.53}$$

The flow velocity of the mixture as whole is defined by the relation:

$$u_i = \frac{1}{\rho} \sum_{s=1}^{N_0} \rho_s\,u_{si} = \frac{1}{\rho} \sum_{s=1}^{N_0} m_s\,n_s\,u_{si} \tag{10.54}$$

The diffusion velocity w_{si} of the sth species in the mixture is

$$w_{si} = u_{si} - u_i \tag{10.55}$$

It is evident that

$$\sum_{s=1}^{N_0} \rho_s \, w_{si} = 0 \tag{10.56}$$

(iii) The pressure, stress tensor, temperature, heat flux etc. are expressed in terms of the average of powers of the random or peculiar velocity components. For a mixture of gases, there are two ways to define the peculair velocity components and then the macroscopic state variables and the stress tensor etc. These two peculiar velocity components are c_{si} and c_{si}^{*} which are defined as follows:

$$q_{msi} = u_i + c_{si} = u_{si} + c_{si}^{*} \tag{10.57}$$

When the diffusion velocity is small, we usually define the pressure, temperature, etc. in terms of mean values of the powers of c_{si} while if the diffusion velocity is large, we should define the pressure, temperature, etc. in terms of the mean value of the powers of c_{si}^{*}. When the mases of various species are of the same order of magnitude, the diffusion velocity is usually small. On the other hand, when the difference of masses between various species is large such as the case between electrons and ordinary atoms or molecules, the diffusion velocity of the lighter particles will be large. Since the form of these definitions with c_{si} is the same as that with c_{si}^{*}, we shall consider the case with c_{si} only.

The pressure tensor is defined as follows:

$$p_{sij} = m_s \int c_{si} \, c_{sj} \, F_s \, \mathrm{d}\vec{p}_s \tag{10.58}$$

The partial pressure of the sth species is then

$$p_s^{*} = \frac{1}{3} \left(p_{s11} + p_{s22} + p_{s33} \right) = \frac{m_s}{3} \overline{c_{si} \, c_{si}} = \frac{m_s}{3} \overline{c_s^2} \tag{10.59}$$

The viscous stress tensor τ_s of the sth species is

$$\tau_{sij} = - p_{sij} + \delta_{ij} \, p_s^{*} \tag{10.60}$$

where $\delta_{ij} = 0$ if $i \neq j$ and $\delta_{ij} = 1$ if $i = j$.

The kinetic temperature T_s of the sth species is defined as

$$T_s = \frac{m_s}{3 \, k \, n_s} \overline{c_s^2} = \frac{m_s}{3k} \int (c_1^2 + c_2^2 + c_3^2) \, F_s \, \mathrm{d}\vec{p}_s \tag{10.61}$$

Comparing equations (10.52 b), (10.59) and (10.61), we have the equation of state of a perfect gas:

$$p_s^{*} = k \, n_s \, T_s \tag{10.62}$$

The total pressure of the mixture is

$$p = \sum_{s=1}^{N_0} p_s^{*} \tag{10.63}$$

The temperature of the mixture as a whole is defined as

$$T = \frac{1}{N} \sum_{s=1}^{N_0} n_s \, T_s \tag{10.64}$$

From equations (10.62) to (10.64), we have the equation of state of the mixture of perfect gases as follows:

$$p = kNT \qquad (10.65)$$

The third moment of the velocity is defined as

$$S_{sijk} = m_s\, n_s\, \overline{c_{si}\, c_{sj}\, c_{sk}} \qquad (10.66)$$

The heat flux by condition and diffusion is then

$$Q_{csi} = S_{sijj} = S_{si11} + S_{si22} + S_{si33} = q_{si}^* + \frac{5}{2}\, p_s^*\, w_{si} \qquad (10.67)$$

The term q_{si}^* is the heat flux by conduction. If the gas consists of only one kind of particles, $w_{si} = 0$, equation (10.67) gives the heat flux by conduction.

With the definition of the macroscopic variables by equations (10.52) to (10.67), we may derive the conservation equations of ordinary gasdynamics from the Boltzmann equation by taking the moments of the Boltzmann equation. Let Q_s^n be a function of power of c_{si}. We take the nth moment of Boltzmann equation by multiplying Q_s^n to the Boltzmann equation and then integrating the resultant equation over the whole momentum space $\mathrm{d}\vec{p}_s$.

(i) Equation of continuity. For the zeroth moment, we multiply equation (10.22) by $Q_s^0 = m_s$ and integrate the resultant equation with respect to p_s. We have then

$$\frac{\partial\, \rho_s}{\partial t} + \frac{\partial\, \rho_s\, u_{si}}{\partial x_i} = \int m_s \left(\frac{\delta\, F_s}{\delta t}\right)_c \mathrm{d}\vec{p}_s = \Delta\, (m_s\, n_s) = \beta_s \qquad (10.68)$$

The term β_s is the mass source per unit volume of the sth species which may be due to chemical reaction or ionization process. In order to evaluate the collision terms, we have to assume certain model for the kinetic picture so that the collision term Δ can be evaluated. For complicated physical phenomena, the collision term is difficult to evaluate. However, since the form of equation (10.68) is the same as that obtained from the macroscopic approach, we have confidance on this equation and we may postulate some relations between the collision term and other physical quantities such as temperature T_s, density n_s etc.

If we sum up all the N_0 equations of continuity (10.68), we have the equation of continuity of the mixture as a whole:

$$\frac{\partial\, \rho}{\partial t} + \frac{\partial\, \rho\, u_i}{\partial x_i} = \sum_{s=1}^{N_0} \rho_s = \beta \cong 0 \qquad (10.69)$$

where β is the rate of transformation of mass to energy by nuclear reaction. In most of the fluid dynamical problems, it is negligible. Hence we may put $\beta = 0$ and equation (10.69) is the equation of continuity used in fluid dynamics [cf. eq. (5.3)].

Equation (10.68) may be written in the following form:

$$\rho \left(\frac{\partial\, k_s}{\partial t} + u_i\, \frac{\partial\, k_s}{\partial x_i}\right) = -\left(\frac{\partial\, k_s\, \rho\, w_{si}}{\partial x_i}\right) + \beta_s \qquad (10.70)$$

where k_s is the mass fraction of the sth species, i.e., $k_s = \rho_s/\rho$. Equation (10.70) is the well known equation of diffusion if we replace the diffusion velocity w_{si} in terms of diffusion coefficients and the gradient of partial pressure and other physical quantities (2, 6). Of course, it is convenient to use the diffusion coefficient when the diffusion velocity is small. If the diffusion velocity is not small in comparison with the flow velocity or sound speed, we should use the equations of motion of the sth species to determine accurately the diffusion velocity instead of using the approximate expression in term of diffusion coefficient. It is especially true for ionized gas in which the amount of electron is large.

(ii) Equations of motion. For the first moment, we multiple equation (10.22) by $Q_s{}^1 = m_s c_{si}$ and integrate the resulting equation with respect to $\vec{p}_s$. We have the equation of motion of the sth species:

$$\rho_s \frac{D u_i}{D t} + \frac{D}{D t}(\rho_s w_{si}) + \rho_s w_{si} \epsilon_0 + \rho_s w_{sk} \frac{\partial u_i}{\partial x_k} + \frac{\partial}{\partial x_k}(p_{sik}) - \rho_s G_i$$

$$= \int m_s c_{si} \left(\frac{\delta F_s}{\delta t}\right)_c d\vec{p}_s = \Delta_{s1i} \tag{10.71}$$

where

$$\frac{D}{D t} = \frac{\partial}{\partial t} + u_k \frac{\partial}{\partial x_k} \;;\; \epsilon_0 = \frac{\partial u_k}{\partial x_k}$$

The term $- \rho_s G_i$ is the body force term which includes the gravitational force and the electromagnetic force etc., as shown in the force term ϕ_{si} of equation (10.22). The collision term Δ_{s1i} gives the ith component of the change of momentum due to collision between species. In general, this term is very complicated, especially when the gas is rarefied. There are terms including heat transfer as well as the difference of velocity (2, 6). If the gas is not rarefied, for first approximation, we have

$$\Delta_{s1i} = - \sum_t K_{st}(w_{si} - w_{ti}) \tag{10.72}$$

where K_{st} is known as the friction coefficient between species s and t. The summation is over all the species.

If we sum up all the equations of motion of all species (10.71), we have the equation of motion of the mixture as a whole:

$$\rho \frac{D u_i}{D t} + \frac{\partial p_{ik}}{\partial x_k} - \rho G_i = \sum \Delta_{s1i} = 0 \tag{10.73}$$

where $p_{ik} = \sum_s p_{sik}$ and $\rho G_i = \sum_s \rho_s G_i$ and the summation of Δ_{s1i} is zero because of the conservation of momentum of the mixture as a whole. Equation (10.73) is the equation of motion which has been used in ordinary gasdynamics [cf. equation (5.4)].

Equation (10.71) with the help of equation (10.73) may be written as follows:

$$\left[\frac{D}{D t}(\rho_s w_{si}) + \rho_s w_{si} \epsilon_0 + \rho_s w_{sk} \frac{\partial u_i}{\partial x_k} - \frac{\partial \tau_{sik}}{\partial x_k} - \frac{\rho_s}{\rho} \frac{\partial \tau_{ik}}{\partial x_k}\right] +$$

$$+ \left(\frac{\partial p_s{}^*}{\partial x_i} - \frac{\rho_s}{\rho} \frac{\partial p}{\partial x_i}\right) - (\rho_s - \rho) G_i = \Delta_{s1i} \tag{10.74}$$

Equation (10.74) is the equation for the diffusion velocity w_{st}. If the diffusion velocity is small, the terms in the square bracket are negligible and we have the usually expression of the diffusion velocity in terms of the pressure gradient and the body forces. For instance, for a mixture of two species and neglecting the thermal diffusion so that equation (10.72) may be used for Δ_{s1t}, we have

$$w_{si} = \frac{1}{K_{st}} \frac{\rho_t}{\rho} \left(\frac{\rho_s}{\rho} \frac{\partial p}{\partial x_i} - \frac{\partial p_s^*}{\partial x} \right) \tag{10.75}$$

Hence for a first approximation the diffusion velocity may be expressed in terms of the gradient of the partial pressures. The friction coefficient K_{st} is inversely proportional to the diffusion coefficient D_{st}. Ordinarily, we further assume that the total pressure of the mixture p as well as the temperature of each species T_s are constant, we have then that the diffusion velocity is proportional to the gradient of the concentration of the species.

(iii) Equation of internal energy. For internal energy, we multiply equation (10.22) by $Q_s{}^{20} = \frac{1}{2} m_s c_s{}^2$ and integrate the resultant equation with respect to $\vec{p}_s$. We have then:

$$\frac{D}{Dt}\left(\frac{3}{2} p_s^* \right) + \frac{5}{2} p_s^* \epsilon_0 - \frac{1}{2} (\tau_{sij}) \epsilon_{ij} + \frac{\partial q_{csi}}{\partial x_i} +$$

$$+ \rho_s \frac{D u_i}{Dt} w_{si} - G_i \rho_s w_{si} = \Delta_{s20} \tag{10.76}$$

where

$$\epsilon_{ij} = \frac{\partial u_j}{\partial x_i} + \frac{\partial u_i}{\partial x_j} - \frac{2}{3} \delta_{ij} \epsilon_0$$

Equation (10.76) is the equation of internal energy of the sth species. The collision term $\Delta_{s\,20}$ will include terms of the difference of temperature $(T_s - T_t)$ because in general the kinetic temperatures of various species may be different. Except the case when the mass of various species differ greatly from one another, the difference of temperatures between species is usually small. We may use the assumption $T_s = T$ for all the species and drop the energy equation of the species (10.76) in the analysis of the flow problem. However for the case of ionized gas, the temperature of the electron may differ considerably from other species, we should use the energy equation of the electrons in addition to the energy equation of the mixture as a whole if accurate results are required, particularly the density of the mixture is low. The energy equation of the mixture as a whole is obtained by summing up all the N_0 equations of (10.76) and is

$$\frac{D}{Dt}\left(\frac{3}{2} p \right) + \frac{5}{2} p - \frac{1}{2} \tau_{vij} \epsilon_{ij} + \frac{\partial q_{ci}}{\partial x_i} = 0 \tag{10.77}$$

(iv) Equation of viscous stresses and equation of heat flow. If we multiply equation (10.22) by $Q_s{}^2 = m_s c_{si} c_{sj}$ and integrate the resultant equation with respect to $\vec{p}_s$, we will have a differential equation for the viscous stress tensor p_{stij}. Similarly, if we multiple equation (10.22) by $Q_s{}^{30} = \frac{1}{2} m_s c_s{}^2 c_{si}$ and integrate the resultant equation with respect to $\vec{p}_s$, we have a differential equation for the

12*

heat flux by conduction q_{sci}.   Hence in general, both the viscous stress tensor and the heat flux by conduction are governed by complicated differential equations, especially when the gas is rarefied.   For rarefied gasdynamics, the fundamental equations are very complicated.   However, if the mean free path of the gas is small, we may derive some simple relations between the viscous stress tensor and the velocity gradient and the coefficient of viscosity $(2,3,5,6)$ from the differential equation of the viscous stress tensor in the same scheme as we derive the relation (10.75) for the diffusion coefficient.   In other words, we may obtain the well known Navier-Stokes relations from the differential equation for the viscous stress tensor as a first approximation, i.e.,

$$\tau_{vij} = \mu \left(\frac{\partial u_i}{\partial x_j} + \frac{\partial u_j}{\partial x_i} \right) + \mu_1 \delta_{ij} \frac{\partial u_k}{\partial x_k} \tag{10.78}$$

where μ and μ_1 are respectively the ordinary coefficient of viscosity and the second coefficient of viscosity and they are functions of the temperature and the composition of the mixture.

Similarly from the differential equation of the heat flux $(2, 6)$ we may obtain the Fourier's law of heat conduction, i.e.,

$$q_{ci} = - \varkappa \frac{\partial T}{\partial x_i} \tag{10.79}$$

In deriving the internal energy equation (10.76), we assume that the particles are monatomic particles and its internal energy is simply $U_{ms} = (^3/_2) \, k \, T/m_s$. Hence we have the simple expression $(3 \, p_s^*/2)$ in the equation.   For complicated molecules, this simple expression of internal energy should be replaced by the correct one including all the other modes of internal energy such as rotational energy, vibrational energy etc.   But the general form of the energy equation is the same.   This shows the modern approach to the study of fluid dynamics for high temperature gas mixture in which both the microscopic and the macroscopic treatments are used.   By the microscopic treatment, we assure us the correct form of the fundamental equations.   For instance, the thermal diffusion (3) was found from the microscopic treatment.   By macroscopic treatment, we may deal with very complicated physical phenomena so that simplified expression for the collision terms may be assumed by the guidance of experimental facts even though they cannot easily derived accurately from the kinetic theory.

7. Radiation stresses and radiation energy density.   In radiation gasdynamics, we derive the fundamental equations from the relativistic Boltzmann equation in a similar manner as we did for the material particles.   However, since we consider only the case that the velocity of the material particles are negligibly small in comparison with the speed of light and the classical mechanics is applicable to the material particles, we need to consider the relativistic mechanics for the photons only.   In other words, we need to define the macroscopic quantities from the distribution function for photons F_R from the four dimensional space point of view.   We define the stress tensor $p_{R\alpha\beta}$ in the four dimensional space as follows:

$$p_{R\alpha\beta} = c \int F_R \frac{p_\alpha \, p_\beta}{|\vec{p}|} \, \mathrm{d}\vec{p} \tag{10.80}$$

where the integration is taken over the whole three-dimensional momentum space, i.e., for p_i varying from minus infinity to plus infinity. In this definition, F_R is assumed to vanish sufficiently rapidly when $|\vec{p}|$ approaches infinity; furthermore, whenever α or β equal to 4, we have $p_4 = i\,|\vec{p}| = + i\,h\,\nu/c$. It is easy to show that the stress tensor $p_{R\alpha\beta}$ is a symmetrical 4-tensor. Since the length of the momentum vector p_α is zero, we have

$$p_{R44} = p_{Rii} \tag{10.81}$$

In the three dimensional space, the components of the 4-dimensional tensor $p_{R\alpha\beta}$ are as follows:

$$p_{Rij} = -\frac{1}{c}\int_0^\infty d\nu \int I_\nu\, n_i\, n_j \, d\omega = \text{radiation stress tensor } ij\text{th} \atop \text{components [cf. eq. (2.18)]} \tag{10.83}$$

$$q_{Ri} = c\, p_{Ri4} = \int_0^\infty d\nu \int I_\nu\, n_i \, d\omega = \text{radiative heat flux} \atop \text{[cf. eq. (2.8)]} \tag{10.83}$$

$$p_{R44} = E_R = p_{Rii} = \frac{1}{c}\int_0^\infty d\nu \int I_\nu \, d\omega = \text{radiation energy density} \atop \text{[cf. eq. (2.14)]} \tag{10.84}$$

It is interesting to find the expression of the components of the radiation stress tensor in the frame such that the mass velocity of the gas is zero. If we denote the value in this frame by an additional subscript 0, and assume that the mass velocity u_i is small negligibly in comparison with the speed of light c, we have for first approximation:

$$p_{Rij0} = p_{Rij} - \frac{1}{c^2}\,(q_{Ri}\,u_j + q_{Rj}\,u_i) \tag{10.85a}$$

$$q_{Ri0} = q_{Ri} - p_{Rij}\,u_j - E_R\,u_i \tag{10.85b}$$

$$E_{R0} = E_R - \frac{2}{c^2}\,q_{Ri}\,u_i \tag{10.85c}$$

and

$$p_{Rij} = p_{Rij0} + \frac{1}{c^2}\,(q_{Ri0}\,u_j + q_{Rj0}\,u_i) \tag{10.86a}$$

$$q_{Ri} = q_{Ri0} + p_{Rij0}\,u_j + E_{R0}\,u_i \tag{10.86b}$$

$$E_R = E_{R0} + \frac{2}{c^2}\,q_{Ri0}\,u_i \tag{10.86c}$$

Usually the terms with the factor $(1/c^2)$ in equations (10.85) and (10.86) are negligible. But the convection radiation energy density $E_R\,u_i$ and the work done by the radiation stress tensor $p_{Rij0}\,u_j$ may not be neglected.

Now we write the collision terms with the following symbols:

$$M_{Ri} = \int\left(\frac{\delta F_R}{\delta t}\right)_c p_i\, d\vec{p} = \frac{1}{c}\int_0^\infty d\nu \int (j_\nu - k_\nu\, I_\nu)\, n_i\, d\omega \tag{10.87}$$

$$N_R = \int \left(\frac{\delta F_R}{\delta t}\right)_c c \left|\vec{p}\right| \mathrm{d}\vec{p} = \int_0^\infty \mathrm{d}\nu \int (j_\nu - k_\nu I_\nu) \, \mathrm{d}\omega \qquad (10.88)$$

Now if we eliminate the collision terms $(\delta F_R/\delta t)_c$ from equation (10.48), equations (10.87) and (10.88) become respectively:

$$M_{Ri} = \frac{\partial}{\partial x_\alpha}(p_{Ri\alpha}) = \frac{\partial}{\partial x_j}(p_{Rij}) + \frac{1}{c^2}\frac{\partial q_{Ri}}{\partial t} \qquad (10.87\,\mathrm{a})$$

$$N_R = c\frac{\partial}{\partial x_\alpha}(p_{R4\alpha}) = \frac{\partial}{\partial x_i}(q_{Ri}) + \frac{\partial E_R}{\partial t} \qquad (10.88\,\mathrm{a})$$

Now we are in a position to derive from the Boltzmann equation the fundamental equations of radiation gasdynamics. For simplicity, we assume that the gas is a mixture of one kind of material particles and photons. Since the change of mass due to the production of photons is negligible, there will be no change in the equation of continuity (10.69). Since there will be no source terms of this gas, our equation of continuity of radiation gasdynamics is equation (10.69). It should be noticed that the random velocity c_i is closely related to the momentum $\vec{p}$ in the Galilean frame in which the mass velocity of the species s vanishes, i.e.,

$$p_{sm0i} = p_{smi} - m_s u_{si} \qquad (10.89)$$

Now we multiply the Boltzmann equation (10.22) by the momentum component p_{smi} and integrate the resultant equation in the three dimensional momentum space $\mathrm{d}p$. We have cf. equation (10.71)

$$\frac{\partial}{\partial t}(\rho_s u_{si}) + \frac{\partial}{\partial x_k}(P_{sik}) = M_{si} + \int F_s \phi_{si} \, \mathrm{d}\vec{p} \qquad (10.90)$$

where

$$P_{sik} = p_{sik} + \rho_s u_{si} u_{sk}, \quad p_{sik} = P_{s0ik}$$

$$M_{si} = \int \left(\frac{\partial F_s}{\partial t}\right)_c p_{smi} \, \mathrm{d}\vec{p} \qquad (10.91)$$

$$\phi_{si} = m_s G_i + e_s E_i + \frac{e_s}{m_s}\left(\vec{p}\times\vec{B}\right)_i$$

$\vec{B}$ is the magnetic induction.

When we multiply p_i to the Boltzmann equation for photons and integrate the resultant equation, we have equation (10.87 a). Adding equations (10.87 a) and (10.90), we have

$$\frac{\partial}{\partial t}(\rho u_i) + \frac{\partial}{\partial x_k}(\rho u_i u_k + p_{sik} + P_{R0ik}) + \frac{1}{c^2}\frac{\partial}{\partial t}(q_{R0i}) +$$

$$+ \frac{1}{c^2}\frac{\partial}{\partial x_k}(q_{R0i} u_k + q_{R0k} u_i) = \rho G_i + \rho_e E_i + \left(\vec{J}\times\vec{B}\right)_i \qquad (10.92)$$

Equation (10.92) is the equation of motion of the mixture of gases and photons, i.e., the equation of motion of radiation gasdynamics. Equation (10.92) is the same as equation (5.4) when the terms with the factor $(1/c^2)$ are neglected and $P_{R0ik} = \tau_{Rik}$, $p_{0ik} = -\tau_{ik} + p\,\delta_{ik}$ and $M_{si} + M_{Ri} = 0$.

Similarly, if we add the internal energy equation (10.76) with the radiation energy equation (10.88a) and use the relation

$$N_s + N_R = \varepsilon_0 \qquad (10.93)$$

where

$$N_s = \int \left(\frac{\delta F_s}{\delta t}\right)_c \left(\frac{1}{2}\frac{|\vec{p}|^2}{m_s} + \chi_s\right) d\vec{p}$$

and ε_0 is the energy released per unit time and per unit volume due to chemical and nuclear reactions and χ_s is the ionization energy for sth species. We have the energy equation of radiation gasdynamics:

$$\frac{\partial}{\partial t}(U_m + E_R + \tfrac{1}{2}\rho\,u^2) + \frac{\partial}{\partial x_i}[(U_m + E_{R0} + \tfrac{1}{2}\rho\,u^2)\,u_i] +$$

$$+ \frac{\partial}{\partial x_i}[(P_{0ik} + P_{R0ik})\,u_k + q_{ci} + q_{R0i}] = \varepsilon_0 + \rho\,G_i\,u_i + J_i\,E_i \quad (10.94)$$

where the therms with factor $(1/c^2)$ are neglected. Equation (10.94) is the same as equation (5.7) with $Q = \varepsilon_0 + J_i\,E_i$.

We have mentioned that P_{R0ik} is equal to the radiation stress tensor τ_{Rik} given by equation (2.18). It is interest to discuss the components of the 4-dimensional tensor $P_R^{\alpha\beta}$. Let us consider material particles with flow velocity v^α of the same order of magnitude as the velocity of light c. Let us denote v^α the 4-velocity vector of the gas:

$$v^\alpha = \left(1 - \frac{v^2}{c^2}\right)^{-\frac{1}{2}}(u_i,\,c); \quad v^\alpha\,v_\alpha = c^2 \qquad (10.95)$$

Following ECKART (11), we introduce the following symbols:

$$S^{\alpha\beta} = \frac{1}{c^2}\,v^\alpha\,v^\beta - g^{\alpha\beta} \qquad (10.96\,\text{a})$$

$$\mathfrak{P}^{\alpha\beta} = S^{\alpha\lambda}\,S^{\beta\mu}\,P_{R\lambda\mu} \qquad (10.96\,\text{b})$$

$$\mathfrak{F}^\alpha = -S^{\alpha\lambda}\,v^\mu\,P_{R\lambda\mu} \qquad (10.96\,\text{c})$$

$$V = \frac{1}{c^2}\,v^\lambda\,v^\mu\,P_{R\lambda\mu} \qquad (10.96\,\text{d})$$

Hence the 4-tensor $P_R^{\alpha\beta}$ becomes

$$P_R^{\alpha\beta} = \mathfrak{P}^{\alpha\beta} + \frac{1}{c^2}(\mathfrak{F}^\alpha\,v^\beta + \mathfrak{F}^\beta\,v^\alpha + V\,v^\alpha\,v^\beta) \qquad (10.97)$$

Both $S^{\alpha\beta}$ and $\mathfrak{P}^{\alpha\beta}$ are symmetrical and the following relations hold:

$$S^{\alpha\beta}\,v_\beta = 0; \quad \mathfrak{P}^{\alpha\beta}\,v_\beta = 0; \quad \mathfrak{F}^\alpha\,v_\alpha = 0$$

$$g_{\alpha\beta}\,\mathfrak{P}^{\alpha\beta} = -V; \quad P_R^{\alpha\beta}\,v_\beta = \mathfrak{F}^\alpha + V\,v^\alpha \qquad (10.98)$$

the unit vector $g_{\alpha\beta}$ is defined as follows:

$$ds^2 = g_{\alpha\beta}\,dx^\alpha\,dx^\beta = -(dx^1)^2 - (dx^2)^2 - (dx^3)^2 + (dx^4)^2 \qquad (10.99)$$

By the orthogonality relations between $\mathfrak{P}^{\alpha\beta}$ and v^α, we have

$$\mathfrak{P}^{i4} = \frac{1}{c}\,\mathfrak{P}^{ij}\,u^j;\quad \mathfrak{P}^{44} = \frac{1}{c^2}\,\mathfrak{P}^{ij}\,u^i\,u^j \tag{10.100}$$

Similarly, from the orthogonality relation between $\mathfrak{F}^\alpha$ and v^α, we have

$$\mathfrak{F}^4 = \frac{1}{c}\,\mathfrak{F}^i\,u^i \tag{10.101}$$

and then equation (10.98) gives

$$V = \left(\delta^{ij} - \frac{1}{c^2}\,u^i\,u^j\right)\mathfrak{P}^{ij} \tag{10.102}$$

In the Galilean frame with a particular event, we have

$$u_0{}^i = 0,\ v_0 = 0,0,0,c$$

and then

$$\mathfrak{P}_0{}^{ij} = P_{R0}^{ij};\quad \mathfrak{F}_0{}^i = q_{R0}^i;\quad V_0 = E_{R0},\quad \mathfrak{P}_0^{\alpha 4} = 0,\quad \mathfrak{F}_0{}^4 = 0 \tag{10.103}$$

Equation (10.103) gives the relations which we use for the photons.

8. Local thermodynamic equilibrium. As we have discussed in chapter IV, section 7, for most of the flow problems, the assumption of local thermodynamic equilibrium is a good approximation for the actual case. Now we are going to discuss the hypotheses of local thermodynamic equilibrium from the relativistic point of view (cf. reference *11*). As we have already discussed in chapter IV, § 7, for local thermodynamic equilibrium, the emission coefficient in the G_0 frame is

$$j_{\nu_0}^0 = \exp\left(-\,h\,\nu_0/k\,T_0\right) k_{\nu_0}^0\, I_{\nu_0}^0 + [1 - \exp\left(-\,h\,\nu_0/k\,T_0\right)]\, k_{\nu_0}^0\, B_{\nu_0}\,(T_0) \tag{10.104}$$

where T_0 is the temperature of the gas defined in the G_0 frame. We may regard it as a 4-invariant and write $T = T_0$. The sub- and superscript 0 refers to the values in the G_0 frame. Now in terms of E, A and F_R defined in equations (10.46) and (10.51), equation (10.104) becomes

$$E_0 = A_0\, F_{R0}\exp\left(-\,c\,|\vec{p}_0\,|/k\,T_0\right) + [1 - \exp\left(-\,c\,|\vec{p}_0\,|/k\,T_0\right)]\,A_0\,g_{R0} \tag{10.105}$$

where

$$g_{R0} = \frac{2}{h^3}\,[\exp\left(c\,|\vec{p}_0|/k\,T_0\right) - 1]^{-1} \tag{10.106}$$

g_{R0} is the distribution function of the photons which corresponds to a black body stationary in G_0. In any G frame, the local thermodynamic equilibrium hypothesis gives the following relation:

$$E = A\,F_R\exp\left(-\,p^\alpha\,v_\alpha/k\,T\right) + [1 - \exp\left(-\,p^\alpha\,v_\alpha/k\,T\right)]\,A\,g_R \tag{10.107}$$

where

$$g_R = \frac{2}{h^3}\,[\exp\left(p^\alpha\,v_\alpha/k\,T\right) - 1]^{-1} \tag{10.108}$$

is the distribution function of the photon gas which corresponds to a black body moving with velocity v with respect to G.

Finally the radiative transfer equation for photons (10.50) in local thermo-dynamic equilibrium condition becomes

$$p^\alpha \frac{\partial F_R}{\partial x^\alpha} = \frac{1}{c}(p^\alpha v_\alpha) K (g_R - F_R) \tag{10.109}$$

where

$$K = \frac{c}{p^\alpha v_\alpha} A [1 - \exp(-p^\beta v_\beta/k T)] \tag{10.110}$$

K is simply the reduced absorption coefficient k_v' [cf. eq. (4.50)], i.e., in the G_0 frame

$$K = k_{v_0}^0 [1 - \exp(-h v_0/k T_0)] = k'^0_{v_0} \tag{10.111}$$

For local thermodynamic equilibrium condition, we may derive the radiation stress tensor P_R for small mean free path of radiation or large value of K as we did in chapter V, § 8. From equation (10.109), we have for first approximation:

$$F_R = g_R - \frac{c}{p^\alpha v_\alpha} \frac{1}{K} p^\beta \frac{\partial g_R}{\partial x^\beta} \tag{10.112}$$

where g_R is a function of x^β because both T and v^α are in general functions of x^β.

If we substitute the expression (10.112) into the radiation stress tensor $p_R^{\alpha\beta}$, i.e., equation (10.80), we may obtain the 4-dimensional stress tensor in the local thermodynamic equilibrium condition. Such a calculation is rather lengthy. We shall calculate only $\mathfrak{P}^{\alpha\beta}$, $\mathfrak{F}^\alpha$ and V in a G_0 frame. In such a frame, we have

$$v_0{}^i = v_{0i} = 0; \quad v_0{}^4 = v_{04} = c \tag{10.113}$$

We have then (cf. reference 12).

$$\mathfrak{P}_0{}^{ij} = \frac{1}{3} a_R T_0{}^4 \delta^{ij} - \frac{4 a_R T_0{}^4}{15 K_R c} \left(\frac{\partial v_0{}^k}{\partial x_0{}^k} \delta^{ij} + \frac{\partial v_0{}^i}{\partial x_0{}^j} + \right.$$
$$\left. + \frac{\partial v_0{}^j}{\partial x_0{}^i} + 5 \frac{c}{T_0} \frac{\partial T_0}{\partial x_0{}^4} \delta^{ij} \right) \tag{10.114}$$

$$\mathfrak{F}_0{}^i = -\frac{4 a_R T_0{}^4}{3 K_R} \left(\frac{\partial u_0{}^i}{\partial x_0{}^4} + \frac{c}{T_0} \frac{\partial T_0}{\partial x_0{}^i} \right) \tag{10.115}$$

$$V_0 = \mathfrak{P}_0{}^{ii} = a_R T_0{}^4 - \frac{4 a_R T_0{}^4}{3 K_R c} \left(\frac{\partial v_0{}^i}{\partial x_0{}^i} + 3 \frac{c}{T_0} \frac{\partial T_0}{\partial x_0{}^4} \right) \tag{10.116}$$

where K_R is the Rosseland mean absorption coefficient equation (5.18) which will be regarded as a 4-invariant.

From equations (10.97) and (10.114) to (10.116), we have the radiation stress $\mathfrak{P}^{\alpha\beta}$, radiation flux $\mathfrak{F}^\alpha$ and radiation energy V in any G frame are respectively:

$$\mathfrak{P}^{\alpha\beta} = p_R \left(\frac{v^\alpha v^\beta}{c^2} - g^{\alpha\beta} \right) + \mu_R \left[\frac{\partial v^\lambda}{\partial x^\lambda} g^{\alpha\beta} + \frac{\partial v^\alpha}{\partial x^\lambda} g^{\lambda\beta} + \right.$$
$$\left. + \frac{\partial v^\beta}{\partial x^\lambda} g^{\alpha\lambda} - \frac{1}{c^2} \frac{\partial}{\partial x^\lambda}(v^\alpha v^\beta v^\lambda) - 5 \frac{v^\lambda}{T} \frac{\partial T}{\partial x^\lambda} \left(\frac{v^\alpha v^\beta}{c^2} - g^{\alpha\beta} \right) \right] \tag{10.117}$$

$$\mathfrak{F}^\alpha = -5\,\mu_R \left[v^\lambda \frac{\partial\,v^\alpha}{\partial\,x^\lambda} + \frac{c^2}{T}\,\frac{\partial\,T}{\partial\,x^\lambda}\left(\frac{v^\alpha\,v^\lambda}{c^2} - g^{\alpha\lambda}\right)\right] \tag{10.118}$$

$$V = 3\,p_R - 5\,\mu_R \left(\frac{\partial\,v^\lambda}{\partial\,x^\lambda} + 3\,\frac{v^\lambda}{T}\,\frac{\partial\,T}{\partial\,x^\lambda}\right) \tag{10.119}$$

where

$$p_R = \frac{1}{3}\,a_R\,T^4 = \text{radiation pressure} \tag{10.120 a}$$

$$\mu_R = \frac{4\,a_R\,T^4}{15\,K_R\,c} = \text{coefficient of radiative viscosity} \tag{10.120 b}$$

The expression of the radiation stress tensor $p_R^{\alpha\beta}$ is then

$$p_R^{\alpha\beta} = p_R\left(4\,\frac{v^\alpha\,v^\beta}{c^2} - g^{\alpha\beta}\right) + \mu_R\left[\frac{\partial\,v^\lambda}{\partial\,x^\lambda}\,g^{\alpha\beta} + \frac{\partial\,v^\alpha}{\partial\,x^\lambda}\,g^{\alpha\lambda} + \frac{\partial\,v^\beta}{\partial\,x^\lambda}\cdot g^{\alpha\lambda} - \right.$$
$$\left. - \frac{6}{c^2}\,\frac{\partial}{\partial\,x^\lambda}(v^\alpha\,v^\beta\,v^\lambda) + \frac{5}{T}\,\frac{\partial\,T}{\partial\,x^\lambda}\left(v^\lambda\,g^{\alpha\beta} + v^\alpha\,g^{\lambda\beta} + v^\beta\,g^{\alpha\lambda} - \frac{6}{c^2}\,v^\alpha\,v^\beta\,v^\lambda\right)\right] \tag{10.121}$$

The above results such as equation (10.121) is relativistically correct. In ordinary radiation gasdynamics, the flow velocity u is much smaller than the velocity of light c and for any macroscopic quantity Q, the modulus of $\partial\,Q/\partial\,x^4$ is much smaller than the modulus of $\partial\,Q/\partial\,x^i$. If we neglect the higher order terms, equation (10.103) gives:

$$p_{R0}^{ij} = \mathfrak{P}^{ij}\,;\quad q_{R0}^{i} = \mathfrak{F}^i\,;\quad E_{R0} = V \tag{10.122}$$

For the case of small mean free path of radiation, we have

$$p_{R0}^{ij} = p_R\,\delta^{ij} - \mu_R\left(\frac{\partial\,u^k}{\partial\,x^k}\,\delta^{ij} + \frac{\partial\,u^i}{\partial\,x^j} + \frac{\partial\,u^j}{\partial\,x^i}\right) \tag{10.123 a}$$

$$q_{R0}^{i} = -5\,\mu_R\left(\frac{c^2}{T}\,\frac{\partial\,T}{\partial\,x^i} + v_\alpha\,\frac{\partial\,u^i}{\partial\,x^\alpha}\right) \tag{10.123 b}$$

$$E_{R0} \doteq p_{R0}^{ii} = 3\,p_R - 5\,\mu_R\,\frac{\partial\,v^\alpha}{\partial\,x^\alpha} \tag{10.123 c}$$

Equations (10.123) reduce to the same relations given in chapter V, § 8 when the terms of the order of u^2/c^2 are neglected.

9. Rarefied radiation gasdynamics. From the results of section 4 to 8, we see that there is a similarity between the stress tensor of ordinary gasdynamics and the radiation stress tensor. In general, both of these stress tensors should be expressed as integrals of momentum components. However in the extreme cases, these integrals may be greatly simplified. The parameters which characterize the conditions are the mean free path of the gas given by equation (10.43) and the mean free path of radiation which is inversely proportional to the mass absorption coefficient K [cf. eqs. (10.110) or (5.12)]. It is possible to divide the flow field into various regimes according to the values of these mean free paths.

We may use the following non-dimensional parameters to define the rarefication of radiation gasdynamics:

$$K_f = \frac{L_f}{L} = \text{Knudsen number of gasdynamics} \qquad (10.124\,\text{a})$$

$$K_r = \frac{L_R}{L} = \text{Radiation Knudsen number} \qquad (10.124\,\text{b})$$

where L_f is the mean free path of the gas (10.43) which may be expressed in terms of the kinematic viscosity coefficient ν_g as follows:

$$\nu_g = \tfrac{1}{2} L_f \bar{c}_a \qquad (10.125\,\text{a})$$

From equation (10.38) and (10.39), we have

$$\bar{c}_a = (8/\pi\,\gamma)^{\frac{1}{2}}\, a \qquad (10.125\,\text{b})$$

Equation (10.124 a) becomes

$$K_f = 1.255\sqrt{\gamma}\; M/R_e \qquad (10.126)$$

where $M = U/a$ is the Mach number of the flow field, U is the typical velocity of the flow field, $R_e = U L/\nu_g$ is the Reynolds number of the flow field and L is the characteristic length of the flow field.

L_R is the mean free path of radiation defined by equation (5.12). In general L_R depends on the frequency of radiation rays. However in defining the radiation Knudsen number K_r some mean value of the mean free path of radiation may be used. Usually if the mean free path of radiation is small in comparison to L, the Rosseland mean value (5.18) is used and if the mean free path of radiation is large in comparison with L, the Planck mean value is usually used (5.36). It should be noticed that the radiation Knudsen number is inversely proportional to the optical thickness. For small radiation Knudsen number, the gas is optically thick while for large Knudsen number, the gas is optically thin.

When the Knudsen number of gasdynamics K_f is negligibly small, the fluid may be considered as a continuous medium and the stress tensor may be expressed by the Navier-Stokes relations in terms of viscosity coefficient. When the Knudsen number K_f is not negligibly small, the stress tensor should be expressed by complicated relations other than the Navier-Stokes relations. In the extreme case, the discrete character of the gas particles should be considered.

The choice of the characteristic length L depends on the problem considered. Hence we may choose the typical dimension of a body in the flow field as L when we study the forces and heat transfer to the body in a gas flow. We may use the boundary layer thickness on the body as L when we are interested in the skin friction and heat transfer through the boundary layer. When we investigate the transition region in a shock wave, the thickness of the shock wave may be used as L. Because of the various choices of the characteristic length L, whether a gas flow should be considered as a rarefied gas or not, it depends on the problem considered. Once the value of L is chosen, the Knudsen number K_f tells us the degree of rarefication.

We may divide the rarefied gas flow into the following regions according to the Knudsen number. Since Knudsen number K_f is a function of the Mach number M and the Reynolds number R_e, it is convenient to divide the flow field into various regimes according to the values of the Mach number and the Reynolds number (10):

Regimes	Range of M and R_e	
Free molecule flow	$M/R_e \gtreqless 3$	
Transition flow region	$M/R_e \lesseqgtr 3$ & $M/\sqrt{R_e} \gtreqless 0.1$	(10.127)
Slip flow	$0.1 \gtreqless M/\sqrt{R_e} \gtreqless 0.01$	
Continuum flow	$M/\sqrt{R_e} \lesseqgtr 0.01$	

When the Knudsen number is large in comparison to unity, the flow is in the free molecule regime in which the gas particles should be considered to be free in motion without the influence of other particles. When the Knudsen number is very small, we have the continuum flow regime in which the Navier-Stokes equations are good description of the actual flow field. The intermediate regimes, transition region and slip flow, of a rarefied gas is rather complicated. One would expect that the flow will gradually change from the continuum flow regime into the free molecule flow region as the value of the Knudsen number increases. The division of the intermediate regimes into several regimes is usually arbitrary. The most common division is shown in equation (10.127). In the slip flow regime, the fluid still behaves as a continuous medium and the Navier-Stokes relations for the viscous stresses and the velocity gradients holds but there is a slip of velocity on the solid boundary and a jump in temperature on the solid boundary due to the rarefied effects. In the transition flow regime, the flow depends greatly on the discrete character of the gas. In general we should use complicated relations, sometimes differential equations, for the relation between the viscous stress tensor and other macroscopic variables as well as for the heat flux. Our present knowledge of the flow in this transition flow region is very limited.

In the present space age, the flow around a body flying at hypersonic speeds under rarefied gas flow conditions is very important. It is advisable to study the rarefied gas flow field in detail. We shall refine the division of the regimes of rarefied gas flow according to Probstein as follows (9):

In order to give a definite example, we consider the stagnation point region of an axisymmetrical blunt body of radius R_b flying at a hypersonic speed. This is also the problem in which the thermal radiation effect is important as we have discussed in chapter IX, § 8. We consider a detached shock in front of a blunt as shown in Fig. 10.2. The mean free path of the gas behind the shock is L_{fs} is much smaller than that in the free stream $L_{f\infty}$. For hypersonic flow at a very high temperature, we may make the following approximations:

$$\frac{L_{fs}}{L_f} \sim \frac{\delta}{R_b} \sim k_\rho \gtreqless \frac{\rho_\infty}{\rho_s} \ll 1 \qquad (10.128)$$

where δ is the shock detached distance. The density ratio k_ρ across a hypersonic shock wave is of the order of 0.1 to 0.07 or less. The basic condition for the continuum flow may be taken as

$$K_f = \frac{L_{fs}}{\delta} \ll 1 \qquad (10.129\,a)$$

Even though the flow may be considered as a continuum flow so that the Navier-Stokes relations of the viscous stresses may be used, the actual solution of the problem may be different from the classical results according to the values of the Knudsen number K_f. The following subdivision may be made:

(i) Vorticity interaction regime. In this regime, we have

$$K_f = \frac{L_{fs}}{\delta} \ll k_\rho \qquad (10.129\,b)$$

This is essentially the classical case of gasdynamics. The thickness of the shock is thin. Immediately behind the shock, the flow may be considered as inviscid and rotational. There is a very thin boundary layer flow near the body surface. This is the case which we discuss in chapter IX, § 8 (i) and (ii).

(ii) Viscous layer regime. In this regime, we have

$$K_f = \frac{L_{fs}}{\delta} \ll \sqrt{k_\rho} \qquad (20.129\,c)$$

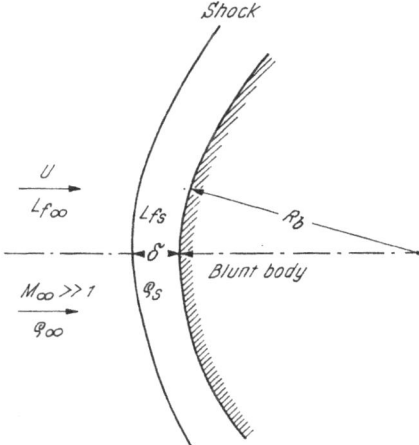

Fig. 10.2. Stagnation point flow of a blunt body in a hypersonic flow

The Reynolds number in this case is smaller than that in case (i). The thickness of the shock is still thin but the flow immediately behind the shock should be considered as a viscous and heat-conducting flow. This is the case which we discussed in chapter IX, § 8 (iii).

(iii) Incipient merged layer regime. In this regime, we have

$$K_f = \frac{L_{fs}}{\delta} \ll 1 \qquad (10.129\,d)$$

The Reynolds number in this case is smaller than that in case (ii). Now the thickness of the shock is no longer thin. In fact, we can not distinguish the shock transition and the viscous region behind the shock.

When the Knudsen number K_{fs} is not negligibly small, we may also define the flow field into three different regimes. The detailed analysis of these regimes should be based on the kinetic theory of gases.

(iv) Fully merged layer regime. In this regime, we have

$$K_f = \frac{L_{fs}}{\delta} \cong 1 \qquad (10.129\,e)$$

In this regime, the shock wave is no longer a discontinuity. A strict treatment of this regime requires the kinetic theory of gases.

(v) Transition layer regime. In this regime, we have

$$K_f = \frac{L_{fs}}{\delta} > 1 \qquad (10.129\,f)$$

The discrete character of the particles in the flow field becomes important.

(vi) First order collision regime. In this regime, we have the local Knudsen number is large but not large enough to reach the free molecule regime. In this analysis, we need to take into account of the collisions between a free stream molecule and a re-emitted molecule.

Finally, when the Knudsen number K_{fs} is much larger than unity, the collision between molecules is negligible and we have the free molecule flow region.

It should be noted that the conductive heat flux is closely related to the viscous stresses as they both are due to the motion of molecules. Hence the conductive heat flux varies similarly as the viscous stresses according to the Knudsen numbers in various regimes discussed above.

As to the radiation stresses and radiative heat flux, we may divide them into three distinguish regimes according to the radiation Knudsen number:

(i) Optically thick case. When the radiation Knudsen number K_s is small, i.e.,

$$K_r = \frac{L_R}{L} \ll 1 \qquad (10.130\,a)$$

Both the radiation stresses and the radiative heat flux may be expressed in terms of the macroscopic variables such as the temperature gradient etc. in the same manner as the corresponding quantities of the flow variables [cf. eqs. (5.15), (5.16) and (5.21)]. In many problems, the radiation pressure number is small, we may consider only the radiative heat flux by increasing the corresponding value of the coefficient of heat conductivity [cf. eq. (8.10)].

(ii) Intermediate case. When the radiation Knudsen number K_r is of the order of unity, i.e.,

$$K_r = \frac{L_R}{L} \cong 1 \qquad (10.130\,b)$$

we have to use the integral expressions for the radiative stresses and heat flux (cf. chapter IX, § 9).

(iii) Optically thin case. When the radiation Knudsen number K_s is large, i.e.,

$$K_r = \frac{L_R}{L} \gg 1 \qquad (10.130\,c)$$

we may use some very simple expressions for the radiation stresses and radiative heat flux as shown in equation (6.35) and similar expressions.

We may also subdivide the above regimes. For instance, when the radiation Knudsen number is extremely small, the no slip condition holds for radiation heat flux as well as the conductive heat flux. If the radiation Knudsen number is small but not negligible, a temperature jump such as in the slip flow regime of rarefied gasdynamics may occur in the radiative heat flux term (cf. chapter VI, § 8).

10. Free molecule flows. Now we are going to discuss the free molecule regime of radiation gasdynamics in some detail. In highly rarefied gas, the heat transfer by conduction and convection in the gas will be small in comparison with that by radiation. Hence thermal radiation becomes a dominant factor in the heat transfer problem. Since the thermal radiation is important, the surface temperature of the body will be greatly influenced by the thermal radiation. The surface temperature has great influence on the free molecule flow. As a result, the thermal radiation would change the free molecule flow significantly and the interaction between the flow variables and the thermal radiation will not be negligible.

In a free molecule flow, the collision between the molecule is much less than the collision between molecules and the surface of the body in the flow field. As a result, we may neglect completely the collisions between the molecules and consider only the interaction between the gas molecules and the surface of the body in the flow field. Thus in the free molecule flow, we may divide the gas molecules into two parts; the incident stream in which the flow velocity and temperature of the gas are known and the molecular distribution function is Maxwellian distribution function (10.34), corresponding to the known flow velocity and temperature; and the reflected stream which composes of the molecules reflected from the surface. The molecular distribution function of the reflected stream could be determined from the known condition of the incident stream and the properties of the surfaces either theoretically or experimentally. Such a determination is rather complicated. However, for engineering problems, it is sufficient to know a thermal accomodation coefficient α_0 and two reflection coefficients σ_f and σ_f' which are defined as follows:

$$\alpha_0 = \frac{dE_i - dE_r}{dE_i - dE_w} \tag{10.131 a}$$

$$\sigma_f = \frac{\tau_i - \tau_r}{\tau_i - \tau_w}, \quad (\tau_w = 0) \tag{10.131 b}$$

$$\sigma_f' = \frac{p_i - p_r}{p_i - p_w} \tag{10.131 c}$$

where subscripts i, r and w refer to the values of the incident stream, the reflected stream and the wall of the body respectively. dE is the energy flux, τ is the tangential momentum on the surface and p is the normal momentum on the surface. The values of α_0, σ_f and σ_f' depend on the properties of the surface as well as the gas and they give some overall phenomenological average properties. The value of α_0 is between 0.87 to 0.97 and the value of σ_f is from 0.79 to 1.00 and no information is available for σ_f'.

There are two special cases which are of interest, i.e., the entirely specular reflection and the entirely diffuse reflection. By specular reflection, we mean that the molecule after collision with the surface has the same tangential velocity component before collision but a reversal of its normal velocity component to the surface. By diffuse reflection we mean that the molecules issue from the surface with a Maxwellian velocity distribution at a temperature not necessarily equal to that of the surface. For entirely specular reflection with vanishing energy exchange, we have $\alpha_0 = \sigma_f = \sigma_f' = 0$ while for entirely diffuse reflection which has been completely accommodated to the wall temperature T_w, we have $\alpha_0 = \sigma_f = \sigma_f' = 1$. Actually, the values of α_0, σ_f and σ_f' are between 0 and 1.

After the values of α_0, σ_f and σ_f' are known, we may calculate the heat transfer and the force on a body from the conditions of the incident stream and that of the surface of the body. In the calculation of the heat transfer the accommodation coefficient α_0 may be different for different modes of energy. For instance, the surface collision will have less effect on the vibrational energy than on the translational energy or rotational energy. But for a first approximation, we may assume that α_0 is the same for all modes of the internal energy.

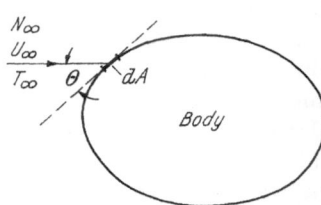

Since it is uncertain for the values of the reflection coefficients, from now on, we consider only the technical surface from which the reflection of the molecules is diffuse but the accommodation coefficient of the surface is a constant different or equal to 1. We consider a uniform free stream with a velocity U_∞ and a temperature T_∞ passing over a convex body so that multi-reflection does not occur. Let us consider the elementary surface on the body dA (Fig. 10.3). We further assume

Fig. 10.3. A body in a free molecule flow

that there is no gain or loss of molecules during the reflection process. The number of the incident molecules crossing a unit area of dA in a unit time is

$$N_i = \frac{N_x c_{m_x}}{2\sqrt{\pi}} \left\{ \exp\left(-S^2 \cos\theta\right) + \sqrt{\pi}\, S \cos\theta \left[1 + \operatorname{erf}\left(S \cos\theta\right)\right] \right\} =$$

$$= N_\infty c_{m_\infty} Z \qquad (10.132)$$

where θ is the local angle of attack and subscript ∞ refers to the values in the free stream, c_m is the most probable speed defined in equation (10.38), $S = U_\infty / c_m$ is known as speed ratio. The parameter Z defined by equation (10.132) is a non-dimensional number flux. The energy fluxes of the incoming stream consist of two parts: One is the translational energy of the molecules E_t and the other is due to the internal degree of freedom E_j. The energy fluxes of these two parts are:

$$E_t = \tfrac{1}{2} m N_\infty c_{m\,\infty}^3 \left\{ \left(\frac{2 + S^2}{\sqrt{2\pi}}\right) \exp\left(-S^2 \sin^2\theta\right) + \left(\frac{5 + 2 S^2}{4}\right) S \cdot \sin\theta \left[1 + \right. \right.$$

$$\left. + \operatorname{erf}\left(\sin\theta\right)\right] \right\} = \tfrac{1}{2} m\, c_{m\,\infty}^2\, N_i\, \varepsilon. \qquad (10.133)$$

$$E_j = \tfrac{1}{2} j\, k\, T_\infty\, N_i \qquad (10.134)$$

where j is the number of internal degrees of freedom that partake in the energy exchange with the surface, k is the Boltzmann constant and ε is the non-dimensional translational energy flux per molecule. Figs. 10.4 and 10.5 show respectively the variation of the non-dimensional number flux Z and the non-dimensional translational energy flux ε.

The energy efflux of the reflected molecules may be expressed in terms of the accommodation coefficient by equation (10.131a). Since we consider the energy flux on a unit area, the expressions of equations (10.133) and (10.134) are actually dE given in equation (10.131). $E_w = (2 + \frac{1}{2}j)\,k\,T_w\,N_i$. Then the reflected energy flux is

$$\frac{E_r}{E_i} = 1 - \alpha\left[1 - \frac{4+j}{2+j}\left(\frac{T_w}{T_\infty}\right)\right] \qquad (10.135)$$

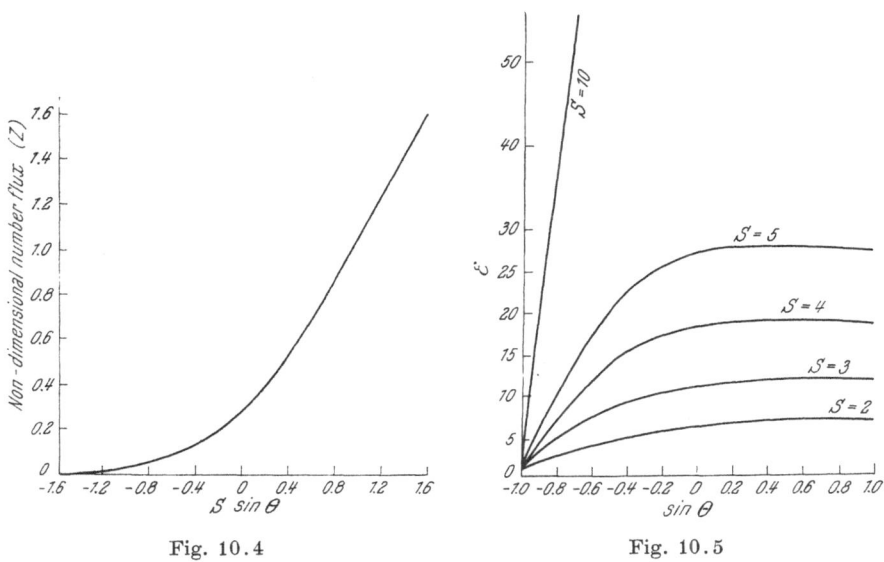

Fig. 10.4   Fig. 10.5

Fig. 10.4. Non-dimensional number Z vs speed ratio perpendicular to the Surface. (Fig. 1 of reference 1 by S. ABARBANEL, Courtesy of AIAA)

Fig. 10.5. Non-dimensional translational energy flux per molecule at various incident angle θ. (Fig. 2 of reference 1 by S. ABARBANEL, Courtesy of AIAA)

Now we may examine the heat transfer in a free molecular flow including the radiation effect. For highly rarefied gas, we may neglect the absorption and emission of the thermal radiation by the gas molecules. Hence we need to consider the emissivity of the surface of the body only. Furthermore, in order to determine the temperature of the surface, the heat conductivity of the solid surface is very important. We shall consider the following three cases:

(i) Insulated surfaces. The surface temperature distribution of a thermally insulated body depends on the local heat fluxes only. The balance of the energy fluxes gives:

$$E_t + E_j = E_r + e_w\,\sigma\,T_n{}^4 \qquad (10.136)$$

In non-dimensional form, equation (10.136) becomes

$$\eta_n{}^4 + \eta_n = H_n \qquad (10.136a)$$

where
$$\eta_n = \left[\frac{2\,e_w}{(4+j)\,\alpha\,k\,N_\infty\,c_{m_\infty}\,Z}\right]^{\frac13} T_n \tag{10.136 b}$$

and
$$H_n = \left(\frac{2\,\varepsilon+j}{4+j}\right)\eta_\infty = \left(\frac{2\,\varepsilon+j}{4+j}\right)\left[\frac{2\,e_w}{(4+)\,\alpha\,k\,N_\infty\,c_{m_\infty}\,Z}\right]^{\frac13} T_\infty \tag{10.136 c}$$

T_n is the temperature of the insulated surface, e_w is the surface emissivity and σ is the Stefan Boltzmann constant for the radiative flux. Since H_n is known for any given problem, equation (10.136 a) gives the surface temperature T_n. For large value of speed ratio S, we have $S \gg 1$ and

$$\eta_n = H_n^{\frac14} \tag{10.137}$$

and
$$T_n = \left[\frac{(2\,\varepsilon+j)\,\alpha\,N_\infty\,k^{\frac32}\,T_\infty^{\frac32}\,Z}{\sqrt{2\,m}\,\,e_w\,\sigma}\right]^{\frac12} \tag{10.137 a}$$

(ii) Perfect conducting surface. For perfect conducting surface, we have to consider the balance of the total energy balance over the whole body, i.e.,

$$\int\int (E_t + E_j)\,\mathrm{d}A = \int\int E_r\,\mathrm{d}A + \int\int e_w\,\sigma\,T_c^4\,\mathrm{d}A \tag{10.138}$$

In non-dimensional form, equation (10.138) becomes

$$\eta_c^4 + \eta_c = H_c \tag{10.138 a}$$

where
$$\eta_c = \left[\frac{2\,e_w\,\sigma\int\int\mathrm{d}A}{(4+j)\,\alpha\,k\,N_\infty\,c_{m_\infty}\int\int Z\,\mathrm{d}A}\right]^{\frac13} T_c \tag{10.137 a}$$

and
$$H_c = \left[\frac{(2\,\varepsilon+j)\int\int Z\,\mathrm{d}A}{(4+j)\int\int Z\,\mathrm{d}A}\right]\left[\frac{2\,e_w\,\sigma\int\int\mathrm{d}A}{(4+j)\,\alpha\,k\,N_\infty\,c_{m_\infty}\int\int Z\,\mathrm{d}A}\right]^{\frac13} T_\infty \tag{10.138 c}$$

T_c is the temperature of the perfect conductor. Since equation (10.138) consists of the following ratios of integrals:

$$N_a = \frac{\int\int Z\,\mathrm{d}A}{\int\mathrm{d}A} \tag{10.139}$$

and
$$F_a = \frac{\int\int(2\,\varepsilon+j)\,Z\,\mathrm{d}A}{(4+j)\int\int Z\,\mathrm{d}A} \tag{10.140}$$

the results of the solution of equation (10.138 a) depend greatly on the geometery of the body considered. In reference 1, Abarbanel calculated the ratios N and F for various simple configurations:

For a flat plate at an angle of attack θ or a symmetrical double wedge with the semi-wedge angle θ, we have

$$N_a = \frac{\exp(-h_0{}^2) + \sqrt{\pi}\, h_0 \operatorname{erf} h_0}{2\sqrt{\pi}}; \quad F_a(4+j) = 2\,S^2 + 5 + j -$$

$$-[1 + \sqrt{\pi}\, h_0 \exp(h_0{}^2 \operatorname{erf} h_0]^{-2} \tag{10.141}$$

where $h_0 = S \sin \theta$. $\operatorname{erf} = $ error function.

For a sphere, we have

$$N_a = \frac{2\exp(-S^2) + \sqrt{\pi}\,(S^{-1} + 2\,S)\operatorname{erf} S}{8\sqrt{\pi}} \tag{10.142 a}$$

$$(4+j)\,F_a = 2\,S^2 + 5 + j - (\tfrac{1}{2}\sqrt{\pi}\operatorname{erf} S)\,[S\exp(-S^2) + \sqrt{\alpha}\,(1+2\,S^2)\operatorname{erf} S]^{-1} \tag{10.142 b}$$

(iii) Body with finite thermal conductivity.

For a body of finite thermal conductivity, we have to solve the heat conduction problem in the body with the boundary condition determined by the heat flux balance on the surface of the body. The fundamental equation of heat conduction for steady case is simply the Laplace equation of temperature T in the body, i.e.,

$$\nabla^2 T = 0 \tag{10.143a}$$

equation (10.143a) should be solved with the following conditions on the external surface of the body:

$$\varkappa \frac{\partial T}{\partial n} + e_w \sigma T^4 + E_r = E_t + E_j \tag{10.143 a}$$

where $\varkappa$ is the coefficient of thermal conductivity of the body and n is the direction of the outward normal of the surface of the body.

In non-dimensional form, equations (10.143) may be written as follows:

$$\nabla^2 \eta = 0 \tag{10.144a}$$

$$\frac{\partial \eta}{\partial n*} + \eta^4 + X\eta = H \quad \text{(on the external surface)} \tag{10.144 b}$$

where
$$n* = \frac{n}{L}; \quad X = \tfrac{1}{2}(4+j)\left(\frac{\alpha\, k\, L\, N_i}{\varkappa}\right);$$

$$H = \left(\frac{e_w \sigma L^4}{4\,\varkappa}\right)^{\frac{1}{3}} \cdot \tfrac{1}{2}(2\varepsilon + j)\, k\, N_i\, T; \quad \eta = (e_w \sigma L \varkappa)^{\frac{1}{3}}\, T$$

For many practical cases, the term $X\eta$ in equation (10.144b) is negligible. Some simple configurations such as the flat plate, solid sphere and thin spherical shell have been studied by ABARBANEL in reference 1. For example, for a conducting flat plate of small thickness so that it is essentially a solid slab with linear temperature distribution, equation (10.144b) may be approximately written as follows:

$$\eta_1{}^4 + \eta_1 - \eta_2 = H_1 \tag{10.145a}$$

$$\eta_2{}^4 + \eta_2 - \eta_1 = H_2 \tag{10.145b}$$

The surface temperature η_1 and η_2 or T_1 and T_2 may be easily determined from equations (10.145).

So far we assume that there is no radiation source in the incoming stream. For space flight, there may be radiation source in the incoming stream such as the solar radiation. E_r may easily take the solar radiation into consideration by adding the solar energy flux on the elementary surface by the amount:

$$E_s = e_s Js \cos \theta_s \qquad (10.146)$$

where J_s is the local solar constant, e_w is the surface absorptivity which is equal to the emissivity and θ_s is the angle between the direction of sun rays and the normal of the surface d A.

References

1. ABARBANEL, S.: Radiative heat transfer in free molecule flow. Jour. Aero. Sci. vol. 28, No. 4, pp. 299–307, April 1961.
2. BURGERS, J. M.: The bridge between particle mechanics and continuum mechanics. Proc. of Plasma Dynamics, Ed. by F. Clauser, Addison Wesley Pub., pp. 119–186, 1960.
3. CHAPMAN, S., and T. G. COWLING: The mathematical Theory of Non-uniform Gases. Cambridge University Press, 1939.
4. HARTNETT, J. P.: A survey of thermal accommodation coefficients. Proc. of 2nd International Symp. on Rarefied Gasdynamics, Academic Press, New York, 1960.
5. HIRSCHFELDER, J. O., C. F. CURTISS, and R. B. BIRD: Molecular Theory of Gases and Liquids. John Wiley & Sons, Inc., New York, 1954.
6. PAI, S. I.: Magnetogasdynamics and Plasma Dynamics. Springer Verlag, Vienna and Prentice Hall Inc., N. J., 1962.
7. PATTERSON, G. N.: Molecular flow of gases. John Wiley & Sons, Inc., New York, 1956.
8. PATTERSON, G. N.: A state of art survey of some aspects of the mechanics of rarefied gases and plasma. UTIAS Review No. 18, Institute of Aerospace Studies, Univ. of Toronto, March 1964.
9. PROBSTEIN, R. F.: Continuum theory and rarefied hypersonic aerodynamics. Rarefied Gas Dynamics, Pergamon Press, pp. 416–431, 1960.
10. SCHAAF, S. A., and P. L. CHAMBRE: Flow of rarefied gases. Sec. H. of Fundamental of Gas Dynamics, vol. III. High Speed Aerodynamics and Jet propulsion, Princeton Univ. Press, 1958.
11. SIMON, R.: The conservation equations of a classical plasma in presence of radiation. A & ES rept. 62-1, Purdue University, 1962.
12. SYNGE, J. L.: The relativistic gas. North-Holland Pub. Co., Amsterdam, 1957.

Chapter XI

Radiative Properties of High Temperature Gases

1. Introduction. One of the most important properties of gases in our previous discussions of radiation gasdynamics is the coefficient of absorption of gas. In our macroscopic analysis of radiation gasdynamics, the absorption coefficient is assumed to be a known function of the state variables: density and temperature of the gas, as well as a function of the composition of the gas mixture. In order to determine the coefficient of absorption, we have to use the microscopic analysis or the experimental method. The absorption coefficient represents an average but complicated physical process in the atoms of the gases due to the interaction of matter and energy. Since they are the physical processes in the atoms, we have to use the quantum theory in order to obtain the accurate results. However, the classical theory of absorption and emission of radiation will give us qualitatively correct results in a rather simple manner. Hence we shall first discuss the classical theory in section 2.

Thermal radiation is simply one form of the electromagnetic radiation. In fact, visible light, thermal radiation, radio waves, x-rays etc., are electromagnetic waves. The only difference between them is their wave lengths. Hence the study of the radiation of electromagnetic energy will give us all the essential features of absorption and emission of thermal radiation. In section 2, we shall consider the case with the classical formulation of an atomic model in order to explain the absorption of electromagnetic radiation; while in section 3, we discuss briefly the quantum theory of radiation. Since the quantum theory of radiation (cf. reference *11*. HEITLER's book) is a very complicated subject, we can only point out a few of the most important results of the theory of radiation. For those readers who want to obtain detailed results, special treatise should be referred to (*11*).

In section 4, we shall describe some important results of spectroscopy of high temperature gases which is useful for our study of the absorption coefficient of thermal radiation. Since one of the most important medium in radiation gasdynamics is air, we shall discuss the absorption and emission of radiation in high temperature air in sections 5 and 7. Another important gas is hydrogen gas which will also be discussed. We are especially interested in the variation of the absorption coefficient with the state variables and their mean values over the entire frequency range such as the Planck mean value and the Rosseland mean value (§ 7). The scattering coefficient will be discussed in § 6.

The processes involved in the radiation by gases are very complicated. Usually the gases of interest, such as air, are mixture of many species, each of which is made of complex molecules. Furthermore, a gas may be contaminated from the pure radiation point of view. The physical processes responsible for absorption and emission of radiation in gases are different in different temperature range. Hence the complete theoretical analysis of high temperature gas or gas mixture is very complicated. Many useful information of absorption coefficient of high temperature gas may be obtained by experiments. In section 8, we shall discuss a few of the experimental technique and results.

In most of the analysis in the previous chapters, we consider only the case of local thermodynamic equilibrium conditions. The theoretical calculation of absorption coefficient of high temperature gas is usually based on the equilibrium condition. In actual flow problems, the thermal radiation may not be in local thermodynamic equilibrium. The study of non-equilibrium radiation in gas-dynamics is still in a very preliminary stage. We shall discuss very briefly some essential points in the non-equilibrium radiation in section 9.

2. Classical theory of absorption and emission of radiation. In the determination of the absorption and emission of radiation in a medium, we have

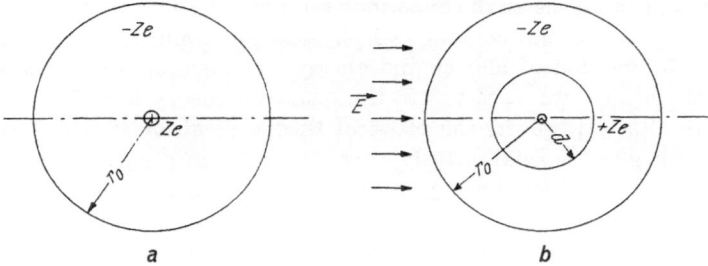

Fig. 11.1. Electric dipole of an atom

to know the structure of the molecules of the medium because the absorption and emission of radiation energy is related with the motion of a system of charges in the medium. For simplicity, we consider the case of a monatomic gas. In the classical theory, the atom is postulated to consist of a concentrated nucleus of charge Ze surrounded by an electron atmosphere of charge $-Ze$ where e is the absolute electric charge of an electron and Z is the atomic number. The atmosphere of electrons may be assumed to confine in a sphere of radius r_0 (Fig. 11.1a). Without external applied electric field, the charge center of the electron atmosphere coincides with the nucleus. With an external applied electric field, the charge center of the electron cloud will displace a distance d (Fig. 11.1b). Hence we say that the external electric field induces in the atom a dipole of moment:

$$M_0 = Zed = \alpha_e E \qquad (11.1)$$

Because the force to produce the displacement is proportional to the external electric field E and the dipole moment is also proportional to the electric field, the proportional factor α_e is known as the electronic polarizability of the atom. Here $\alpha_e = 4\pi\epsilon_0 r_0^3$ and ϵ_0 is the dielectric constant in free space.

Now if we assume that the external field, i.e., the electric field of the incident radiation wave, is varying with time, the charge center will follow this oscillation and energy will be absorbed by the atom. Let us consider a simple case in which the incident electric field has only one component in the z-direction and varies sinusoidally in time, i.e.,

$$\vec{E} = \vec{i_z}\, E_{z0} \exp\,(i\,w\,t) \tag{11.2}$$

where E_{z0} is a constant and $w = 2\,\pi\,\nu$ and ν is the frequency of the wave. We assume that the motion of an electron may be a simple oscillator with a damping force proportional to its velocity and restoring force proportional to its distance from equilibrium condition. The equation of motion of an electron under the influence of the external electric field E is then

$$z'' + \gamma_a\,z' + w_0{}^2\,z = \frac{e\,E_{z0}}{m}\exp\,(i\,w\,t) = \frac{e\,E}{m} \tag{11.3}$$

where m is the mass of an electron, $g = m\,\gamma_a$ is the proportional constant of the damping force and $K_a = m\,w_0{}^2$ is the proportional constant of the restoring force. The resonance frequency of the oscillator is $w_0 = 2\,\pi\,\nu_0$. The distance z is measured from the equilibrium position where $\vec{E} = 0$. We need to consider the particular solution of z of equation (11.3) because the complementary solution represents the transient effects when the radiation first strikes the atom. The particular solution is

$$z = \frac{\dfrac{e}{4\,\pi^2\,m}\,E_{z0}\exp\,(2\,\pi\,i\,\nu\,t)}{\nu_0{}^2 - \nu^2 + i\,\gamma_a\,\dfrac{\nu}{2\,\pi}} \tag{11.4}$$

The distance z is the dipole distance d shown in Fig. 11.1 b.

For more exact analysis of absorption of a gas, we should use the local electric field $\vec{E'}\,(t)$ instead of the external electric field $\vec{E}$ in the equation of motion (11.3). The relation between the local electric field $\vec{E'}$ and the external electric field $\vec{E}$ is given by the CLAUSIUS-MOSOTTI-LORENTZ equation (21):

$$\vec{E'} = \vec{E} + \frac{\vec{P}}{3\,\epsilon_0} \tag{11.5}$$

where P is the polarization vector, which is

$$\vec{P} = N e \vec{z} \tag{11.6}$$

where we assume that there are N oscillators per unit volume of the gas, each having a dipole moment ez. Substituting equations (11.5) and (11.6) into equation (11.3), we have a differential equation for the polarization vector P as follows:

$$P'' + \gamma_a\,P' + w_e{}^2\,P = \frac{N\,e^2}{m}\,E = \frac{N\,e^2}{m}\,E_{z0}\exp\,(i\,w\,t)\cdot \tag{11.7}$$

where

$$w_e{}^2 = w_0{}^2 - \frac{N\,e^2}{3\,m\,\epsilon_0} \tag{11.8}$$

Except the difference of some constants, equation (11.7) is identical to equation (11.3). Hence the steady state solution for P is

$$P = \frac{\dfrac{Ne^2}{4\pi^2 m} E_{z0} \exp(2\pi i \nu t)}{\nu_e^2 - \nu^2 + i\gamma_a \dfrac{\nu}{2\pi}} \tag{11.9}$$

If there are more than one type of oscillators, the polarization vector is equal to the sum of the polarization vectors due to all the oscillators. We shall consider only one type of oscillators here in order to illustrate the principle (cf. reference 21).

The dielectric constant of the medium or the permeability of the medium is

$$\epsilon = 1 + \frac{P}{E\,\epsilon_0} = 1 + \frac{\dfrac{Ne^2}{4\pi^2 m\,\epsilon_0}}{\nu_e^2 - \nu^2 + i\nu\dfrac{\gamma_a}{2\pi}} \tag{11.10}$$

The complete index of refraction may be written as follows:

$$n - i\varkappa_a = \sqrt{\epsilon} = 1 + \frac{\dfrac{Ne^2}{8\pi^2 m\,\epsilon_0}}{\nu_e^2 - \nu^2 + i\nu\dfrac{\gamma_a}{2\pi}} + \text{ higher order terms} \tag{11.11}$$

If we neglect the higher order terms, we have

$$n = \text{ ordinary index of refraction } = 1 +$$

$$+ \frac{Ne^2(\nu_e^2 - \nu^2)}{8\pi^2 m\,\epsilon_0 \left[(\nu_e^2 - \nu^2) + \nu^2 \left(\dfrac{\gamma_a}{2\pi}\right)^2 \right]} \tag{11.12}$$

and $\qquad \varkappa_a = \text{ absorptivity } = \dfrac{Ne^2\nu(\gamma_a/2\pi)}{8\pi^2 m\,\epsilon_0 \left[(\nu_e^2 - \nu^2)^2 + \nu^2\left(\dfrac{\gamma_a}{2\pi}\right) \right]} \tag{11.13}$

For many practical problems, the effective resonance frequency w_e is very close to the resonance frequency w_0. If we are interested in the frequency range ν close to the resonance frequency ν_0, we have the following approximate formula for the index of refraction n and the absorptivity:

$$n = 1 + \frac{Ne^2(\nu_0 - \nu)}{16\pi^2 m\,\epsilon_0\nu\left[(\nu_0 - \nu)^2 + \left(\dfrac{\gamma_a}{2\pi}\right)^2 \right]} \tag{11.14}$$

$$\varkappa_a = \frac{Ne^2(\gamma_a/2\pi)}{32\pi^2 m\,\epsilon_0\nu\left[(\nu_0 - \nu)^2 + \left(\dfrac{\gamma_a}{2\pi}\right)^2 \right]} \tag{11.15}$$

Equation (11.14) shows that $(n-1)$ changes sign at $\nu = \nu_0$ while equation (11.15) show that $\varkappa_\alpha$ reaches a maximum at $\nu = \nu_\alpha$. Fig. 11.2 shows the variation of the absorptivity with frequency.

We are specially interested in the relation between the absorptivity $\varkappa$ and the coefficient of absorption of radiation k_ν. It is well known that the electric field E is perpendicular to the direction of propagation. We may assume that the direction of propagation of radiation is in the x-direction. If the phase velocity of the radiation wave is v which is given by the formula:

$$v = \frac{c}{n - i\varkappa_a} \tag{11.16}$$

The z-component of the electric field travelling in the positive x-direction is then

$$E = E_{z0} \exp\left[2\pi i\nu\left(t - \frac{x}{v}\right)\right] \tag{11.17}$$

The specific intensity of radiation is proportional to the square of the amplitude of the electric field. Hence we have

$$I = I_0 \exp\left(-4\pi\varkappa_\alpha\nu x/c\right) = I_0 \exp\left(-k_\nu\rho x\right) \tag{11.18}$$

where we assume that the variation of the coefficient of absorption with x is small and equation (11.18) is essentially the same as that given in equation (3.2). From equation (11.18), we have immediately the expression for the coefficient of the absorption of radiation as a function of the frequency ν and other properties of the atoms, i.e.,

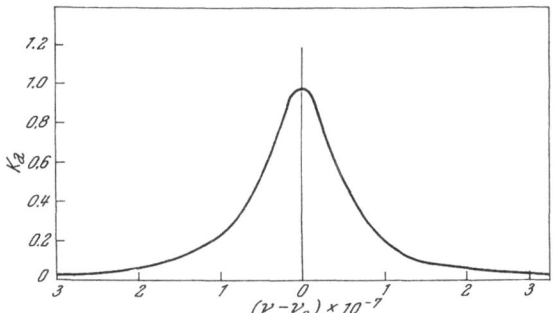

Fig. 11.2.   Variation of absorptivity $\varkappa_a$ with frequency

$$\rho k_\nu = \frac{4\pi\varkappa_a\nu}{c} = \frac{N e^2 (\gamma_a/2\pi)}{8\pi m\epsilon_0 c\left[(\nu_0 - \nu)^2 + \left(\frac{\gamma_a}{4\pi}\right)^2\right]} \tag{11.19}$$

It is interesting to notice that the more accurate formula of quantum theory has the same form as equation (11.19) except that the number density N is replaced by Nf where f is known as the oscillator strength and the damping factor γ_α is replaced by another factor of different meaning.

Even though the classical theory gives qualitatively correct formula for the coefficient of absorption, there are many basic defects in the classical theory. The classical theory does not explain the existence of different natural frequencies in the atom and it does not predict the correct width of the spectral lines. In order to get the correct results of the coefficient of absorption, we have to use the quantum theory.

The complementary solution of equation (11.3) gives the emission of radiation of a free atom. This solution may be written as follows:

$$z = z_0 \exp\left(-\tfrac{1}{2}\gamma_\alpha t\right) \cdot \cos\left(w'' t\right) \tag{11.20}$$

where

$$w''^2 = w_0^2 - (\tfrac{1}{2}\gamma_\alpha)^2 \tag{11.21}$$

The intensity of radiation decreases exponentially with time, i.e.,

$$\frac{I}{I_0} = \exp\left(-\gamma_a t\right) \tag{11.22}$$

If we express the amplitude of the oscillator in Fourier integral, we have from equation (11.20)

$$z = \frac{1}{\sqrt{\pi}} \int_0^\infty a\,(w)\ \cos wt\ dw + \frac{1}{\sqrt{\pi}} \int_0^\infty b\,(w)\ \sin wt\ dw \tag{11.23}$$

where

$$a\,(w) = \frac{z_0}{2\sqrt{\pi}} \left[\frac{\frac{1}{2}\gamma_a}{(\frac{1}{2}\gamma_a)^2 + (w - w_0)^2} + \frac{\frac{1}{2}\gamma_a}{(w + w_0)^2 + (\frac{1}{2}\gamma_a)^2} \right]$$

$$b\,(w) = \frac{z_0}{2\sqrt{\pi}} \left[\frac{w - w_0}{(\frac{1}{2}\gamma_a)^2 + (w - w_0)^2} + \frac{w + w_0}{(w + w_0)^2 + (\frac{1}{2}\gamma_a)^2} \right]$$

The intensity of radiation I is then

$$I = C\,[a^2\,(w) + b^2\,(w)] = \frac{C}{4} \left\{ \frac{1}{(\frac{1}{2}\gamma_a)^2 + (w - w_0)^2} + \right.$$

$$\left. + \frac{1}{(\frac{1}{2}\gamma_a)^2 + (w + w_0)^2} + \frac{2\,[(\frac{1}{2}\gamma_a)^2 + (w + w_0)\,(w - w_0)]}{[(\frac{1}{2}\gamma_a)^2 + (w + w_0)^2]\,[(\frac{1}{2}\gamma_a)^2 + (w - w_0)^2]} \right\} \tag{11.24}$$

where C is a proportional constant.

Near the resonance frequency w_0, only the first term of the right hand side of equation (11.24) is important and we have then

$$\frac{I}{I_0} = \frac{1}{1 + \left(\dfrac{w - w_0}{\frac{1}{2}\gamma_a}\right)^2} \tag{11.25}$$

where $w - w_0 = \pm \frac{1}{2}\gamma_a$, $I/I_0 = \frac{1}{2}$. Hence γ_a characterises the half width of the spectral line.

3. The quantum theory of radiation (*11*). The accurate theory of radiation must be based on the quantum mechanics. In fact, the quantum was discovered by Planck in his study of thermal radiation. We do not plan to review the quantum mechanics and assume that the readers already know the basic concepts of quantum mechanics. Here we shall only discuss those points of quantum theory which are important in evaluation of the absorption coefficient and emission coefficient of thermal radiation.

Since radiation represents the interaction of electromagnetic field with the electrons in the molecules. We have to have a good description of the structure of the molecules before we can postulate a good theory of radiation.

An atom or a monatomic molecule consists of a heavy nucleus of positive charge around which the electrons revolve. For neutral gas, the net electric charge is zero. The positive charge of the nucleus is equal and opposite to the total negative change of electrons around it. A molecule consists of a number of heavy nuclei, the atomic nuclei of the atoms or ions which form the molecule;

round these nuclei the electrons revolve. For atoms or monatomic molecules, the kinetic energy due to the rotational motion of the particles is negligible small; while for a molecule, the kinetic energy due to rotational motion of the molecule is of the same order of magnitude as that of translational motion. In equilibrium condition, we have an energy of $\frac{1}{2} k T$ per degree of freedom whether it is due to translational motion or rotational motion where k is the Boltzmann constant and T is the temperature.

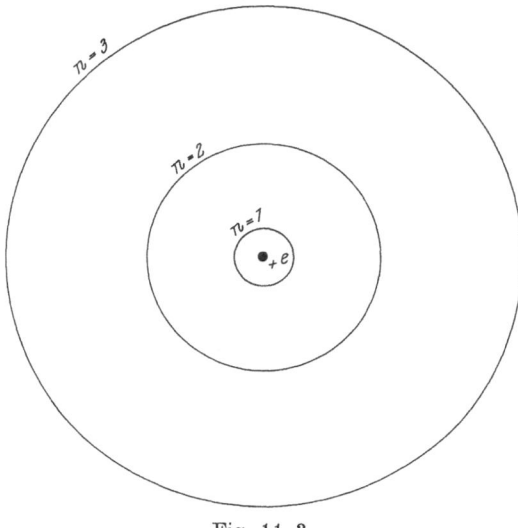

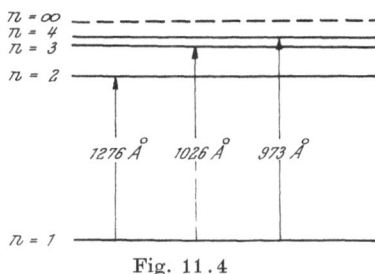

Fig. 11.3   Fig. 11.4

Fig. 11.3. Orbits of an electron around a nucleus of a hydrogen atom
Fig. 11.4. Energy levels and the most important lines of the hydrogen spectrum

For molecules, the relative motion of the nuclei, i.e., the vibrational mode of internal energy, may be also of the same order of magnitude as that of the translational motion. These vibrational energies are important at temperature higher than 2,000 °K.

There is a given amount of energy associated with each electron revolving around the nuclei. In order to see clearly such energy associated with an electron, we consider the simplest case: the Bohr's hydrogen atom which consists of a small heavy nucleus with an electron moving as a satellite in a circular orbit about the nucleus. The nucleus has a mass approximately equal to that of the hydrogen atom and a positive electric charge equal in magnitude to the charge of electron. According to the quantum theory of atom, the radii "a" of the orbit of the electron are limited to a number of possible stable states which is given by the formula:

$$a = a_1 \frac{n^2}{Z} = \frac{n^2 h^2}{4 \pi^2 m Z e^2} \qquad (11.26)$$

where a_1 is the Bohr radius and Ze is the positive electric charge of the nucleus and n is known as the first quantum number which may be any positive integers, i.e., 1, 2, 3, ... For every integer n, we have a possible orbit (Fig. 11.3) and the energy level associated with this orbit is E_n which is given by the formula:

$$E_n = - \frac{Z e^2}{2 a} = - \frac{Z^2 a^2 m 4 \pi^2}{2 n^2 h^2} \qquad (11.27)$$

where e is the absolute charge of an electron and m is the mass of an electron. The energy scale (Fig. 11.4) is so normalized that its zero point corresponds to $n \to \infty$. Hence $-E_n$ is the work needed to remove the electron from the state n to rest at infinity and it is the ionization energy from the state n. Normally, the electron is in the ground state which corresponds to $n = 1$ and the radius of this grounds state is $a_1 = h^2/(4\,\pi^2\,m e^2) = 0.528$ Å which is generally known as the "Bohr radius." The energy $-E_1$ is the ionization energy from the ground state which is usually expressed in electron volts and is referred to as ionization potential. If an atom receives an amount of energy greater than its ionization potential energy from the ground state, one of the electron will move away from the influence of the nucleus and the atom becomes an ion. Ordinarily, the electron is in the ground state. If the electron is in a state other than the ground state, the atom is said in an excited state. For a transition from a state E_n to another state E_m, the energy emitted or absorbed is according to the formula:

$$E_n - E_m = h\,\nu \tag{11.28}$$

where h is the Planck constant and ν is the frequency of the photon emitted.

The hydrogen atom is the simplest type of atom of which the nucleus is a singly charged particle known as a proton. For other heavy atoms, the nucleus consists of a number of protons equal to its atomic number Z which is roughly one half the atomic weight. The balance of the atomic weight is made up of the electrically neutral particles, called neutrons, having the same mass as the protons. Thus, helium, whose atomic number is two (2) and atomic weight four (4) has two protons and two neutrons in its nucleus. In isotopes of an element, the number of protons is the same for all isotopes of that element but the number of neutrons is different so that various isotopes have different masses. For a given element, there are ordinarily Z number of electrons revolving around its nucleus. Each of the electrons has its own state, which is specified by various quantum numbers. No two electrons in an atom can have the same values of all the five quantum numbers. The electrons are normally in various shells according to the total quantum number n. For $n = 1$, the shell is called K-shell in which no more than two electrons may exist. The electron in hydrogen atom and the electrons in the helium atom are usually in the K-shell. For $n = 2$, the shell is known as L-shell which will be filled by 8 electrons. For elements, if its electrons fill the shell, it is an inert element; while if the outmost shell has less electrons than the number required to fill it, the element is chemically active. Hence hydrogen atom is a chemically active element while helium is an inert gas. If the electron of an atom is not in the lowest state, the atom is excited. If one of the electrons moves away from the orbit $n \to \infty$, the atom is singly charged, i.e., ionized. If more than one electrons move away, the atom is multiplied charged.

When two or more atoms form a molecule, there are binding force between their nuclei and some energy, the chemical energy, may release during the formation. On the other hand, if a molecule receives an amount of energy, the atoms in it move away from one another so that they will not behave as a single particle. Such a process is called dissociation. For high temperature gas flow, the processes of dissociation and recombination are very important.

The main feature of the quantum theory of absorption and emission of radiation by atoms is that the electrons in the atom can exist only at certain energy levels. Hence emission of energy occurs when an electron passes from a higher level to a lower level according to equation (11.28). Thus the set of allowed light frequencies is discrete and is characteristic of the atom type. Similarity, absorption of energy occurs when an electron passes from a lower level to a higher level also according to equation (11.28). If the energy transferred is larger than the ionization energy, an electron may move from a bound state to a free state. We refer to these photon-ionization as bound-free transition. For a free electron, the energy may be emitted or absorbed continuously as in the case of classical theory. We refer to these case as free-free transition.

Since in the determination of absorption coefficient of a gas, we consider a large number of atoms and molecules in various states, we have to use statistical approach. In other words, we have to find the probability of the various transitions. The most successful treatment is the use of Einstein's probability coefficients (cf. chapter III, §§ 2 and 3). These Einstein's probability coefficients are:

(i) Spontaneous emission probability coefficient A_{nm} is the probability of an atom in the excited state n to emit a quantum of energy $h \nu_{nm}$ in the absence of external radiation field and in the direction of the wave within an elementary solid angle $d\omega$ and in the time interval dt.

(ii) Absorption probability coefficient B_{mn} is the probability of an atom in a state m exposed to radiation of frequency ν_{mn} absorbing a quantum $h \nu_{mn}$ in time dt.

(iii) Stimulated or induced emission probability coefficient B_{nm} is the probability of the emission of a quantum $h \nu_{nm}$ induced by an external radiation field of frequency ν_{nm}.

It should be noticed that these Einstein probability coefficients are atomic constants which may be determined experimentally or theoretically by quantum mechanics. There are definite relations between these coefficients. Let us consider the case of thermal equilibrium conditions. Under this condition, the relative population of atoms of two levels n and m is given by the Boltzmann formula:

$$N_n g_m = N_m g_n \exp\left(-h \nu_{nm}/k T\right) \tag{11.29}$$

where N is the number density of the atoms, g is the statistical weight of a level and subscript n and m refer to the state n and m respectively. In the thermal equilibrium condition, the variation of specific intensity is zero and from equation (3.34) we have

$$N_n \left(A_{nm} + B_{nm} I_\nu\right) = N_m B_{mn} I_\nu \tag{11.30}$$

where we write ν for ν_{nm} for simplicity. The specific intensity in the thermal equilibrium condition is the Planck radiation function B_ν of equation (4.22). Hence equation (11.30) becomes

$$g_n A_{nm} = g_m \frac{2 h \nu^3}{c^2} B_{mn} \left\{ \frac{\exp\left(\dfrac{h \nu}{k T}\right) - (g_n B_{nm}/g_m B_{mn})}{\exp\left(h \nu/k T\right) - 1} \right\} \tag{11.31}$$

Since Einstein probability coefficients are constants of atoms only and independent of the temperature, we should have the following relations from equation (11.31), i.e.,

$$g_n B_{nm} = g_m B_{mn} = g_n A_{nm} \frac{c^2}{2 h \nu^3_{nm}} \qquad (11.32)$$

If we know one of the Einstein probability coefficients, we may obtain the rest of the coefficients from equation (11.32). In the study of the absorption coefficient by quantum theory, attempt has been made so that similar formula to the classical theory may be found. In doing so, it was found that the Einstein absorption probability coefficient B_{mn} may be expressed in terms of oscillator strength f_{mn} which is usually referred to as f-value, i.e., $f_{mn} = f$. We have:

$$B_{mn} = \frac{\pi e^2}{h \nu m c} f_{mn} \qquad (11.33)$$

For simple atoms, the f-value may be easily calculated by the quantum mechanics. But for complex molecules, the calculation of the f-value is very tedious and only approximate results can be obtained. Hence one of the current research problems in the high temperature gas is the determination of the gf-value of the gas experimentally.

After the f-value is obtained, the absorption coefficient based on the quantum theory is expressed by a similar formula (11.19) except that the number density N is replaced by Nf and the damping coefficient γ_α by a new constant Γ, i.e.,

$$\rho k_\nu = \frac{Nf e^2 \Gamma}{4 \pi m \epsilon_0 c [4 \pi^2 (\nu_0 - \nu)^2 + (\frac{1}{2} \Gamma)^2]} \qquad (11.34)$$

4. Spectroscopy of high temperature gas. Because of the bound-bound transition, each type of atom will emit light at various definite frequencies. These characteristic wave lengths (6) may be photographed on ordinary films such as the one shown in Fig. 11.5. These prominant lines are commonly called the spectral lines of the substance. The spectral lines depend on the composition of the gas mixture and the temperature. For a given composition and temperature, the location of the spectral lines depends on the atomic constant. One of the interesting research work of spectroscopy is to determine the atomic constants such as the gf-values, line strength etc., from the spectrum diagram of a gas of known composition and temperature.

Even though we say spectral lines, they are not really lines in the strict sense, i.e., a line without width. Actually, the lines have shapes such as that sketched in Fig. 11.6. There are many physical processes which spread the spectral line energy out. We shall not discuss them here (6, 22). But empirically, we may use a shape function $P(\nu)$ to take into account of the line broadening. The shape function satisfies the condition:

$$\int_0^\infty P(\nu) \, d \nu = 1 \qquad (11.35)$$

In equation (3.34), we neglect the line broadening effect. If we consider the line broadening effect with the help of the shape function $P(\nu)$, equation (3.34) should be written as follows:

$$\frac{1}{c}\frac{\partial I_\nu}{\partial t} + \frac{\partial I_\nu}{\partial s} = h\nu\, P(\nu)\,[N_n A_{nm} - I_\nu (N_m B_{mn} - N_n B_{nm})] \quad (11.36)$$

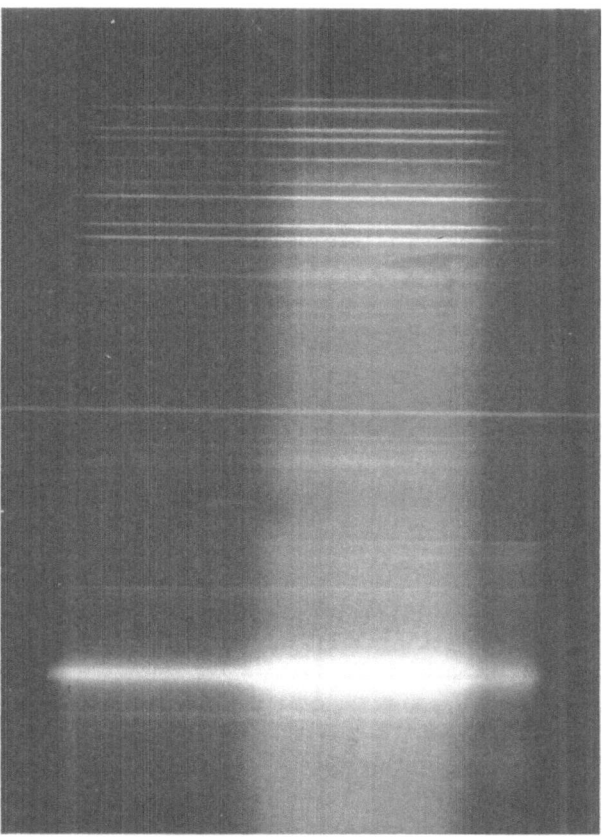

Fig. 11.5. Time-resolved spectrogram of visible light emitted near the end wall of a gas-driven shock tube. (Photograph courtesy of M. H. MILLER, T. D. WILKERSON, G. CHARATIS and P. MURLHY, Institute for Fluid Dynamics and Applied Mathematics, University of Maryland)

From equation (11.36), we see that the absorption coefficient k_ν is

$$\rho\, k_\nu = N_m B_{mn}\, h\, \nu\, P(\nu) \quad (11.37)$$

Hence the absorption coefficient k_ν depends on the shape function, i.e., on the line broadening. Hence the study of the line broadening parameters is very important in the study of the radiative properties of a medium.

Empirically the shape function $P(\nu)$ may be adequately expressed by (i) a Gaussian function

$$P(\nu) = \exp\{-k_1 (\nu - \nu_0)^2\} \quad (11.38)$$

or (ii) and Agnesi function

$$P(\nu) = \frac{k_1}{k_1 + (\nu - \nu_0)^2} \qquad (11.39)$$

where k_1 is a constant.

Special treatise (6, 22) should be consulted for the detailed discussion of the problems of line broadening.

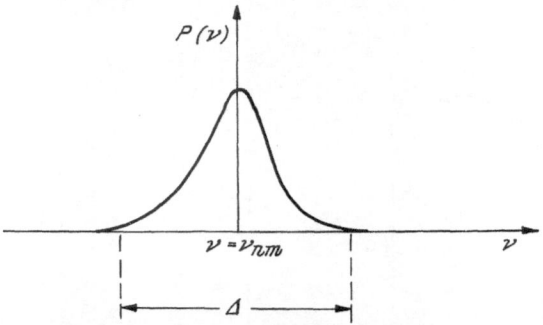

Fig. 11.6. Shape function $P(\nu)$ of a spectral line at $\nu = \nu_{mn}$

5. The absorption coefficient of high temperature gases. Before we make any calculation of the flow problem of radiation gasdynamics, we should know the absorption coefficient of the gas as a function of frequency ν, temperature T and density ρ of the gas. The most important medium for practical interest is air. Hence considerable amount of research work have been done to determine the absorption coefficient of air. The absorption of radiation in earth atmosphere has been extensively investigated by meteorologists because thermal radiation is one of the most important factor in the heat transfer processes in atmosphere. The importance of the variation of the absorption or emission of radiation with frequency has long been recognized. In meteorology, it has been found the necessity of separate treatment of the short-wave radiation from a hot body like sun and the long-wave radiation from cold bodies like the earth and the atmospheric constituent gases (8). For meteorological studies, the range of temperature of air considered is rather small and the absorption of air at low temperature, say below 1,000 °K, is well understood. But in radiation gasdynamics which is greatly stimulated by very high temperature flow of gases such as in re-entry problem, radiation from nuclear detonation, etc., the range of temperature of the gas considered is very large from low temperature of atmospheric air to high temperature of several million degrees. For such a large range of temperature, chemical reactions and ionization take place and the composition of the air will change greatly. The problem of determination of the absorption of air becomes very complicated. Hence the important processes which determine the absorption coefficient of air change with the temperature. At low temperature, the air consists essentially of molecular nitrogen and molecular oxygen. Hence the molecular transition determines the absorption coefficient of air. At very high temperature, the molecules dissociate into atoms, and ionization may occur. Hence the transition processes of atoms, ions and free electrons dominate the absorption of radiation. It is customary to divide the temperature range into several divisions and in each division only the most important processes are considered. The division of temperature range is purely empirical. For instance, in reference *16*, MAGEE and HIRSCHE-FELDER divided the temperature range into four divisions while in reference *2*, Armstrong et al. divided it into two divisions.

Another important point is that when the temperature and density of the gas is high, especially for highly ionized gas, absorption may occur by excitation of collective modes involving many particles simultaneously. No complete theoretical calculation of absorption coefficient of air has been performed by taking the interaction into account yet. However for most practical impor-

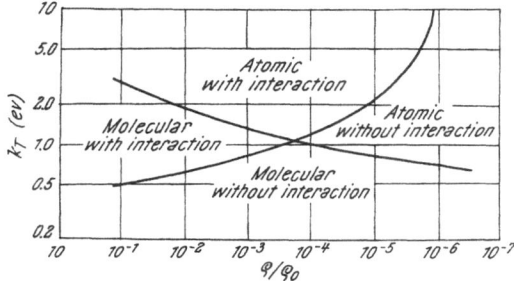

Fig. 11.7. Temperature-density division of important radiative process. (Fig. 1 of reference 2 by B. H. ARMSTRONG, J. SOKOLOFF, R. W. NICHOLLS, D. H. HOLLAND and R. E. MEYEROTT, Courtesy of Pergamon Press, Ltd.)

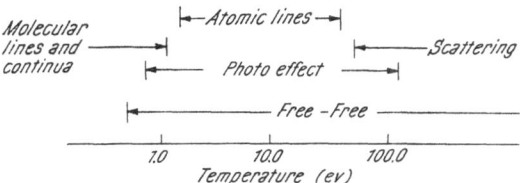

Fig. 11.8. Important radiative processes in various temperature range. (Fig. 2 of reference 2 by B. H. ARMSTRONG, J. SOKOLOFF, R. W. NICHOLLS, D. H. HOLLAND and R. E. MEYEROTT, Courtesy of Pergamon Press, Ltd.)

tant cases at present time, the interaction is negligible and the absorption of radiation in a gas is due to the interaction of individual particles and the photons. Hence most of the calculations such as those given in references *2* and *12* are based only on the elementary interaction involving a single gas particle. Fig. 11.7 shows some empirical divisions of temperature and density in which the atomic process or the molecular process is important as well as the cases in which the interaction between particles are important. Fig. 11.8 shows important radiative processes in various temperature ranges.

If we assume that the interaction between particles is not important, we see from equation (11.19) or (11.34), the absorption coefficient is proportional to the number density N_m. We may call the rest of the factor in the absorption coefficient besides N_m, the cross section for absorption of a photon of frequency ν [cf. eq. (11.37)]. For air, the particles should be divided not only by the state of the atom m but also by the types of particles such as N_2, O_2, N, O, ions, electrons etc. If we refer to the type of the particles by the subscript s, the cross section of absorption depends on the type s as well as the initial state m and the final state n. The total absorption coefficient of the air is the sum of the absorption by all the particles. Symbolically, we may write the absorption coefficient of the air as follows:

$$\rho \, k_\nu = \sum_{s,\,m,\,n} N_{sm} \, \sigma_{smn} \, (\nu) \qquad (11.40)$$

where N_{sm} is the number density of the s-type particles in the state m and σ_{smn} is the cross section of a s-type particle in a transition from the initial state m to the final state n. The summation is extended over all the types of particles in the air and all their initial and final states.

The number density N_{sm} may be expressed in terms of the partial number density of s-type particles N_s or the total number density of the air N as follows:

$$N_{sm} = N_s \, P_{sm} = N \, P_s \, P_{sm} \qquad (11.41)$$

where P_{sm} is the probability that a particle of type s is in the state m and P_s is the probability that a particle in the air chosen at random is of the type s.

For a given temperature and density, we may calculate the probabilities P_s and P_{sm} from the statistical mechanics without considering the radiation field. The cross section of absorption σ_{smn} should be computed by the quantum theory of radiation without references to the number density. Such computations are rather lengthy and we will not discuss them. References 2 and 12 may be referred to for the detailed discussion of the computation. It would be of interest to know the most important processes at various temperature range without reference to the density dependence of the effects (Fig. 11.8). At present time, we have good experimental information for temperature below one electron volt (ev) and good theoretical information for temperature above 500 ev. Research is needed for the intermediate temperature range. Since in this intermediate temperature range, the theoretical computations are very tedious, experimental research in the determination of the absorption coefficient should be of special interest. Fig. 11.9 shows a typical case of absorption coefficient of air at various photon energies, i.e., frequencies ν.

At very high temperature, hydrogen will be the most important gas in radiation gasdynamics. For instance, at extreme high altitude, the atmosphere of earth is composed mainly of hydrogen and in fusion process hydrogen is the main medium. By hydrogen gas, we mean a mixture of H_2, H, H^+, H^-, electrons and various quantum states of H. In reference 13, the absorption coefficient and thermal properties of hydrogen gas have been calculated from the pressure range of one to one thousand atmospheres and a temperature range from 3,000 °R to 20,000 °R. The spectrum absorption coefficient of hydrogen gas is tabulated for 33 wave numbers between 1,000 to 400,000 cm^{-1}. Some typical spectral absorption coefficients of hydrogen gas at a pressure of 10 atmospheres are shown in Fig. 11.10.

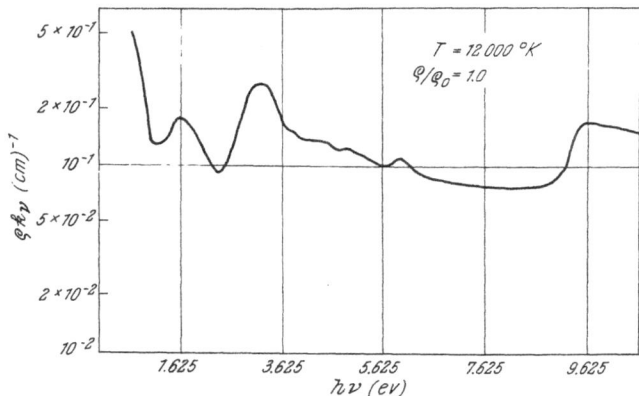

Fig. 11.9. Absorption coefficient *vs* photon energy of air. (Fig. 4 of reference 2 by B. H. ARMSTRONG, J. SOKOLOFF, R. W. NICHOLLS, D. H. HOLLAND and R. E. MEYEROTT, Courtesy of Pergamon Press, Ltd.)

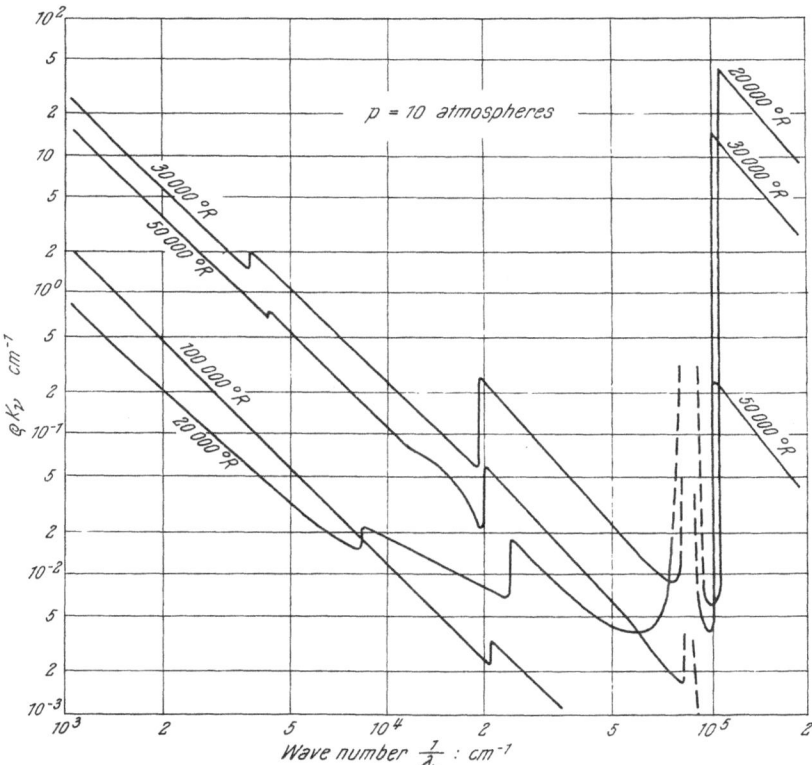

Fig. 11.10. Absorption coefficient of hydrogen gas as a function of wave number at a pressure of 10 atmospheres. (Fig. 4 of reference 13 by N. L. KRASCELLA, Courtesy of NASA)

6. Scattering coefficient of radiation. Even though the scattering phenomena may be neglected in many flow problems of radiation gasdynamics during the preliminary study of the thermal radiation effects, scattering is one of the most important phenomena in radiative processes. In many physical phenomena involving the radiative processes such as those in physical meteorology and astrophysical problems, scattering should be studied in detail. We shall not discuss them here. Special treatizes should be referred to (*1, 4, 8.*) What we are going to discuss in this section is the effect of the scattering on the flow problems of radiation gasdynamics.

First we shall state some terminologies about scattering coefficient. In chapter III, we use the absorption coefficient k_ν as the sum of the true absorption coefficient k_{ν_t} and the scattering coefficient k_{ν_s} [cf. eq. (3.5)]. In some literature (*13*), the absorption coefficient k_ν is called the extinction coefficient and the true absorption coefficient k_{ν_t} is simply called the absorption coefficient. The name of the scattering coefficient k_{ν_s} is unchanged. As long as we notice the distinction of these names, there should not be any misunderstanding of these terms. In this section, we consider the scattering coefficient k_{ν_s} only.

Scattering takes place when the dimensions of the optical obstacles are of the same order of magnitude as the wave length of the radiation rays and may thus be caused by molecules of the gas and by small liquid or solid particles, in the gas. What we have considered are the cases where the amount of foreign particles in the gaseous medium is negligibly small and where the scattering by molecules of the gas itself is considered as a part of the true absorption coefficient (see Fig. 11.8). Hence we may neglect the scattering phenomena in our analysis of the flow problem of radiating gases. If the foreign particles in the gaseous medium are not negligible, we have to consider both the true absorption coefficient and the scattering coefficient due to these foreign particles.

One of the important cases where the scattering process cannot be neglected is the radiative processes in atmosphere. From meteorological study (*8*), it is well known that scattering of radiation from the sun by water drops and dust particles in the atmosphere plays a very important role. For instance, with a clear sky and at high solar altitude, the major portion of short wave radiation, about 85% comes directly from the sun and the rest by scattering. On the other hand, with sun only 10° above the horizon and with haze and smoke in the air, the percentage of direct short wave from the sum may be less than the scattering radiation.

In section 5, we considered the dry air only. In the atmosphere of earth, the water vapor plays a decisive role in the determination of both the true absorption coefficient of the atmosphere and the scattering coefficient of the air. The water vapor has a broad absorption band in the near infrared region which appears in all the radiation from sun at the bottom of the atmosphere. Hence the absorption by water vapor in a solar beam is greater than that due to all other atmospheric gases. By water drops and dust particles whose size is large in comparison with the wave length of radiation rays, diffuse reflection is produced. It affects the scattering strongly. The diffuse reflection from the tops of cloud represents the largest single item of loss from the total radiation energy that enters the earth atmosphere.

Scattering phenomenon has engineering application too. Small solid particles may be used as seeding agents to control radiative heat transfer in gaseous nuclear rocket engines. In gaseous core nuclear rocket, energy generated by nuclear fission is transferred by thermal radiation to the surrounding propelleant such as hydrogen. The propellant may be sufficient opaque to permit absorption of the energy in the thrust chamber and to prevent damage of the rocket walls by excessive radiative heat transfer. However, hydrogen below a temperature of 6,000 °K at a high pressure of 1,000 atmospheres is transparent to thermal radiation. Hence it is desirable to increase the opacity by seeding the hydrogen

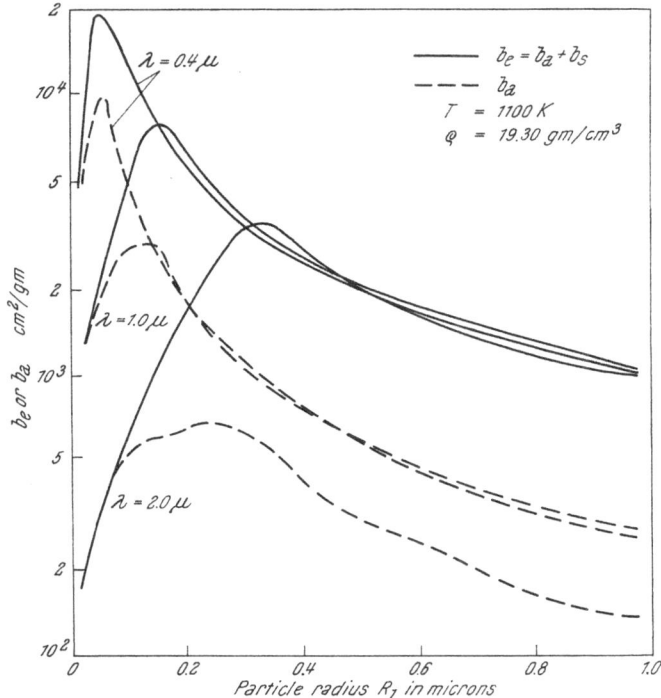

Fig. 11.11. Effect of radius on the extinction and absorption parameters of spherical tungsten particles. (Fig. 31 of reference 14 by N. L. KRASCELLA, Courtesy of United Aircraft Co. Research Laboratories)

with high melting-and boiling-point temperature solids and liquids in the form of small particles. In references *14* and *17*, both theoretical and experimental investigations of the absorption and scattering characteristics of small particles in hydrogen have been carried out. The seeding agents should have the following properties: non-reactivity with high temperature hydrogen; high melting or boiling point temperature and high opacity per unit mass. It was found that tungsten, rhenium and tantalium satisfy these requirements. Fig. 11.11 shows some typical curves of variation of extinction and absorption parameters of spherical tungsten particles where

$$b_a = \sigma_a/\rho \; V = \text{absorption parameter}$$

$$b_e = \sigma_s/\rho \; V = \text{extinction parameter} \qquad (11.42)$$

$$b_s = b_e - b_a = \text{scattering parameter}$$

where ρ is the mass density and V is the volume of the spherical particle of radius R_1. The absorption cross section is σ_a and the extinction cross section is σ_s.

7. Planck and Rosseland mean absorption coefficients of air and hydrogen. From Fig. 11.9 we see that the absorption coefficient k_ν is a complex function of the frequency ν. For very accurate computations, we should use such functional form of absorption coefficient k_ν of the frequency ν in evaluating the radiative integrals. But in many flow problems of radiation gasdynamics as we have discussed in previous chapters, we can only solve the fundamental equations of radiation gasdynamics approximately, hence it is sufficient to use some mean value of absorption coefficient over the whole frequency range in the fundamental equation coefficient over the whole frequency range in the fundamental equations. Two of the mean absorption coefficients are of special interest. They are the Planck mean absorption coefficient which is very useful in the optically thin case and which is defined by equation (5.36), i.e.,

$$K_p = \frac{\int\limits_0^\infty k_\nu' \, B_\nu \, d\nu}{\int\limits_0^\infty B_\nu \, d\nu} = \frac{\pi \int\limits_0^\infty k_\nu' \, B_\nu \, d\nu}{\sigma T^4} \tag{11.43}$$

where σ in the Stefan-Boltzmann constant with respect to radiative flux defined by equation (4.28) and the Rosseland mean absorption coefficient which is very useful in the optically thick case and which is defined by equation (5.18), i.e.,

$$K_R = \left(\int\limits_0^\infty \frac{\partial B_\nu}{\partial T} d\nu \right) \Big/ \left(\int\limits_0^\infty \frac{1}{k_\nu'} \frac{\partial B_\nu}{\partial T} d\nu \right) \tag{11.44}$$

where B_ν is the Planck's radiation function defined by equation (4.22) and k_ν' is the reduced absorption coefficient defined by equation (4.50).

If we know the exact function of the absorption coefficient k_ν in terms of the frequency ν, it is a straight forward procedure to compute both the Planck's mean value K_p and the Rosseland mean value K_R. In reference 2, these mean values have been given as functions of temperature and density of the air which are reproduced in Fig. 11.12 and 11.13. From these figures, some general conclusions may be drawn as follows:

(i) Even though the general shapes of these two mean absorption coefficients with temperature and density are not exactly the same, qualitatively there are some similarities between these two mean values.

(ii) Both the Planck mean and the Rosseland mean absorption coefficients increase approximately linear with the density of the gas.

(iii) At low temperature side, both of these mean absorption coefficients increase with the temperature; in the intermediate temperature range both of them are approximately independent of the temperature and in the high temperature range, both of them decrease with the increase of temperature.

For engineering problems, it is advisable to use some simple formulas for the mean values of the absorption coefficient of the medium. Similar to the case of the coefficient of viscosity, it is suggested that a power law for the mean value of the absorption coefficient or rather for the mean value of the mean free path of radiation which is the reciprocal of the mean mass absorption coefficient would be useful in the analysis of practical flow problems. The following power law has been used in literature:

$$\frac{L_R}{L_{R0}} = \left(\frac{T_0}{T}\right)^{m_1} \left(\frac{\rho_0}{\rho}\right)^{m_2} \tag{11.45}$$

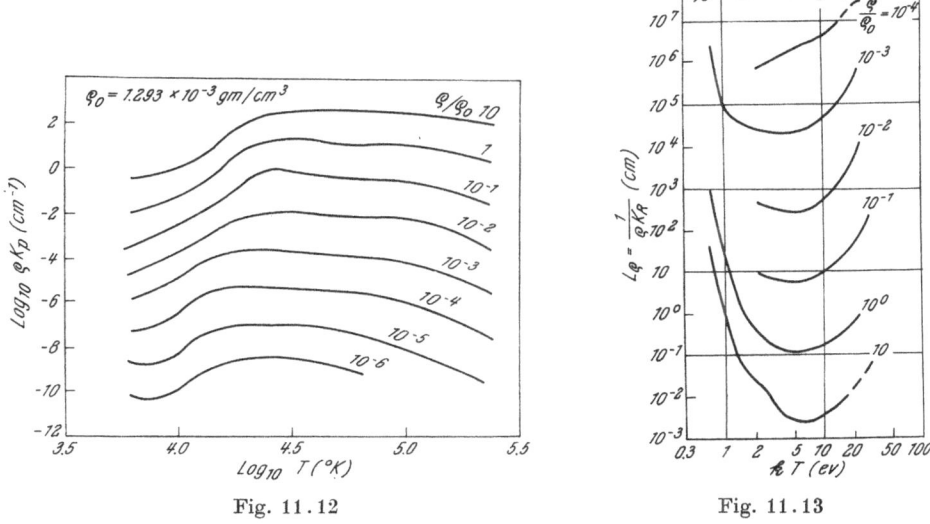

Fig. 11.12   Fig. 11.13

Fig. 11.12. Planck mean absorption coefficient of air. (Fig. 11 of reference 2 by B. H. ARM-STRONG, J. SOKOLOFF, R. W. NICHOLLS, D. H. HOLLAND and R. E. MEYEROTT, Courtesy of Pergamon Press, Ltd.)

Fig. 11.13. Rosseland mean free path of radiation $L_R = 1/(\rho\,K_R)$ of air. (Fig. 12 of reference 2 by B. H. ARMSTRONG, J. SOKOLOFF, R. W. NICHOLLS, D. H. HOLLAND and R. E. MEYEROTT, Courtesy of Pergamon Press, Ltd.)

where $L_R = 1/(\rho\,K_p)$ or $1/(\rho\,K_R)$ is the corresponding mean value of the mean free path of radiation. The subscript 0 refers to the values at some reference conditions and m_1 and m_2 are real numbers, positive or negative and not necessarily integers, chosen to give the best overall fit with the opacity data such as those shown in Figs. 11.12 and 11.13. Strictly speaking, we should have different values of m_1 and m_2 for the Planck mean value from those for Rosseland mean value. From Figs. 11.12 and 11.13, we see clearly that the formula (11.42) is not able to cover the whole range of temperature. Hence the values of m_1 and m_2 change with temperature. However for a given temperature range we make use one set of constant values for m_1 and m_2 in the analysis of the flow problems of radiation gasdynamics. For instance, for air in the temperature range of 7,000 °K to 12,000 °K, the following values of m_1 and m_2 may be used for the Planck mean value:

$$m_1 = 4.4; \; m_2 = 1.0$$

with $\rho_0 = 1.23 \times 10^{-3}$ gr./cc, $T_0 = 10,000\,°K$ and $L_{R0} = 50$ cm.   In general the value $m_2 = 1$ may be used for most of the cases of both the Planck mean and the Rosseland mean values while the value of m_1 changes considerably with the temperature.   In the neighborhood of 20,000 °K, the value of m_1 should be

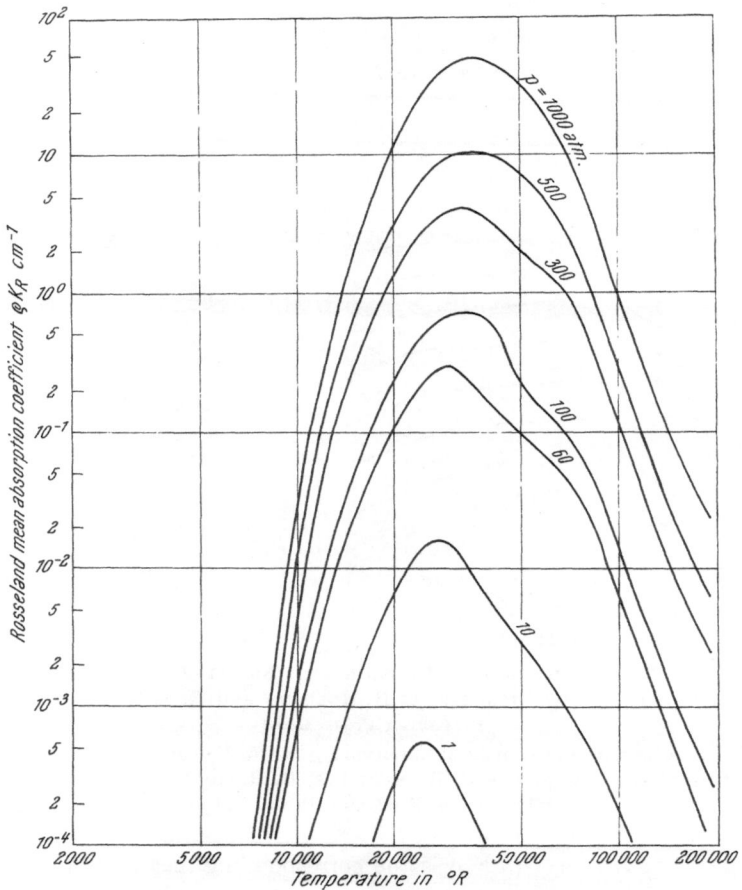

Fig. 11.14. Rosseland mean absorption coefficient for hydrogen gas.
(Fig. 6 of reference 13 by N. L. KRASCELLA, Courtesy of NASA)

of 2.5 in order to be good approximately.   As the temperature increases, the value of m_1 decreases.   At very high temperature, m_1 may be of negative value.

For problems with wide range of temperature, we should use better approximation of the absorption coefficient than that given by equation (11.45).   We have already discussed them in chapter VII, section 5.   For instance equations (7.50) and (7.52) give better approximations.

Fig. 11.14 shows the Rosseland mean absorption coefficient for hydrogen gas.   The general trend of variation of this mean absorption coefficient with temperature and density is the same as that for air.   Hence the approximate formula such as equation (11.45) for the variation of mean value of the mean free path of radiation with temperature and density may be used for hydrogen gas if the proper values of m_1, L_{R0} etc. are chosen.

8. Some experimental investigations of opacity of gases. The experimental investigation on the radiative transfer in gases has been carried out by physicists over a long time and considerable data have been published in literature. However, most of the data of the laboratory experiments are in the lower temperature range, say below 10,000 °K or 1 ev., by using the arc or flame as light source. The good experimental informations of radiative transfer properties in gases at higher temperature than 10,000 °K are still meager. However the use of the shock tube as a spectroscopic source is very helpful for the study of radiative properties of gases at higher temperature. Considerable informations as to the transition probabilities and line broadening coefficient have been obtained recently (22). For instance Fig. 11.5 shows a time-resolved spectrogram of visible light emitted near the end wall of a gas-driven shock tube. Using a test gas of 3/10% methane (CH$_4$) in (1/50) atmosphere of neon, various neutral carbon lines are seen between the orange-red neon lines (above) and the H_β line for the Balmer series at 4861 Å (below). Time runs from left to right, and the film exposure begins with 120 microseconds of steady emission behind the reflected shock. We note the Stark-broadening of H_β due to ionization and free charged particles in this gas. The gas temperature here is roughly 1 ev. (= 11.606 °K) and the incident shock Mach number 7.4 + 0.1. Later shock reflections between the end wall and the driving interface give rise to higher temperatures, degrees of ionization and line intensities. The luminosity is finally quenched by mixing with the cooler, expanded driver gas. Such films, when suitably calibrated and combined with other data, enable the "f-values" or "line strengths" to be measured for many types of atoms and molecules. Other radiation data can also be colected for these gas conditions, which are similar to those for flows in reaction motors and plasma jets and around re-entrant missibles.

In most of the spectroscopical study, the temperature of the hot gas can be determined accurately if the gas is optically thin and if reliable transition probability is available. Such determination of temperature without disturbing the medium is very useful for diagnostic studies of plasma physics and astrophisics. However, if accurate value of transition probability is not available, serious errors may occur in the temperature determination. Another method known as the "reversal temperature" method has been used in many experimental situations in the determination of the temperature which is independent of the previous knowledge concerning the transition probabilities or the optical thickness of the gas. Both the ordinary spectroscopic study in a shock tube and the reversal temperature method have been carried out at the Institute of Fluid Dynamics and Applied Mathematics, University of Maryland by Drs. Wilkerson and Charatis and their associates. We shall not discuss these results which belong to the physics of radiation rather than radiation gasdynamics. For readers who are interested in these experimental technique and results, reference 22 should be referred to.

9. Non-equilibrium radiation. In all of our previous discussions, we consider only the case of local thermodynamic equilibrium conditions because it is a good approximation for many practical problems such as steady flow problems in which a local temperature can be defined and it is also the case where reasonable good theoretical analysis can be made. Because our present knowledge of radiation gasdynamics is still meager, it is advisable to study the local thermodynamic equilibrium case in detail in order to have better understandings on the thermal radiation effects on the flow problems.

At present time, little has been done on the non-equilibrium radiation field in the analysis of the flow problem of radiation gasdynamics. It would be expected that in the near future, the case of non-equilibrium radiation field will be extensively studied. In reference 7, some preliminary study on the non-equilibrium radiation field has been made by dividing the frequence range into three division and found the source function in these range separately. The resultant equation has been used in the prediction of the precursor effects in shock tube. Since the final results were not successful, we are not going to discuss this analysis here.

Another way to study the non-equilibrium radiation field is to apply similar method used in the study of non-equilibrium flow in a chemically reacting medium. The local thermodynamic equilibrium case is similar to the equilibrium flow case in a dissociating flow, i.e., the degree of dissociation is consistent to the local kinetic temperature of the gas. In the non-equilibrium flow, we may define a dissociation or vibrational temperature which is different from the kinetic temperature and which determine the degree of dissociation or vibrational energy. Thus in the non-equilibrium thermal radiation case, we may define a radiation temperature T_R which is different from the local kinetic temperature (10) and which determines the value of the absorption coefficient k_v' as well as the Planck's radiation function B_v. As the time increases, the radiation temperature will reach the local kinetic temperature as a limit. An equation between the radiation temperature T_R and the local kinetic temperature T should be postulated by considering the radiative collision process just as the corresponding chemical reaction equation. Probably some relaxation time of radiative transfer should be introduced in such analysis.

References

1. ALLER, L. H.: Astrophysics: The atmospheres of the sun and stars. Second edition. The Ronald Press Co., New York, 1963.

2. ARMSTRONG, B. H., J. SOKOLOFF, R. W. NICHOLLS, D. H. HOLLAND and R. E. MEYEROTT: Radiative properties of high temperature air. Jour. Quant. Spect. Rad. Transf. Vol. 1, pp. 143–162, Pergamon Press, 1961.

3. BORN, M.: Atomic Physics. Hafner Publishing Co., 1946.

4. CHANDRASEKHAR, S.: Radiative Transfer. Dover Publications, New York, 1960.

5. CHAPIN, C., Jr.: Physics of Hydrogen Radiation. Report A & ES 62-12, School of Aero. & Eng. Sci., Purdue Univ., 1962.

6. CONDON, E. U., and G. H. SHORTLEY: The Theory of Atomic Spectra. Cambridge Univ. Press, 1953.

7. FARRARI, G., and J. H. CLARKE: Photoionization upstream of a strong shock wave. Report CM-1020, Div. of Eng., Brown Univ., 1963.

8. GODSKE, C. L., T. BERGERON, J. BERKNES, and R. C. BUNDGAARD: Dynamic Meteorology and Weather Forecasting. American Meteorological Soc. and Carnegie Institution of Washington, 1957.

9. GOULARD, R.: Fundamental equations of radiation gasdynamics. Report A & ES 62-4, School of Aero. & Eng. Sci., Purdue Univ., 1962.

10. HANSON, C. F.: A radiation model for non-equilibrium molecular gases. AIAA Jour. Vol. 2, No. 4, pp. 611–616, April 1964.

11. HEITLER, W.: The Quantum Theory of Radiation. Oxford Univ. Press, 3rd Ed., 1954.

12. KIVEL, B., and K. BAILEY: Tables of radiation from high temperature air. Res. Report 21, AVCO Res. Lab., 1957.

13. KRASCELLA, N. L.: Tables of the composition, opacity and thermodynamics properties of hydrogen at high temperature. NASA SP-3005, 1963.

14. KRASCELLA, N. L.: Theoretical Investigation of the absorption and scattering characteristics of small particles. United Aircraft Corp. Res. Lab. report C-910092-1, Sept. 1964.

15. LIGHTHILL, M. J.: Dynamics of a dissociating gas. pt. 2 Quasiequilibrium transfer theory. Jour. Fluid Mech. vol. 8, pt. 2, pp. 161–182, 1960.

16. MAGEE, J. L., and J. O. HIRSCHFELDER: Thermal radiation phenomena. Chap. 3 of report on blast wave. Report LA-2000, Los Alamos Sci. Lab., 1947.

17. MARTENEY, P. J.: Experimental investigation of the opacity of small particles. United Aircraft Corp. Res. Lab. report C-910092-2, Sept. 1964.

18. MENSEL, D. H.: Selected papers on physical processes in ionized plasma. Dover Publications, Co., 1962.

19. PAI, S. I.: Thermal Radiation effects in hypersonic flow field Proc. on Non-linear Engineering Problems. Academics Press, pp. 163–183, 1964.

20. PLANCK, M.: The Theory of Heat Radiation. Dover Publication, 1959.

21. VON HIPPEL, A. R.: Dielectrics and Waves. John Wiley and Sons, Inc., New York, 1954.

22. WILKERSON, T.: The use of the shock tube as a spectroscopic source with the application to the measurement of the gf-value for lines of neutral and singly ionized chromium. Ph. D. Thesis, Univ. of Michigan, 1961, also Tech. report AFOSR 1151, University of Michigan, 1961.

A List of Important Symbols

Since Radiation Gasdynamics is a combined subject of gasdynamics and radiative transfer, many well established symbols in gasdynamics may have entirely different meaning in radiative transfer textbook. For instance, the letter "a" is usually used as sound speed in gasdynamics but it is used for Stefan-Boltzmann constant in radiative transfer analysis. We would like to use those well established symbols in both gasdynamics and radiative transfer as much as possible if there is no confusion. However, if confusion may occur, we shall put a subscript in one of them to distinguish the meanings of these symbols. For instance, we shall use "a" for sound speed but "a_R" for Stefan-Boltzmann constant. In the following, we list those important symbols used in this book which occur in more than one sections. Those less important symbols which occur only in one section are not listed but are explained in the text.

Symbols:	Meanings:
a	sound speed
A	area
$a_R = 7.67 \times 10^{-15} \dfrac{\text{ergs}}{\text{cm}^3 - {}^\circ K^4}$	Stefan-Boltzmann constant defined by equation (4.31)
a_ν	hemispherical absorptivity coefficient [see equation (6.24)]
$a_{\theta', \nu}$	directional absorptivity (6.15)
A_{nm}	Einstein coefficient of spontaneous emission (3.25)
b	impact parameter (10.23)
$B(T)$ (or B) $= \dfrac{\sigma}{\pi} T^4$	Planck integrated radiation function defined by equation (4.27)
$\vec{B}(B_x, B_y, B_z) = \mu_e \vec{H}$	magnetic induction
B_ν	Planck radiation function defined by equation (4.22)
B_{mn}	Einstein coefficient of absorption (3.6)
B_{nm}	Einstein coefficient of induced emission (3.25)
c	speed of light
$\vec{c}_a$ (c_i, etc.)	random or peculiar velocity (10.36)
C_p	specific heat at constant temperature
$Cp*$	effective specific heat at constant pressure defined in equation (8.9)
C_{PR}	Radiation effective specific heat at constant pressure defined by equation (9.60a)
c_v	specific heat at constant volume
D_R	Rosseland diffusion coefficient of radiation (5.20)
e	absolute electric charge of an electron
E (E_a, E_e, etc.)	energy
$\vec{E}$ (magnitude E)	electric field

Symbols: Meanings:

$\bar{e}_m = U_m + \frac{1}{2} q^2 + \phi + E_R/\rho$ total energy of a gas per unit mass

E_R radiation energy density defined by equation (2.14)

e_w coefficient of emissivity of a wall

$f(\eta)$ (or f)   streamfunction of similarity solution defined by equation (9.75c)

f_{mn} (or f)   oscillator strength, equation (11.33)

$\vec{F}$ (F)   body force vector

$\vec{F}_e$ electromagnetic force vector

$\vec{F}_g$ non-electromagnetic force

$\vec{F}_R$ (F_{Ri})   radiation body force (2.21)

F_R distribution function of photons

F_s (F_r or F)   molecular distribution function, equation (10.1)

$\vec{g}$ (or g)   gravitational acceleration

g_n (or g)   statistical weight

$h = 6.62 \times 10^{-27}$ ergs/sec. Planck constant

$\vec{h}$ (h_x, h_y, h_z)   perturbed magnetic field

H Boltzmann H-function defined in equation (10.27)

$\vec{H}$ magnetic field strength

$i = \sqrt{-1}$ imaginary number

$\vec{i}, \vec{j}, \vec{k}$ unit vector along x-, y- and z-axis respectively

I integrated intensity defined by equation (2.3)

I_ν specific intensity defined by equation (2.1)

$\vec{J}$ (J_x, J_y, J_z)   electric current density

j_ν radiation emission coefficient (3.29)

J_ν source function of radiation (3.33)

$k = 1.379 \times 10^{-16}$ ergs/°K   Boltzmann constant

K_p Planck mean absorption coefficient defined by equation (5.36)

$K_r = L_R/L$ Knudsen number of radiation

K_R Rosseland mean absorption coefficient defined by equation (5.18)

$k_s = \rho_s/\rho$ mass fraction of sth species

k_ν (k_{ν_t}, k_{ν_s})   absorption coefficient (3.1)

k_ν' reduced absorption coefficient defined by equation (4.50)

l, m, n directional cosines

L typical length

L_R ($L_{R\nu}$)   mean free path of radiation or Rosseland mean value of mean free path of radiation

L_{Rp} Planck mean value of mean free path of radiation

Symbols:	Meanings:
m	mass
M	Mach number
n	index of refraction
$\vec{n}$ (n_i or n_j)	unit normal and its components
N	number density
p (P or p^*)	pressure
p_g	gas pressure
p_R	radiation pressure defined by equation (2.19)
P_r	Prandtl number (7.29)
$\vec{p}_s$	momentum vector (10.1)
$\vec{q}$ (q_i, q)	flow velocity vector
$\vec{q}_R$ (q_{Ry} etc.)	Radiation heat flux defined by equation (2.9)
r	radial distance or simply distance between two points
R	gas constant (5.2)
r_w (r, r_v etc.)	reflection coefficient of a wall
R_b	nose radius
$R_e = UL\,\rho/\mu$	Reynolds number
R_E	electric field number (9.43c)
R_F	radiative flux number (7.41)
R_h	Hartmann number (9.47)
R_p	Radiation pressure number (7.18)
R_r	relativistic parameter (7.17)
R_t	time parameter (7.15)
R_σ	magnetic Reynolds number (9.43c)
s	distance along a radiation ray
S	entropy
t	time
T	temperature
u, v, w	x-, y- and z-component of flow velocity q respectively
U	typical velocity
U_m	internal energy
U_ν	spectral energy density of radiation defined by equation (2.13)
U_λ	wave length energy density of radiation defined by equation (4.8)
V	volume
V_x	x-wise Alfven wave speed (8.13)
V_y	y-wise Alfven wave speed (8.13)
w_s	diffusion velocity of sth species
x, y, z	Cartesian coordinates

Symbols: Meanings:

α shock angle

γ ratio of specific heats (7.25)

$\delta^{ij} = 0$ if $i \neq j$; $\delta^{ij} = 1$ if $i = j$

ε_n exponential integral (5.38)

θ angle

$\varkappa$ coefficient of heat conductivity

$\varkappa_R$ coefficient of heat conductivity due to radiation (5.22)

λ wave length (4.9)

μ coefficient of viscosity

μ_e magnetic permeability

ν (ν_{mn}) frequency

ν_g coefficient of kinematic viscosity

ν_H magnetic diffusivity

ρ density

ρ_e excess electric charge

$\sigma = 5.68 \times 10^{-5}$ erg. cm^{-2} °K^{-4} sec^{-1} Stefan-Boltzmann constant for radiative flux (4.28)

σ_e electric conductivity

σ_m cross section of absorption (3.15)

σ_0 small area (2.1)

τ (τ_ν) optical thickness defined by equation (3.3) or similar equation

τ_R (τ_{Rij}) radiation stress tensor (2.18)

τ_s (τ_{sij}) viscous stress tensor (5.5)

ψ streamfunction

ω solid angle

$$\nabla = \vec{i}\,\frac{\partial}{\partial x} + \vec{j}\,\frac{\partial}{\partial y} + \vec{k}\,\frac{\partial}{\partial z}$$ gradient operator

The meaning of most subscripts has been described in the text but the following are some of the common ones:

Subscript i or j on a vector refers to its ith or j-th component.

Subscript x, y, or z also refers to its corresponding component.

Author Index

Abarbanel, S., 194, 196
Allen, R. A., 6, 162
Aller, L. H., 6, 14, 34, 217
Ambartsumyan, V. A., 6, 14
Armstrong, B. H., 6, 80, 217

Bailey, K., 49, 64, 81, 218
Baldwin, B. S. jr., 97, 110, 120, 121
Bergeron, T., 14, 23, 34, 162, 217
Bethe, H. A., 120
Bird, R. B., 196
Bjerknes, J., 14, 23, 34, 162, 217
Born, M., 34, 64, 217
Bridgeman, P. W., 80
Bungaard, R. C., 14, 23, 34, 162, 217
Burgers, J. M., 196

Camm, J. C., 6
Chambre, P. L., 196
Chandrasekhar, S., 6, 14, 23, 26, 34, 49, 64, 217
Chapin, C. jr., 217
Chapman, D. R., 113, 121, 163
Chapman, S., 196
Clarke, J. F., 107, 110, 120
Clarke, J. H., 217
Condon, E. U., 217
Cowling, T. G., 196
Curtiss, C. F., 196

DeSilva, C. N., 149, 163

Edwards, D. K., 50, 65, 163
Einstein, T. H., 162
Emde, E., 49

Farrari, G., 217
Fay, J. A., 6
Fuchs, K., 120

Garden, R., 64
Geiger, R. E., 163
Georgiev, S., 162
Godske, C. L., 14, 23, 34, 162, 217
Goulard, M., 49, 64
Goulard, R., 6, 49, 64, 80, 120, 162, 218
Grosh, R. J., 24, 163
Guess, A. W., 105, 121

Hanson, H. C., 218
Hartnett, J. P., 196
Hayes, W., 162
Heaslet, M. A., 110, 120
Heitler, W., 218
Hirschfelder, J. D., 7, 120, 196, 218
Holland, H. D., 6, 80, 217
Hoshizaki, H., 121
Howe, J. T., 162

Jahnke, E., 49
Jakob, M., 34, 124, 162
Johnson, J. C., 14, 24, 34, 64, 162

Kennet, H., 163
Kivel, B., 6, 49, 64, 81, 163, 218
Koh, J. C. Y., 149, 163
Kornowski, E. T., 163
Kourganoff, V., 14, 24, 50, 64
Krascella, N. L., 218

Li, T. Y., 163
Lighthill, M. J., 218

Magee, J. L., 7, 120, 218
Maghreblian, R. V., 7
Marshak, R. L., 120
Marteney, P. J., 218
Mensel, D. H., 218
Meyerott, R. E., 6, 80, 217
Mitchner, M., 110, 120
Moffat, W. C., 6
Mueller, H. G., 163

Nemchinov, I. V., 163
Nicholls, R. W., 6, 80, 217

Olfe, D. B., 120
Oppenheim, A. K., 163

Pai, S. I., 7, 14, 24, 50, 64, 120, 163, 196, 218
Patterson, G. N., 196
Peierls, R. E., 120
Penner, S. S., 65
Planck, M., 1, 7, 14, 24, 29, 34, 50, 65, 218
Pomerantz, J., 7, 121
Probstein, R. F., 6, 65, 162, 196
Prokof'ev, V. A., 92, 110, 120

Ratcliffe, J. A., 163
Romishevskii, Ye, A., 7, 14, 24, 50, 81, 121
Rose, P. H., 6, 7
Rosseland, S., 7, 14, 50, 65

Sachs, R. G., 121
Sampson, D. H., 7, 50, 65, 80, 81, 111, 121, 157, 163
Scala, S. M., 7, 50, 65, 80, 81, 111, 121, 157, 163
Schaaf, S. A., 196
Sen, H. K., 105, 121
Sforza, P. M., 163
Shortley, G. H., 217
Simons, R., 196
Sokoloff, J., 6, 80, 217
Speth, A. I., 120
Strack, S. L., 163
Synge, J. L., 196

Teare, J. D., 7, 162
Tellep, D. M., 50, 65, 163

Traugott, S., 18, 111, 121

Ünsold, A., 7

Vertushkin, V. K., 7, 14, 24, 50, 81, 121
Viegas, J. R., 162
Vincenti, W. G., 121
Vinokur, V., 110, 120
Viskanta, R., 24, 163
von Hippel, A. R., 218
von Karman, Th., 2, 7
von Neumann, J., 120

Wick, B. H., 163
Wilkerson, T. D., 218
Wilson, K. H., 121
Wolf, E., 64

Yoshikawa, K. K., 113, 121, 163

Zel'dovich, Ia, B., 110, 121
Zhigulev, V. N., 7, 14, 24, 50, 81, 120, 121

Subject Index

Absorbing medium 125
absorption 14, 198
absorption coefficient 15, 22, 57, 78, 207
absorptivity 58, 200
accomodation coefficient 191
accounting method 124
adiabatic change 32
Agnesi function 207
albedo 19
Alfven's wave 86
angle factor 123
atmosphere 130
atomic number 198
atomic structure 203
average quantity 165

Binary collision 170
black body 9, 16, 27, 123
black surface 27, 59
Blasius solution 147
blunt body 112, 150
Bohr radius 203
Boltzmann constant 36
Boltzmann distribution 29, 205
Boltzmann equation 170, 174
Boltzmann H-theorem 171
Boltzmann law 29
boundary condition 51, 52
boundary layer flow 141, 146, 150, 155
boundary layer thickness 73

Calculus method 124
classical theory of radiation 198
Clausius-Mosotti-Lorentz equation 199
coefficient of absorption 15, 207
coefficient of emission 20
coefficient of extinction 6, 211
coefficient of reflection 56, 59, 191
coefficient of viscosity 36
collision cross section 173
collision term 170
compressibility factor 69
configuration factor 123
conservation equations 175
conservation of electric charge 83
continuum flow 188
Couette flow 135
cross section of absorption 18, 209

Damping factor 201
degeneracy 30
density 114
detached shock 116
detached shock distance 164
dielectric constant 200
diffuse reflection 56
diffusion coefficient of radiation 10, 14, 67
diffusion equation 156, 178
diffusion velocity 175, 178
dimensional analysis 66
dipole moment 198
directional absorprivity 56
dispersed shock 110
dispersion relation 85, 87
D-layer 132

Effective heat conductivity 85
effective Prandtl number 144
effective specific heat 85
Einstein coefficients 17, 20, 205
Einstein formula 169
E-layer 132
electric conductivity 83
electric current density 135
electric field 37, 82
electric field number 136
electromagnetic energy flux 83
electromagnetic force 83
electromagnetic radiation 197
electromagnetic waves, 86, 197
emission coefficient 6, 20, 58, 198
emissivity 57, 61, 123
energy levels 203
enthalpy of air 113
entropy 32
equation of continuity 36, 70, 177
equation of energy 37, 70, 179
equation of heat flow 180
equation of motion 36, 70, 178
equation of radiative transfer 21, 23, 37, 70
equation of state 36, 70
equation of viscous stress 179
Euler-Mascheroni constant 49
exosphere 132
experimental investigation of opacity 216
exponential approximation 129

exponential integral 47
extinction coefficient 6, 211

Fast wave 90
film coefficient 74
finite mean free path of radiation 41, 92, 106, 145
first law of thermodynamics 32
first order collision region 190
F-layer 132
flow velocity 166, 175
fluorescence 6
flux of radiation 9, 40
Fokker-Planck collision term 171
four dimensional space 166
free molecule flow 188, 191
frequency 8
friction coefficient 178
fully dispersed shock 107
fully merge layer 189
fundamental equations of radiation gas-dynamics 35

Galilean transformation 167
gas constant 36
gas pressure 32, 176
Gaussian function 207
generalized Ohm's law 83
geometrical factor 123
gf-value 206
gravitational constant 36
gray gas 39, 44
gray surface 59

Half-width of spectral line 202
Hartmann number 137
heat balance in atmosphere 133
heat conduction 1, 126
heat conductivity 41, 85
heat convection 1
heat transfer 1, 122
heat wave 87
heterosphere 131
H-function 171
high temperature gas 197, 207
homosphere 130
hydrogen atom 203
hydrogen gas 215
hypersonic flow 149

ICBM 4
incipient merger layer 189
index of refraction 53, 200
induced emission 20, 205
inductive capacity 82
initial condition 51
injection of foreign gas 159

inspection analysis 66
insulated surface 193
integral equation method 125, 127
integrated intensity 4, 9
internal energy 2, 179
inviscid flow field 112
inviscid shock layer 150
inviscid tail of shock 107
ionosphere 131
isotropic scattering 19
iterative methods 127

Jet mixing 143

Kinematic viscosity 73
kinetic theory of gasdynamics 164
Kirchhoff law of net work 124
Kirchhoff law of radiation 25
Kirchhoff law of solid surface 58
Knudsen number 76, 187
Knudsen number of radiation 75, 187

Laminar jet mixing 143
lapse rate 131
Legendre polynomial 19
linear absorption coefficient 18
linearization 126
local thermodynamic equilibrium 34, 184
longitudinal waves 87
Lorentz invariant 167
Lorentz transformation 167

Mach number 71, 72
magnetic diffusivity 86
magnetic field 82
magnetic permeability 82
magnetic pressure number 135
magnetic Reynolds number 136
magnetogasdynamics 133
Mars probe 4
Maxwell equations of electromagnetic fields 82
Maxwellian distribution 172
mean free path 76, 173
mean free path of radiation 16, 38, 79
mean molecular velocity 172
mesopause 131
mesophere 131
metasphere 132
Milne-Eddington approximation 110, 129
molecular distribution function 165
molecular velocity 165
monatomic gas 202
most probable speed 172

Network method 124
Newton's second law of motion 169

no slip condition 52
non-absorbing medium 123
non-equilibrium radiation 216
nose radius 4, 160
number density 165, 175
Nusselt number 74, 161

Oblique shock 117
Ohm's law 84
one dimensional approximation 45
optical depth 15, 43
optical thickness 15, 43
optically rough surface 55
optically smooth surface 53
optically thick case 39, 82, 98, 103, 153, 190
optically thin case 60, 129, 154, 190
oscillator strength 206

Parallel plates 45, 59, 128, 133
partly dispersed shock 107
Peclet number 76
peculiar velocity 172
perfect conducting surface 194
permeability 200
perturbation technique 172
phase function 19
phosphorescence 6
photons 5, 174
Planck constant 30
Planck mean absorption coefficient 44, 78, 115, 159, 213
Planck radiation law 29
Pi (π) theorem 66
plane Couette flow 135
plane Poiseuille flow 140
polarization 18
polarization vector 199
Prandtl number 71, 73, 144
pressure 67, 176
probability 209
protosphere 132

Quantum 12, 30
quantum theory of radiation 202

Radiation 1
radiation body force 13
radiation decay length 115
radiation diffusion coefficient 10, 41, 67
radiation dispersed shock 110
radiation energy 8
radiation energy density 2, 11, 40, 180
radiation gasdynamics 1, 35, 66
radiation heat conduction 41
radiation heat wave 88
radiation Knudsen number 71, 75
radiation pressure 2, 12, 40, 134, 180

radiation pressure number 69, 75
radiation rays 5, 8
radiation resisted shock 107
radiation slip 63
radiation sound speed 87
radiation stress 2, 12, 40, 180
radiation temperature 217
radiative equilibrium 35
radiative flux number 77
radiative heat flux 2, 4, 9, 40
radiative heat transfer 1, 10, 123
radiative processes in atmosphere 130
radiative properties of high temperature gases 197
radiative viscosity 186
radiative transfer 8, 15
radiative transfer equation 21, 23, 37
radiosity 57, 123
random velocity 172
Rankine-Hugoniot relations of shock wave 99
rarefied radiation gasdynamics 186
ratio of specific heats 70, 72, 101
Rayleigh-Jeans radiation law 29
Rayleigh phase function 19
reduced absorption coefficient 22, 34
re-entry 4
reflection coefficient 57, 59, 191
reflection distribution function 57
reflectivity 56
relativistic Boltzmann equation 175
relativistic mechanics 166
relativistic parameter 69, 75, 166
relativity 166
relaxation length 107
relaxation time 217
resonance frequency 202
Reynolds number 71, 73
Rosseland approximation 40, 128
Rosseland mean absorption coefficient 40, 78, 213
rough surface 55, 59

Scattering coefficient 18, 21, 210
scattering medium 161
Schmidt number 74
shape function 207
shock layer 116, 149
shock standoff distance 164
shock wave 83, 98, 112, 149
shock wave structure 103, 106, 149
similarity parameter 66
similarity solution 147
simultaneous heat conduction and radiation 126
slip flow 188
slow wave 90
small mean free path of radiation 39, 143

smooth surface 53
Snell-Fresnel law 53
solar radiation 196
solar spectra 9
sound speed 86
source function of radiation 22
specific heat 33, 85
specific intensity 8
spectroscopy 206
specular reflection 56
speed ratio 193
spontaneous emission 20, 205
stagnation point 149
standard atmosphere 131
Stanton number 76
state, equation of 36
statistical weight 205
Stefan-Boltzmann constant 31
Stefan-Boltzmann law of radiation 32
stimulated emission 205
stratopause 131
stratosphere 131
streamfunction 142, 147
stress tensor 176
structure of molecule 202
summation invariant 172

Temperature 176
thermal diffusivity 73
thermal radiation 4

thermosphere 131
time parameter 71, 72
transition flow region 188
transition layer 190
transmissivity 125
transparency coefficient 57
transverse wave 85
tropopause 131
troposphere 130
true absorption coefficient 17

Ultra-violet catastrophe 29
unsteady flow 162

Velocity, flow 67, 167, 175
velocity of light 27
viscosity coefficient 36
viscous layer region 189
viscous shock layer 158
viscous solution of shock 111
viscous stress 36
viscous wave 86
vorticity interaction region 189

Water vapor 131
waves 5, 83
wave length 16, 27
wave of small amplitude 82, 92
wedge flow 117
Wien's displacement law 27

Druck von Adolf Holzhausens Nfg., Wien